The Haynes
Holley
Carburetor
Manual

by Mark Ryan
and John H Haynes
Member of the Guild of Motoring Writers

Models covered:

One-barrel models: 1904, 1908, 1920, 1940, 1945, 1946, 1949, 1960, 6145, 6146, 6149

Two-barrel models: 2100, 2110, 2210, 2211, 2245, 2280, 2300, 2305, 2360, 2380, 5200, 5210, 5220, 6280, 6360, 6500, 6510, 6520

Three-barrel models: 3150, 3160

Four-barrel models: 4010, 4011, 4150, 4160, 4165, 4175, 4180, 4190, 4360, 4500

(10225 - 5AA6)

Haynes Group Limited
Haynes North America, Inc.
www.haynes.com

Acknowledgements

This book was written with the gracious help and cooperation of the Holley Replacement Parts Division, 11955 East Nine Mile Road, Warren, Michigan. Thanks are also due to Bob Oliver of Competition Carburetion in Sparks, Nevada, who provided technical assistance, and Mike McCurdy of Conejo Valley Carburetors, who provided technical assistance and many of the carburetors you see photographed in this manual.

© **Haynes North America, Inc. 1993**

With permission from Haynes Group Limited

A book in the Haynes Automotive Repair Manual Series

ISBN-10: 1-56392-069-7

ISBN-13: 978-1-56392-069-1

Library of Congress Catalog Card Number 93-79156

Disclaimer

There are risks associated with automotive repairs. The ability to make repairs depends on individual skill, experience and proper tools. Individuals should act with due care and acknowledge and assume the risk of making automotive repairs. While every attempt is made to ensure that the information in this manual is correct, no liability can be accepted by the authors or publishers for loss, damage or injury caused by any errors in, or omissions from, the information given.

Contents

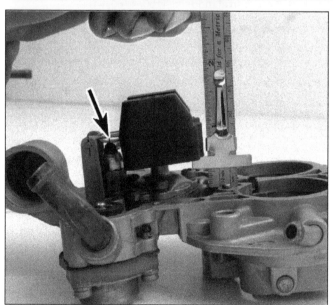

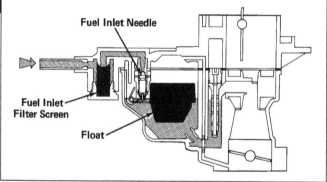

Fuel Inlet Needle

Fuel Inlet
Filter Screen

Float

1 Introduction

About this manual

The goal of this manual is to help you obtain the very best your Holley carburetor has to offer, whether it be performance or economy. Since these carburetors are fine precision instruments, the manual assumes that you already have a basic understanding of how an engine works and are fairly competent with the use of tools and safe working procedures.

For many, the carburetor represents a portion of a vehicle that just shouldn't be tampered with, and because of this repairs or modifications to the carburetor are usually left to someone else. A considerable amount of money can be saved and much knowledge can be gained by taking on these tasks yourself. This will help you better understand what your vehicle is doing under certain operating conditions, and you'll be able to tune your carburetor to correct certain problems and extract the maximum performance from you engine.

There are literally millions of Holley carburetors in existence. The company was started by George Holley and his brother Earl in 1902, originally to build racing engines for motorcycles. They also produced motorcycles and some automobiles, but eventually began concentrating on fuel and ignition systems.

Since that time, Holley carburetors have appeared on just about every type of vehicle you can think of. They've been used as original equipment by every major American automobile manufacturer at one time or another, and have been used extensively in high-performance aftermarket applications. Racers, boaters, off-roaders and hot-rodders have all discovered the merits of the Holley carburetor. Many farmers and ranchers are familiar with Holley carburetors as well - they've been used on tractors and other farm implements, as well as gasoline-powered generators and pumps, etc. As carburetors go, they have no equal.

Of the millions of these carburetors still in existence, all of them will need attention sooner or later. All of the carburetors built by Holley are extremely well made, are relatively simple to overhaul and easy to tune (although tuning a carburetor is usually a trial-and-error proposition). This manual unlocks the "mystery" of the Holley carburetor. Included are chapters on tools, fundamentals of operation, carburetor identification, troubleshooting, removal and installation, overhaul and adjustments, selection and modifications and general specifications.

At the end of the manual there's a glossary of terms used throughout the book. In short, everything you need to know to successfully deal with your Holley carburetor and realize its full potential!

How to use this repair manual

The manual is divided into Chapters. Each chapter is subdivided into sections, some of which consist of consecutively numbered paragraphs (usually referred to as "Steps", since they're normally part of a procedure). If the material is basically informative in nature, rather than a step-by-step procedure, the paragraphs aren't numbered.

The term "see illustration" (in parentheses), is used in the text to indicate that a photo or drawing has been included to make the information easier to understand (the old cliché "a picture is worth a thousand words" is especially true when it comes to how-to procedures). Also, every attempt is made to position illustrations on the same page as the corresponding text to minimize confusion. The two types of illustrations used (photographs and line drawings) are referenced by a number preceding the caption. Illustration numbers denote chapter and numerical sequence within the chapter (i.e. 3.4 means Chapter 3, illustration number 4 in order).

The terms **"Note"**, **"Caution"** and **"Warning"** are used throughout the book with a specific purpose in mind - to attract the reader's attention. A **"Note"** simply provides information required to properly complete a procedure or information which will make the procedure easier to understand. A **"Caution"** outlines a special procedure or special steps which must be taken when completing the procedure where the Caution is found. Failure to pay attention to a Caution can result in damage to the component being repaired or the tools being used. A **"Warning"** is included where personal injury can result if the instructions aren't followed exactly as described.

Even though extreme care has been taken during the preparation of this manual, neither the publisher nor the author can accept responsibility for any errors in, or omissions from, the information given.

Automotive chemicals and lubricants

A wide variety of automotive chemicals and lubricants - ranging from cleaning solvents and degreasers to lubricants and protective sprays for rubber, plastic and vinyl - are available.

Cleaners

Brake system cleaner (sometimes used in place of carburetor cleaner)

Brake system cleaner removes grease and brake fluid from brake parts like disc brake rotors, where a spotless surfaces is essential. It leaves no residue and often eliminates brake squeal caused by contaminants. Because it leaves no residue, brake cleaner is often used for cleaning engine parts as well.

Carburetor and choke cleaner

Carburetor and choke cleaner is a strong solvent for gum, varnish and carbon. Most carburetor cleaners leave a dry-type lubricant film which will not harden or gum up. So don't use carburetor cleaner on electrical components.

Degreasers

Degreasers are heavy-duty solvents used to remove grease from the outside of the engine and from chassis components. They're usually sprayed or brushed on. Depending on the type, they're rinsed off either with water or solvent.

Demoisturants

Demoisturants remove water and moisture from electrical components such as alternators, voltage regulators, electrical connectors and fuse blocks. They are non-conductive, non-corrosive and non-flammable.

Electrical cleaner

Electrical cleaner removes oxidation, corrosion and carbon deposits from electrical contacts, restoring full current flow. It can also be used to clean spark plugs, carburetor jets, voltage regulators and other parts where an oil-free surface is necessary.

Lubricants

Assembly lube

Assembly lube is a special extreme pressure lubricant, usually containing moly, used to lubricate high-load parts (such as main and rod bearings and cam lobes) for initial start-up of a new engine. The assembly lube lubricates the parts without being squeezed out or washed away until the engine oiling system begins to function.

Graphite lubricants

Graphite lubricants are used where oils cannot be used due to contamination problems, such as in locks. The dry graphite will lubricate metal parts while remaining uncontaminated by dirt, water, oil or acids. It is electrically conductive and will not foul electrical contacts in locks such as the ignition switch.

Heat-sink grease

Heat-sink grease is a special electrically non-conductive grease that is used for mounting electronic ignition modules where it is essential that heat is transferred away from the module.

Penetrating oil

Penetrating oil loosens and lubricates frozen, rusted and corroded fasteners and prevents future rusting or freezing.

Silicone lubricants

Silicone lubricants are used to protect rubber, plastic, vinyl and nylon parts.

White grease

White grease is a heavy grease for metal-to-metal applications where water is present. It stays soft under both low and high temperatures (usually from -100 to +190-degrees F), and won't wash off or dilute when exposed to water. Another good "glue" for holding parts in place during assembly.

Sealants

Anaerobic sealant

Anaerobic sealant is much like RTV in that it can be used either to seal gaskets or to form gaskets by itself. It remains flexible, is solvent resistant and fills surface imperfections. The difference between an anaerobic sealant and an RTV-type sealant is in the curing. RTV cures when exposed to air, while an anaerobic sealant cures only in the absence of air. This means that an anaerobic sealant cures only after the assembly of parts, sealing them together.

RTV sealant

RTV sealant is one of the most widely used gasket compounds. Made from silicone, RTV is air curing, it seals, bonds, waterproofs, fills surface irregularities, remains flexible, doesn't shrink, is relatively easy to remove, and is used as a supplementary sealer with almost all low and medium temperature gaskets.

Thread and pipe sealant

Thread and pipe sealant is used for sealing hydraulic and pneumatic fittings and vacuum lines. It is usually made from a teflon compound, and comes in a spray, a paint-on liquid and as a wrap-around tape.

Chemicals

Anaerobic locking compounds

Anaerobic locking compounds are used to keep fasteners from vibrating or working loose and cure only after installation, in the absence of air. Medium strength locking compound is used for small nuts, bolts and screws that may be removed later. High-strength locking compound is for large nuts, bolts and studs which aren't removed on a regular basis.

Anti-seize compound

Anti-seize compound prevents seizing, galling, cold welding, rust and corrosion in fasteners. High-temperature anti-seize, usually made with copper and graphite lubricants, is used for exhaust system and exhaust manifold bolts.

Gas additives

Gas additives perform several functions, depending on their chemical makeup. They usually contain solvents that help dissolve gum and varnish that build up on carburetor, fuel injection and intake parts. They also serve to break down carbon deposits that form on the inside surfaces of the combustion chambers. Some additives contain upper cylinder lubricants for valves and piston rings, and others contain chemicals to remove condensation from the gas tank.

Safety first!

Regardless of how enthusiastic you may be about getting on with the job at hand, take the time to ensure that your safety is not jeopardized. A moment's lack of attention can result in an accident, as can failure to observe certain simple safety precautions. The possibility of an accident will always exist, and the following points should not be considered a comprehensive list of all dangers. Rather, they are intended to make you aware of the risks and to encourage a safety conscious approach to all work you carry out on your vehicle.

Essential DOs and DON'Ts

DON'T rely on a jack when working under the vehicle. Always use approved jackstands to support the weight of the vehicle and place them under the recommended lift or support points.

DON'T attempt to loosen extremely tight fasteners (i.e. wheel lug nuts) while the vehicle is on a jack - it may fall.

DON'T start the engine without first making sure that the transmission is in Neutral (or Park where applicable) and the parking brake is set.

DON'T attempt to drain the engine oil until you are sure it has cooled to the point that it will not burn you.

DON'T touch any part of the engine or exhaust system until it has cooled sufficiently to avoid burns.

DON'T siphon toxic liquids such as gasoline, antifreeze and brake fluid by mouth, or allow them to remain on your skin.

DON'T allow spilled fluids or lubricants to remain on the floor - wipe it up before someone slips on it.

DON'T use loose fitting wrenches or other tools which may slip and cause injury.

DON'T push on wrenches when loosening or tightening nuts or bolts. Always try to pull the wrench toward you. If the situation calls for pushing the wrench away, push with an open hand to avoid scraped knuckles if the wrench should slip.

DON'T rush or take unsafe shortcuts to finish a job.

DON'T allow children or animals in or around the vehicle while you are working on it.

DO wear eye protection when using power tools such as a drill, sander, bench grinder, etc. and when working under a vehicle.

DO keep loose clothing and long hair well out of the way of moving parts.

DO make sure that any hoist used has a safe working load rating adequate for the job.

DO get someone to check on you periodically when working alone on a vehicle.

DO carry out work in a logical sequence and make sure that everything is correctly assembled and tightened.

DO keep chemicals and fluids tightly capped and out of the reach of children and pets.

DO remember that your vehicle's safety affects that of yourself and others. If in doubt on any point, get professional advice.

Batteries

Never create a spark or allow a bare light bulb near a battery. They normally give off a certain amount of hydrogen gas, which is highly explosive.

Always disconnect the battery ground (-) cable at the battery before working on the fuel or electrical systems.

Fire

We strongly recommend that a fire extinguisher suitable for use on fuel and electrical fires be kept handy in the garage or workshop at all times. Never try to extinguish a fuel or electrical fire with water. Post the phone number for the nearest fire department in a conspicuous location near the phone.

Fumes

Certain fumes are highly toxic and can quickly cause unconsciousness and even death if inhaled to any extent. Gasoline vapor falls into this category, as do the vapors from some cleaning solvents. Any draining or pouring of such volatile fluids should be done in a well ventilated area.

When using cleaning fluids and solvents, read the instructions on the container carefully. Never use materials from unmarked containers.

Never run the engine in an enclosed space, such as a garage. Exhaust fumes contain carbon monoxide, which is extremely poisonous. If you need to run the engine, always do so in the open air, or at least have the rear of the vehicle outside the work area.

Household current

When using an electric power tool, inspection light, etc., which operates on household current, always make sure that the tool is correctly connected to its plug and that, where necessary, it is properly grounded. Do not use such items in damp conditions and, again, do not create a spark or apply excessive heat in the vicinity of fuel or fuel vapor.

Keep it clean

Get in the habit of taking a regular look around the shop to check for potential dangers. Keep the work area clean and neat. Sweep up all debris and dispose of it as soon as possible. Don't leave tools lying around on the floor.

Be very careful with oily rags. Spontaneous combustion can occur if they're left in a pile, so dispose of them properly in a covered metal container.

Check all equipment and tools for security and safety hazards (like frayed cords). Make necessary repairs as soon as a problem is noticed - don't wait for a shelf unit to collapse before fixing it.

Accidents and emergencies

Shop accidents range from minor cuts and skinned knuckles to serious injuries requiring immediate medical attention. The former are inevitable, while the latter are, hopefully, avoidable or at least uncommon. Think about what you would do in the event of an accident. Get some first aid training and have an adequate first aid kit somewhere within easy reach.

Think about what you would do if you were badly hurt and incapacitated. Is there someone nearby who could be summoned quickly? If possible, never work alone just in case something goes wrong.

If you had to cope with someone else's accident, would you know what to do? Dealing with accidents is a large and complex subject, and it's easy to make matters worse if you have no idea how to respond. Rather than attempt to deal with this subject in a superficial manner, buy a good First Aid book and read it carefully. Better yet, take a course in First Aid at a local junior college.

2 Tools and equipment

Introduction

For some home mechanics, the idea of using the correct tool is completely foreign. They'll cheerfully tackle the most complex procedures with only a set of cheap open-end wrenches of the wrong type, a single screwdriver with a worn tip, a large hammer and an adjustable wrench. Though they often get away with it, this cavalier approach is stupid and dangerous. It can result in relatively minor annoyances like stripped fasteners, or cause catastrophic consequences like blown engines. It can also result in serious injury.

An assortment of good tools is a given for anyone who plans to work on cars. If you don't already have most of the tools listed below, the initial investment may seem high, but compared to the spiraling costs of routine maintenance and repairs, it's a deal. Besides, you can use a lot of the tools around the house for other types of mechanical repairs. We've included a list of the tools you'll need and a detailed description of what to look for when shopping for tools and how to use them correctly.

Buying tools

There are two ways to buy tools. The easiest and quickest way is to simply buy an entire set. Tool sets are often priced substantially below the cost of the same individually priced tools - and sometimes they even come with a tool box. When purchasing such sets, you often wind up with some tools you don't need or want. But if low price and convenience are your concerns, this might be the way to go. Keep in mind that you're going to keep a quality set of tools a long time (maybe the rest of your life), so check the tools carefully; don't skimp too much on price, either. Buying tools individually is usually a more expensive and time-consuming way to go, but you're more likely to wind up with the tools you need and want. You can also select each tool on its relative merits for the way you use it.

You can get most of the hand tools on our list from the tool department of any large department store or hardware store chain that sells hand tools. Blackhawk, Craftsman, KD, Proto and SK are fairly inexpensive, good-quality choices. Specialty tools are available from mechanics' tool companies such as Snap-on, Mac, Matco, Cornwell, Kent-Moore, Lisle, OTC, etc. These companies also supply the other tools you need, but they'll probably be more expensive.

Also consider buying second-hand tools from garage sales or used tool outlets. You may have limited choice in sizes, but you can usually determine from the condition of the tools if they're worth buying. It's a cheap way of putting a basic tool kit together.

Until you're a good judge of the quality levels of tools, avoid mail order firms (excepting Sears and other name-brand suppliers), flea markets and swap meets. Some of them offer good value for the money, but many sell cheap, imported tools of dubious quality. Like other consumer products counterfeited in the Far East, these tools run the gamut from acceptable to unusable.

If you're unsure about how much use a tool will get, the following approach may help. For example, if you need a set of combination wrenches but aren't sure which sizes you'll end up using most, buy a cheap or medium-priced set (make sure the jaws fit the fastener sizes marked on them). After some use over a period of time, carefully examine each tool in the set to assess its condition. If all the tools fit well and are undamaged, don't bother buying a better set. If one or two are worn, replace them with high-quality items - this way you'll end up with top-quality tools where they're needed most and the cheaper ones are sufficient for occasional use. On rare occasions you may conclude the whole set is poor quality. If so, buy a better set, if necessary, and remember never to buy that brand again.

In summary, try to avoid cheap tools, especially when you're purchasing high-use items like screwdrivers, wrenches and sockets. Cheap tools don't last long. Their initial cost plus the additional expense of replacing them will exceed the initial cost of better-quality tools.

Hand tools

A list of general-purpose hand tools you need for general carburetor work

Allen wrench set (1/8 to 3/8-inch)
Ball peen hammer - 12 oz (any steel hammer will do)
Box-end wrenches
Brushes (various sizes, for cleaning small passages)
Combination (slip-joint) pliers - 6-inch
Center punch
Cold chisels - 1/4 and 1/2-inch
Combination wrench set (1/4 to 1-inch)
Extensions - 1- and 6-inch
E-Z out (screw extractor) set
Feeler gauge set
Files (assorted)
Gasket scraper
Hacksaw and assortment of blades
Locking pliers
Magnet
Phillips screwdriver (no. 2 x 6-inch)
Phillips screwdriver (no. 3 x 8-inch)
Phillips screwdriver (stubby - no. 2)
Pin punches (1/16, 1/8, 3/16-inch)
Pliers - needle-nose
Pliers - locking (Vise-grip)
Pliers - diagonal cutters
Ratchet
Razor blades (industrial)
Scribe
Socket set
Standard screwdriver (1/4-inch x 6-inch)
Standard screwdriver (5/16-inch x 6-inch)
Standard screwdriver (5/16-inch - stubby)
Steel ruler - 6-inch
Tap and die set
Thread gauge
Torx drivers (only necessary on some later models)
Universal joint

What to look for when buying hand tools and general purpose tools

Wrenches and sockets

Wrenches vary widely in quality. One indication of their quality is their cost: The more they cost, the better they are. Buy the best wrenches you can afford. You'll use them a lot.

Wrenches are similar in appearance, so their quality level can be difficult to judge just by looking at them. There are bargains to be had, just as there are overpriced tools with well-known brand names. On the other hand, you may buy what looks like a reasonable value set of wrenches only to find they fit badly or are made from poor-quality steel.

With a little experience, it's possible to judge the quality of a tool by looking at it. Often, you may have come across the brand name before and have a good idea of the quality. Close examination of the tool can often reveal some hints as to its quality. Prestige tools are usually polished and chrome-plated over their entire surface, with the working faces ground to size. The polished finish is largely cosmetic, but it does make them easy to keep clean. Ground jaws normally indicate the tool will fit well on fasteners.

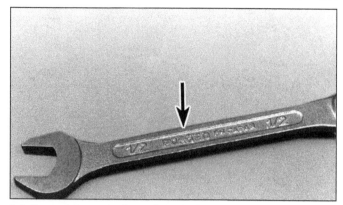

2.1 One quick way to determine whether you're looking at a quality wrench is to read the information printed on the handle - if it says "chrome vanadium" or "forged," it's made out of the right material

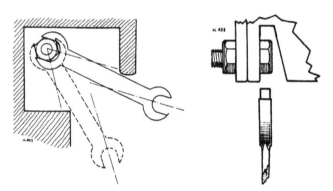

2.2 Open-end wrenches can do several things other wrenches can't - for example, they can be used on bolt heads with limited clearance (above) and they can be used in tight spots where there's little room to turn a wrench by flipping the offset jaw over every few degrees of rotation

A side-by-side comparison of a high-quality wrench with a cheap equivalent is an eye opener. The better tool will be made from a good-quality material, often a forged/chrome-vanadium steel alloy **(see illustration)**. This, together with careful design, allows the tool to be kept as small and compact as possible. If, by comparison, the cheap tool is thicker and heavier, especially around the jaws, it's usually because the extra material is needed to compensate for its lower quality. If the tool fits properly, this isn't necessarily bad - it is, after all, cheaper - but in situations where it's necessary to work in a confined area, the cheaper tool may be too bulky to fit.

Open-end wrenches

Because of its versatility, the open-end wrench is the most common type of wrench. It has a jaw on either end, connected by a flat handle section. The jaws either vary by a size, or overlap sizes between consecutive wrenches in a set. This allows one wrench to be used to hold a bolt head while a similar-size nut is removed. A typical fractional size wrench set might have the following jaw sizes: 1/4 x 5/16, 3/8 x 7/16, 1/2 x 9/16, 5/8 x 11/16 and so on.

Typically, the jaw end is set at an angle to the handle, a feature which makes them very useful in confined spaces; by turning the nut or bolt as far as the obstruction allows, then turning the wrench over so the jaw faces in the other direction, it's possible to move the fastener a fraction of a turn at a time

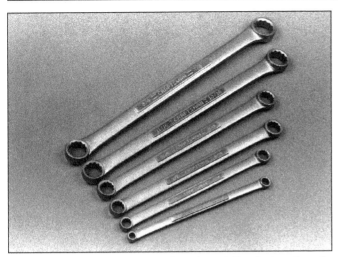

2.3 Box-end wrenches have a ring-shaped "box" at each end - when space permits, they offer the best combination of "grip" and strength

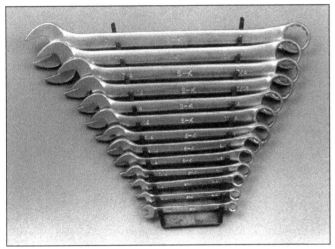

2.5 Buy a set of combination wrenches from 1/4 to 1-inch

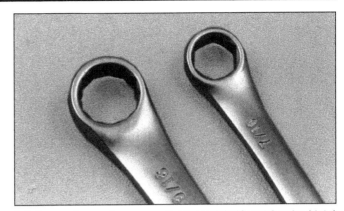

2.4 Box-end wrenches are available in 12 (left) and 6-point (right) openings; even though the 12-point design offers twice as many wrench positions, buy the 6-point first - it's less likely to strip off the corners of a nut or bolt head

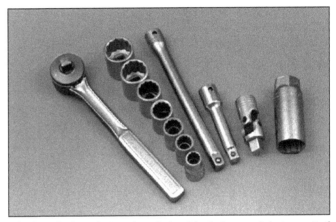

2.6 A typical ratchet and socket set includes a ratchet, a set of sockets, a long and a short extension, a universal joint and a spark plug socket

(see illustration). The handle length is generally determined by the size of the jaw and is calculated to allow a nut or bolt to be tightened sufficiently by hand with minimal risk of breakage or thread damage (though this doesn't apply to soft materials like brass or aluminum).

Common open-end wrenches are usually sold in sets and it's rarely worth buying them individually unless it's to replace a lost or broken tool from a set. Single tools invariably cost more, so check the sizes you're most likely to need regularly and buy the best set of wrenches you can afford in that range of sizes. If money is limited, remember that you'll use open-end wrenches more than any other type - it's a good idea to buy a good set and cut corners elsewhere.

Box-end wrenches

Box-end wrenches (see illustration) have ring-shaped ends with a 6-point (hex) or 12-point (double hex) opening (see illustration). This allows the tool to fit on the fastener hex at 15 (12-point) or 30-degree (6-point) intervals. Normally, each tool has two ends of different sizes, allowing an overlapping range of sizes in a set, as described for open-end wrenches.

Although available as flat tools, the handle is usually offset at each end to allow it to clear obstructions near the fastener, which is normally an advantage. In addition to normal length wrenches, it's also possible to buy long handle types to allow more leverage (very useful when trying to loosen rusted or seized nuts). It is, however, easy to shear off fasteners if not careful, and sometimes the extra length impairs access.

As with open-end wrenches, box-ends are available in varying quality, again often indicated by finish and the amount of metal around the ring ends. While the same criteria should be applied when selecting a set of box-end wrenches, if your budget is limited, go for better-quality open-end wrenches and a slightly cheaper set of box-ends.

Combination wrenches

These wrenches (see illustration) combine a box-end and open-end of the same size in one tool and offer many of the advantages of both. Like the others, they're widely available in sets and as such are probably a better choice than box-ends only. They're generally compact, short-handled tools and are well suited for tight spaces where access is limited.

Ratchet and socket sets

Ratcheting socket wrenches (see illustration) are highly versatile. Besides the sockets themselves, many other interchangeable accessories - extensions, U-drives, step-down

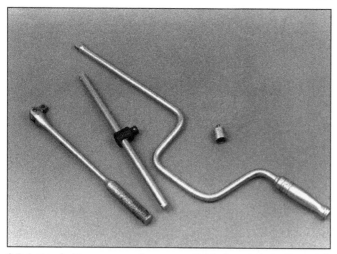

2.7 Lots of other accessories are available for ratchets: From left to right, a breaker bar, a sliding T-handle, a speed handle and a 3/8-to-1/4-inch adapter

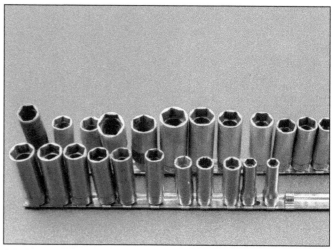

2.8 Deep sockets enable you to loosen or tighten a nut with a long bolt protruding from it

adapters, screwdriver bits, Allen bits, crow's feet, etc. - are available. Buy six-point sockets - they're less likely to slip and strip the corners off bolts and nuts. Don't buy sockets with extra-thick walls - they might be stronger but they can be hard to use on recessed fasteners or fasteners in tight quarters.

A 3/8-inch drive set is the most versatile, but for carburetor work you may want to consider a 1/4-inch drive set.

Interchangeable sockets consist of a forged-steel alloy cylinder with a hex or double-hex formed inside one end. The other end is formed into the square drive recess that engages over the corresponding square end of various socket drive tools.

The most economical way to buy sockets is in a set. As always, quality will govern the cost of the tools. Once again, the "buy the best" approach is usually advised when selecting sockets. While this is a good idea, since the end result is a set of quality tools that should last a lifetime, the cost is so high it's difficult to justify the expense for home use.

As far as accessories go, you'll need a ratchet, at least one extension (buy a three or six-inch size), a spark plug socket and maybe a T-handle or breaker bar. Other desirable, though less essential items, are a speed handle, a U-joint, extensions of various other lengths and adapters from one drive size to another **(see illustration)**. Some of the sets you find may combine drive sizes; they're well worth having if you find the right set at a good price, but avoid being dazzled by the number of pieces.

Above all, be sure to completely ignore any label that reads "86-piece Socket Set," which refers to the number of pieces, not to the number of sockets (sometimes even the metal box and plastic insert are counted in the total!).

Apart from well-known and respected brand names, you'll have to take a chance on the quality of the set you buy. If you know someone who has a set that has held up well, try to find the same brand, if possible. Take a pocketful of nuts and bolts with you and check the fit in some of the sockets. Check the operation of the ratchet. Good ones operate smoothly and crisply in small steps; cheap ones are coarse and stiff - a good basis for guessing the quality of the rest of the pieces.

One of the best things about a socket set is the built-in facility for expansion. Once you have a basic set, you can purchase extra sockets when necessary and replace worn or

2.9 Standard and Phillips bits, Allen-head and Torx drivers will expand the versatility of your ratchet and extensions even further

damaged tools. There are special deep sockets for reaching recessed fasteners or to allow the socket to fit over a projecting bolt or stud **(see illustration)**. You can also buy screwdriver, Allen and Torx bits to fit various drive tools (they can be very handy in some applications) **(see illustration)**.

Using wrenches and sockets

Although you may think the proper use of tools is self-evident, it's worth some thought. After all, when did you last see instructions for use supplied with a set of wrenches?

Which wrench?

Before you start tearing things apart, figure out the best tool for the job; in this instance the best wrench for a hex-head fastener. Sit down with a few nuts and bolts and look at how various tools fit the bolt heads.

A golden rule is to choose a tool that contacts the largest area of the hex-head. This distributes the load as evenly as possible and lessens the risk of damage. The shape most closely resembling the bolt head or nut is another hex, so a 6-point socket or box-end wrench is usually the best choice **(see illustration)**. Many sockets and box-end wrenches have double hex (12-point) openings. If you slip a 12-point box-end wrench over a nut, look at how and where the two are in contact. The corners of the nut engage in every other point of the

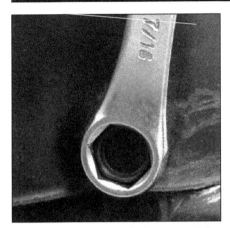

2.10 Try to use a six-point box wrench (or socket) whenever possible - its shape matches that of the fastener, which means maximum grip and minimum slip

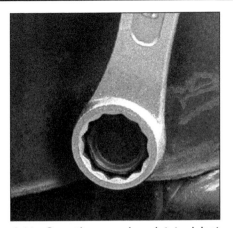

2.11 Sometimes a six-point tool just doesn't offer you any grip when you get the wrench at the angle it needs to be in to loosen or tighten a fastener - when this happens, pull out the 12-point sockets or wrenches - but remember: they're much more likely to strip the corners off a fastener

2.12 Open-end wrenches contact only two sides of the fastener and the jaws tend to open up when you put some muscle on the wrench handle - that's why they should only be used as a last resort

wrench. When the wrench is turned, pressure is applied evenly on each of the six corners **(see illustration)**. This is fine unless the fastener head was previously rounded off. If so, the corners will be damaged and the wrench will slip. If you encounter a damaged bolt head or nut, always use a 6-point wrench or socket if possible. If you don't have one of the right size, choose a wrench that fits securely and proceed with care.

If you slip an open-end wrench over a hex-head fastener, you'll see the tool is in contact on two faces only **(see illustration)**. This is acceptable provided the tool and fastener are both in good condition. The need for a snug fit between the wrench and nut or bolt explains the recommendation to buy good-quality open-end wrenches. If the wrench jaws, the bolt head or both are damaged, the wrench will probably slip, rounding off and distorting the head. In some applications, an open-end wrench is the only possible choice due to limited access, but always check the fit of the wrench on the fastener before attempting to loosen it; if it's hard to get at with a wrench, think how hard it will be to remove after the head is damaged.

The last choice is an adjustable wrench or self-locking pliers/wrench (Vise-Grips). Use these tools only when all else has failed. In some cases, a self-locking wrench may be able to grip a damaged head that no wrench could deal with, but be careful not to make matters worse by damaging it further.

Bearing in mind the remarks about the correct choice of tool in the first place, there are several things worth noting about the actual use of the tool. First, make sure the wrench head is clean and undamaged. If the fastener is rusted or coated with paint, the wrench won't fit correctly. Clean off the head and, if it's rusted, apply some penetrating oil. Leave it to soak in for a while before attempting removal.

It may seem obvious, but take a close look at the fastener to be removed before using a wrench. On many mass-produced machines, one end of a fastener may be fixed or captive, which speeds up initial assembly and usually makes removal easier. If a nut is installed on a stud or a bolt threads into a captive nut or tapped hole, you may have only one fastener to deal with. If, on the other hand, you have a separate nut and bolt, you must hold the bolt head while the nut is removed.

In most cases, a fastener can be removed simply by placing the wrench on the nut or bolt head and turning it. Occasionally, though, the condition or location of the fastener may make things more difficult. Make sure the wrench is square on the head. You may need to reposition the tool or try another type to obtain a snug fit. Make sure the component you're working on is secure and can't move when you turn the wrench. If necessary, get someone to help steady it for you. Position yourself so you can get maximum leverage on the wrench.

If possible, locate the wrench so you can pull the end towards you. If you have to push on the tool, remember that it may slip, or the fastener may move suddenly. For this reason, don't curl your fingers around the handle or you may crush or bruise them when the fastener moves; keep your hand flat, pushing on the wrench with the heel of your thumb. If the tool digs into your hand, place a rag between it and your hand or wear a heavy glove.

If the fastener doesn't move with normal hand pressure, stop and try to figure out why before the fastener or wrench is damaged or you hurt yourself. Stuck fasteners may require penetrating oil, heat or an impact driver or air tool.

Using sockets to remove hex-head fasteners is less likely to result in damage than if a wrench is used. Make sure the socket fits snugly over the fastener head, then attach an extension, if needed, and the ratchet or breaker bar. Theoretically, a ratchet shouldn't be used for loosening a fastener or for final tightening because the ratchet mechanism may be overloaded and could slip. In some instances, the location of the fastener may mean you have no choice but to use a ratchet, in which case you'll have to be extra careful.

Never use extensions where they aren't needed. Whether or not an extension is used, always support the drive end of the breaker bar with one hand while turning it with the other. Once the fastener is loose, the ratchet can be used to speed up removal.

Pliers

Some tool manufacturers make 25 or 30 different types of pliers. You only need a fraction of this selection **(see illustra-**

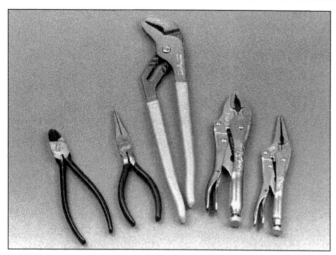

2.13 A typical assortment of the types of pliers you need to have in your box - from the left: diagonal cutters (dikes), needle-nose pliers, Channel-lock pliers, Vise-Grip pliers, needle-nose Vise-Grip pliers

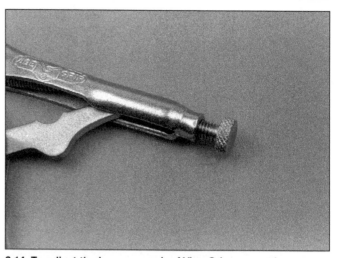

2.14 To adjust the jaws on a pair of Vise-Grips, grasp the part you want to hold with the jaws, tighten them down by turning the knurled knob on the end of one handle and snap the handles together - if you tightened the knob all the way down, you'll probably have to open it up (back it off) a little before you can close the handles

tion). Get a good pair of slip-joint pliers for general use. A pair of needle-nose pliers is handy for reaching into hard-to-get-at places. A set of diagonal wire cutters (dikes) is essential for electrical work and pulling out cotter pins. Vise-Grips are adjustable, locking pliers that grip a fastener firmly - and won't let go - when locked into place. Parallel-jaw, adjustable pliers have angled jaws that remain parallel at any degree of opening. They're also referred to as Channel-lock (the original manufacturer) pliers, arc-joint pliers and water pump pliers. Whatever you call them, they're terrific for gripping a big fastener with a lot of force.

Slip-joint pliers have two open positions; a figure eight-shaped, elongated slot in one handle slips back-and-forth on a pivot pin on the other handle to change them. Good-quality pliers have jaws made of tempered steel and there's usually a wire-cutter at the base of the jaws. The primary uses of slip-joint pliers are for holding objects, bending and cutting throttle wires and crimping and bending metal parts, not loosening nuts and bolts.

Arc-joint or "Channel-lock" pliers have parallel jaws you can open to various widths by engaging different tongues and grooves, or channels, near the pivot pin. Since the tool expands to fit many size objects, it has countless uses for engine and equipment maintenance. Channel-lock pliers come in various sizes. The medium size is adequate for general work; small and large sizes are nice to have as your budget permits. You'll use all three sizes frequently.

Vise-Grips (a brand name) come in various sizes; the medium size with curved jaws is best for all-around work. However, buy a large and small one if possible, since they're often used in pairs. Although this tool falls somewhere between an adjustable wrench, a pair of pliers and a portable vise, it can be invaluable for loosening and tightening damaged fasteners .

The jaw opening is set by turning a knurled knob at the end of one handle. The jaws are placed over the head of the fastener and the handles are squeezed together, locking the tool onto the fastener **(see illustration)**. The design of the tool allows extreme pressure to be applied at the jaws and a variety of jaw designs enable the tool to grip firmly even on damaged heads. Vise-Grips are great for removing fasteners that have been rounded off by badly-fitting wrenches.

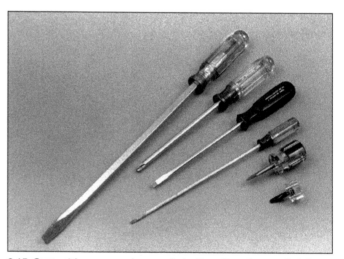

2.15 Screwdrivers come in a myriad of lengths, sizes and styles

As the name suggests, needle-nose pliers have long, thin jaws designed for reaching into holes and other restricted areas. Most needle-nose, or long-nose, pliers also have wire cutters at the base of the jaws.

Look for these qualities when buying pliers: Smooth operating handles and jaws, jaws that match up and grip evenly when the handles are closed, a nice finish and the word "forged" somewhere on the tool.

Screwdrivers

Screwdrivers **(see illustration)** come in a wide variety of sizes and price ranges, but stay away from cheap screwdriver sets at discount tool stores. Even if they look exactly like more expensive brands, the metal tips and shafts are made with inferior alloys and aren't properly heat treated. They usually bend the first time you apply some serious torque.

A screwdriver consists of a steel blade or shank with a drive tip formed at one end. The most common tips are standard (also called straight slot and flat-blade) and Phillips. The other end has a handle attached to it. Traditionally, handles

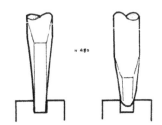

Misuse of a screwdriver – the blade shown is both too narrow and too thin and will probably slip or break off

The left-hand example shows a snug-fitting tip. The right-hand drawing shows a damaged tip which will twist out of the slot when pressure is applied

2.16 Standard screwdrivers - wrong size (left), correct fit in screw slot (center) and worn tip (right)

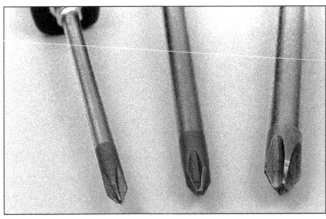

2.17 The tip size on a Phillips screwdriver is indicated by a number from 1 to 4, with 1 the smallest (left - No. 1: center - No. 2; right - No. 3)

were made from wood and secured to the shank, which had raised tangs to prevent it from turning in the handle. Most screwdrivers now come with plastic handles, which are generally more durable than wood.

The design and size of handles and blades vary considerably. Some handles are specially shaped to fit the human hand and provide a better grip. The shank may be either round or square and some have a hex-shaped bolster under the handle to accept a wrench to provide more leverage when trying to turn a stubborn screw. The shank diameter, tip size and overall length vary too.

If access is restricted, a number of special screwdrivers are designed to fit into confined spaces. The "stubby" screwdriver has a specially shortened handle and blade. There are also offset screwdrivers and special screwdriver bits that attach to a ratchet or extension.

The important thing to remember when buying screwdrivers is that they really do come in sizes designed to fit different size fasteners. The slot in any screw has definite dimensions - length, width and depth. Like a bolt head or a nut, the screw slot must be driven by a tool that uses all of the available bearing surface and doesn't slip. Don't use a big wide blade on a small screw and don't try to turn a large screw slot with a tiny, narrow blade. The same principles apply to Allen heads, Phillips heads, Torx heads, etc. Don't even think of using a slotted screwdriver on one of these heads! And don't use your screwdrivers as prybars, chisels or punches! This kind of abuse turns them into very bad screwdrivers.

Standard screwdrivers

These are used to remove and install conventional slotted screws and are available in a wide range of sizes denoting the width of the tip and the length of the shank (for example: a 3/8 x 10-inch screwdriver is 3/8-inch wide at the tip and the shank is 10-inches long). You should have a variety of screwdrivers so screws of various sizes can be dealt with without damaging them. The blade end must be the same width and thickness as the screw slot to work properly, without slipping. When selecting standard screwdrivers, choose good-quality tools, preferably with chrome moly, forged steel shanks. The tip of the shank should be ground to a parallel, flat profile (hollow ground) and not to a taper or wedge shape, which will tend to twist out of the slot when pressure is applied **(see illustration)**.

All screwdrivers wear in use, but standard types can be reground to shape a number of times. When reshaping a tip,

start by filing the very end flat at right angles to the shank. Make sure the tip fits snugly in the slot of a screw of the appropriate size and keep the sides of the tip parallel. Remove only a small amount of metal at a time to avoid overheating the tip and destroying the temper of the steel.

Phillips screwdrivers

Phillips screws are sometimes installed during initial assembly with air tools and are next to impossible to remove later without ruining the heads, particularly if the wrong size screwdriver is used. And don't use other types of cross-head screwdrivers (such as Posi-drive, etc.) on Phillips screws - they won't work.

The only way to ensure the screwdrivers you buy will fit properly is to take a couple of screws with you to make sure the fit between the screwdriver and fastener is snug. If the fit is good, you should be able to angle the blade down almost vertically without the screw falling off the tip. Use only screwdrivers that fit exactly - anything else is guaranteed to chew out the screw head instantly.

The idea behind all cross-head screw designs is to make the screw and screwdriver blade self-aligning. Provided you aim the blade at the center of the screw head, it'll engage correctly, unlike conventional slotted screws, which need careful alignment. This makes the screws suitable for machine installation on an assembly line (which explains why they're sometimes so tight and difficult to remove). The drawback with these screws is the driving tangs on the screwdriver tip are very small and must fit very precisely in the screw head. If this isn't the case, the huge loads imposed on small flats of the screw slot simply tear the metal away, at which point the screw ceases to be removable by normal methods. The problem is made worse by the normally soft material chosen for screws.

To deal with these screws on a regular basis, you'll need high-quality screwdrivers with various size tips so you'll be sure to have the right one when you need it. Phillips screwdrivers are sized by the tip number and length of the shank (for example: a number 2 x 6-inch Phillips screwdriver has a number 2 tip - to fit screws of only that size recess - and the shank is 6-inches long). Tip sizes 1, 2 and 3 should be adequate for engine repair work **(see illustration)**. If the tips get worn or damaged, buy new screwdrivers so the tools don't destroy the screws they're used on **(see illustration)**.

Here's a helpful hint that may come in handy when using Phillips screwdrivers - if the screw is extremely tight and the tip

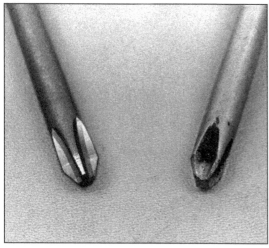 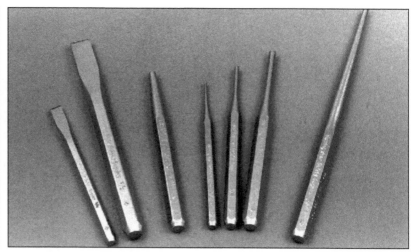

2.18 New (left) and worn (right) Phillips screwdriver tips

2.19 Cold chisels, center-punches, pin punches and line-up punches (left-to-right) will be needed sooner or later for many jobs

tends to back out of the recess rather than turn the screw, apply a small amount of valve lapping compound to the screwdriver tip so it will grip the screw better.

Punches and chisels

Punches and chisels **(see illustration)** are used along with a hammer for various purposes in the shop. Punches are available in various shapes and sizes and a set of assorted types will be very useful. One of the most basic is the center punch, a small cylindrical punch with the end ground to a point. It'll be needed whenever a hole is drilled. The center of the hole is located first and the punch is used to make a small indentation. The indentation acts as a guide for the drill bit so the hole ends up in the right place. Without a punch mark the drill bit will wander and you'll find it impossible to drill with any real accuracy. You can also buy automatic center punches. They're spring loaded and are pressed against the surface to be marked, without the need to use a hammer.

Pin punches are intended for removing items like roll pins (hollow spring steel pins that fit tightly in their holes). Pin punches have other uses, however. You may occasionally have to remove rivets by cutting off the heads and driving out the shanks with a pin punch. They're also very handy for aligning holes in components while bolts or screws are inserted.

Of the various sizes and types of metal-cutting chisels available, a simple cold chisel is essential in any mechanic's workshop. One about 6-inches long with a 1/2-inch wide blade should be adequate. The cutting edge is ground to about 80-degrees **(see illustration)**, while the rest of the tip is ground to a shallower angle away from the edge. The primary use of the cold chisel is rough metal cutting - this can be anything from removing concealment plugs to cutting off the heads of seized or rusted bolts or splitting nuts. A cold chisel is also useful for turning out screws or bolts with messed-up heads.

All of the tools described in this section should be good quality items. They're not particularly expensive, so it's not really worth trying to save money on them. More significantly, there's a risk that with cheap tools, fragments may break off in use - a potentially dangerous situation.

Even with good-quality tools, the heads and working ends will inevitably get worn or damaged, so it's a good idea to maintain all such tools on a regular basis. Using a file or bench grinder, remove all burrs and mushroomed edges from around

the head. This is an important task because the build-up of material around the head can fly off when it's struck with a hammer and is potentially dangerous. Make sure the tool retains its original profile at the working end, again, filing or grinding off all burrs. In the case of cold chisels, the cutting edge will usually have to be reground quite often because the material in the tool isn't usually much harder than materials typically being cut. Make sure the edge is reasonably sharp, but don't make the tip angle greater than it was originally; it'll just wear down faster if you do.

The techniques for using these tools vary according to the job to be done and are best learned by experience. The one common denominator is the fact they're all normally struck with a hammer. It follows that eye protection should be worn. Always make sure the working end of the tool is in contact with the part being punched or cut. If it isn't, the tool will bounce off the surface and damage may result.

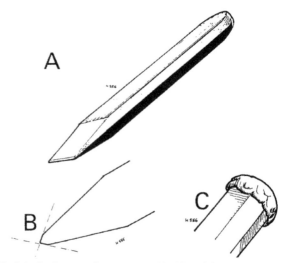

2.20 A typical general purpose cold chisel (A) - note the angle of the cutting edge (B), which should be checked and sharpened on a regular basis; the mushroomed head (C) is dangerous and should be filed to restore it to its original shape

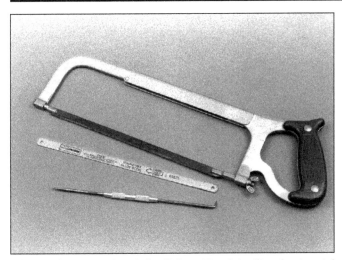

2.21 Hacksaws are handy for little cutting jobs like sheet metal and rusted fasteners

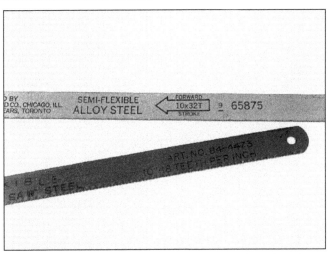

2.22 Hacksaw blades are marked with the number of teeth per inch (TPI) - use a relatively course blade for aluminum and thicker items such as bolts or bar stock; use a finer blade for materials like thin sheet steel

When cutting thin materials, check that at least three teeth are in contact with the workpiece at any time. Too coarse a blade will result in a poor cut and may break the blade. If you do not have the correct blade, cut at a shallow angle to the material

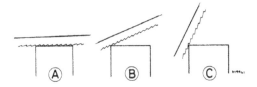

The correct cutting angle is important. If it is too shallow (A) the blade will wander. The angle shown at (B) is correct when starting the cut, and may be reduced slightly once under way. In (C) the angle is too steep and the blade will be inclined to jump out of the cut

2.33 Correct procedure for use of a hacksaw

2.24 Good quality hacksaw blades are marked like this

Hacksaws

A hacksaw **(see illustration)** consists of a handle and frame supporting a flexible steel blade under tension. Blades are available in various lengths and most hacksaws can be adjusted to accommodate the different sizes. The most common blade length is 10-inches.

Most hacksaw frames are adequate. There's little difference between brands. Pick one that's rigid and allows easy blade changing and repositioning.

The type of blade to use, indicated by the number of teeth per inch (TPI) **(see illustration)**, is determined by the material being cut. The rule of thumb is to make sure at least three teeth are in contact with the metal being cut at any one time **(see illustration)**. In practice, this means a fine blade for cutting thin sheet materials, while a coarser blade can be used for faster cutting through thicker items such as bolts or bar stock. When cutting thin materials, angle the saw so the blade cuts at a shallow angle. More teeth are in contact and there's less chance of the blade binding and breaking, or teeth breaking.

When you buy blades, choose a reputable brand. Cheap,

unbranded blades may be perfectly acceptable, but you can't tell by looking at them. Poor quality blades will be insufficiently hardened on the teeth edge and will dull quickly. Most reputable brands will be marked "Flexible High Speed Steel" or a similar term, to indicate the type of material used **(see illustration)**. It is possible to buy "unbreakable" blades (only the teeth are hardened, leaving the rest of the blade less brittle).

Sometimes, a full-size hacksaw is too big to allow access to a frozen nut or bolt. On most saws, you can overcome this problem by turning the blade 90-degrees. Occasionally you may have to position the saw around an obstacle and then install the blade on the other side of it. Where space is really restricted, you may have to use a handle that clamps onto a saw blade at one end. This allows access when a hacksaw frame would not work at all and has another advantage in that you can make use of broken off hacksaw blades instead of throwing them away. Note that because only one end of the blade is supported, and it's not held under tension, it's difficult to control and less efficient when cutting.

Before using a hacksaw, make sure the blade is suitable for the material being cut and installed correctly in the frame

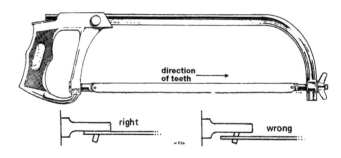

2.25 Correct installation of a hacksaw blade - the teeth must point away from the handle and butt against the locating lugs

(see illustration). Whatever it is you're cutting must be securely supported so it can't move around. The saw cuts on the forward stroke, so the teeth must point away from the handle. This might seem obvious, but it's easy to install the blade backwards by mistake and ruin the teeth on the first few strokes. Make sure the blade is tensioned adequately or it'll distort and chatter in the cut and may break. Wear safety glasses and be careful not to cut yourself on the saw blade or the sharp edge of the cut.

Files

Files **(see illustration)** come in a wide variety of sizes and types for specific jobs, but all of them are used for the same basic function of removing small amounts of metal in a controlled fashion. Files are used by mechanics mainly for deburring, marking parts, removing rust, filing the heads off rivets, restoring threads and fabricating small parts.

File shapes commonly available include flat, half-round, round, square and triangular. Each shape comes in a range of sizes (lengths) and cuts ranging from rough to smooth. The file face is covered with rows of diagonal ridges which form the cutting teeth. They may be aligned in one direction only (single cut) or in two directions to form a diamond-shaped pattern (double-cut) **(see illustration)**. The spacing of the teeth determines the file coarseness, again, ranging from rough to smooth in five basic grades: Rough, coarse, bastard, second-cut and smooth.

You'll want to build up a set of files by purchasing tools of the required shape and cut as they're needed. A good starting point would be flat, half-round, round and triangular files (at least one each - bastard or second-cut types). In addition, you'll have to buy one or more file handles (files are usually sold without handles, which are purchased separately and pushed over the tapered tang of the file when in use) **(see illustration)**. You may need to buy more than one size handle to fit the various files in your tool box, but don't attempt to get by without them. A file tang is fairly sharp and you almost certainly will end up stabbing yourself in the palm of the hand if you use a file without a handle and it catches in the workpiece during use. Adjustable handles are also available for use with files of various sizes, eliminating the need for several handles **(see illustration)**.

Exceptions to the need for a handle are fine swiss pattern files, which have a rounded handle instead of a tang. These small files are usually sold in sets with a number of different shapes.

The correct procedure for using files is fairly easy to master. As with a hacksaw, the work should be clamped securely in a vise, if needed, to prevent it from moving around while being worked on. Hold the file by the handle, using your free hand at the file end to guide it and keep it flat in relation to the surface being filed. Use smooth cutting strokes and be careful not to rock the file as it passes over the surface. Also, don't slide it diagonally across the surface or the teeth will make grooves in the workpiece. Don't drag a file back across the workpiece at the end of the stroke - lift it slightly and pull it back to prevent damage to the teeth.

Files don't require maintenance in the usual sense, but they should be kept clean and free of metal filings. Steel is a

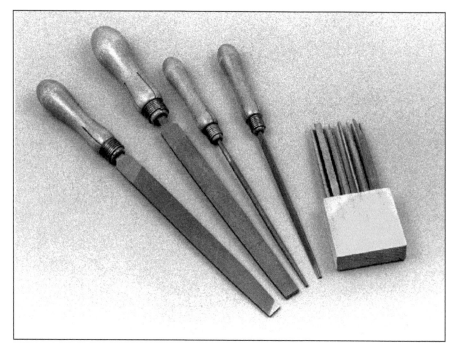

2.26 Get a good assortment of files - they're handy for deburring, marking parts, removing rust, filing the heads off rivets, restoring threads and fabricating small parts

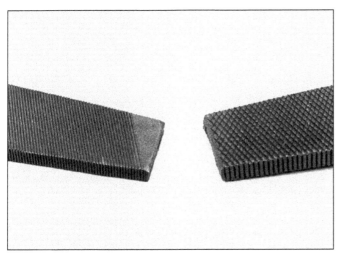

2.27 Files are either single-cut (left) or double-cut (right) - generally speaking, use a single-cut file to produce a very smooth surface; use a double-cut file to remove large amounts of material quickly

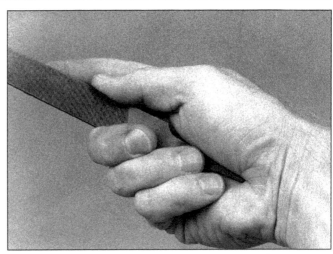

2.28 Never use a file without a handle - the tang is sharp and could puncture your hand

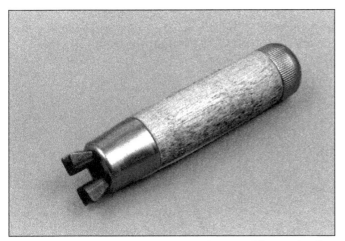

2.29 Adjustable handles that will work with may different size files are also available

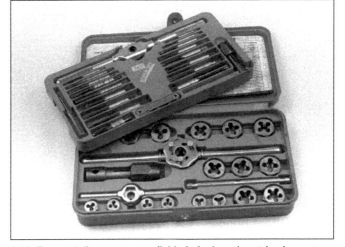

2.30 Tap and die sets are available in inch and metric sizes - taps are used for cutting internal threads and cleaning and restoring damaged threads; dies are used for cutting, cleaning and restoring external threads

reasonably easy material to work with, but softer metals like aluminum tend to clog the file teeth very quickly, which will result in scratches in the workpiece. This can be avoided by rubbing the file face with chalk before using it. General cleaning is carried out with a file card. If kept clean, files will last a long time - when they do eventually dull, they must be replaced; there is no satisfactory way of sharpening a worn file.

Taps and dies

Taps

Tap and die sets **(see illustration)** are available in inch and metric sizes. Taps are used to cut internal threads and clean or restore damaged threads. A tap consists of a fluted shank with a drive square at one end. It's threaded along part of its length - the cutting edges are formed where the flutes intersect the threads **(see illustration)**. Taps are made from hardened steel so they will cut threads in materials softer than what they're made of.

Taps come in three different types: Taper, plug and bottoming. The only real difference is the length of the chamfer on the cutting end of the tap. Taper taps are chamfered for the first 6 or 8 threads, which makes them easy to start but pre-

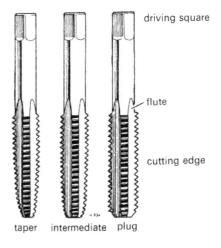

driving square

flute

cutting edge

taper   intermediate   plug

2.31 Taper, plug and bottoming taps (left-to-right)

2.32 If you need to drill and tap a hole, the drill bit size to use for a given bolt (tap) size is marked on the tap

2.34 Hex-shaped dies are especially handy for mechanic's work because they can be turned with a wrench

2.35 A bench vise is one of the most useful pieces of equipment you can have in the shop - bigger is usually better with vises, so get a vise with jaws that open at least four inches

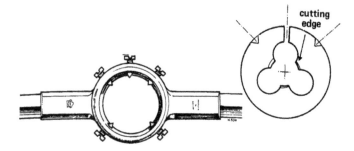

2.33 A die (right) is used for cutting external threads (this one is a split-type/adjustable die) and is held in a tool called a die stock (left)

good when used on threaded holes in aluminum engines. The alternative is to buy high-quality taps if and when you need them, even though they aren't cheap, especially if you need to buy two or more thread pitches in a given size. Despite this, it's the best option - you'll probably only need taps on rare occasions, so a full set isn't absolutely necessary.

Taps are normally used by hand (they can be used in machine tools, but only in machine shop applications). The square drive end of the tap is held in a tap wrench (an adjustable T-handle). For smaller sizes, a T-handled chuck can be used. The tapping process starts by drilling a hole of the correct diameter. For each tap size, there's a corresponding twist drill that will produce a hole of the correct size. This is important; too large a hole will leave the finished thread with the tops missing, producing a weak and unreliable grip. Conversely, too small a hole will place excessive loads on the hard and brittle shank of the tap, which can break it off in the hole. Removing a broken off tap from a hole is no fun! The correct tap drill size is normally marked on the tap itself or the container it comes in **(see illustration)**.

Dies

Dies are used to cut, clean or restore external threads. Most dies are made from a hex-shaped or cylindrical piece of hardened steel with a threaded hole in the center. The threaded hole is overlapped by three or four cutouts, which equate to the flutes on taps and allow metal waste to escape during the threading process. Dies are held in a T-handled holder (called a die stock) **(see illustration)**. Some dies are split at one point, allowing them to be adjusted slightly (opened and closed) for fine control of thread clearances.

Dies aren't needed as often as taps, for the simple reason it's normally easier to install a new bolt than to salvage one. However, it's often helpful to be able to extend the threads of a bolt or clean up damaged threads with a die. Hex-shaped dies are particularly useful for mechanic's work, since they can be turned with a wrench **(see illustration)** and are usually less expensive than adjustable ones.

The procedure for cutting threads with a die is broadly similar to that described above for taps. When using an adjustable die, the initial cut is made with the die fully opened, the adjustment screw being used to reduce the diameter of successive cuts until the finished size is reached. As with taps, a cutting lubricant should be used, and the die must be backed off every few turns to clear swarf from the cutouts.

Bench vise

The bench vise **(see illustration)** is an essential tool in a shop. Buy the best quality vise you can afford. A good vise is

vents them from cutting threads close to the bottom of a hole. Plug taps are chamfered up about 3 to 5 threads, which makes them a good all around tap because they're relatively easy to start and will cut nearly to the bottom of a hole. Bottoming taps, as the name implies, have a very short chamfer (1-1/2 to 3 threads) and will cut as close to the bottom of a blind hole as practical. However, to do this, the threads should be started with a plug or taper tap.

Although cheap tap and die sets are available, the quality is usually very low and they can actually do more harm than

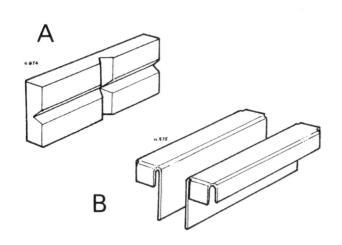

2.36 Sometimes, the parts you have to jig up in the vise are delicate, or made of soft materials - to avoid damaging them, get a pair of fiberglass or plastic "soft jaws" (A) or fabricate your own with 1/8-inch thick aluminum sheet (B)

2.37 Although it's not absolutely necessary, an air compressor can make many jobs easier and produce better results, especially when air powered tools are available to use with it

2.38 Electric drills can be cordless (above) or 115-volt, AC-powered (below)

2.39 Get a set of good quality drill bits for drilling holes and wire brushes of various sizes for cleaning up metal parts - make sure the bits are designed for drilling in metal

expensive, but the quality of its materials and workmanship are worth the extra money. Size is also important - bigger vises are usually more versatile. Make sure the jaws open at least four inches. Get a set of soft jaws to fit the vise as well - you'll need them to grip parts that could be damaged by the hardened vise jaws **(see illustration)**.

Power tools

Really, the only power tool you may need is an electric

drill. But if you have an air compressor and electricity, there's a wide range of pneumatic and electric hand tools to make all sorts of jobs easier and faster.

Air compressor

An air compressor **(see illustration)** makes most jobs easier and faster. Drying off parts after cleaning them, blowing out passages, running power tools - the list is endless. Once you buy a compressor, you'll wonder how you ever got along without it.

Electric drills

A drill motor with a 3/8-inch chuck (drill bit holder) will handle most jobs **(see illustration)**. Collect several different wire brushes to use in the drill and make sure you have a complete

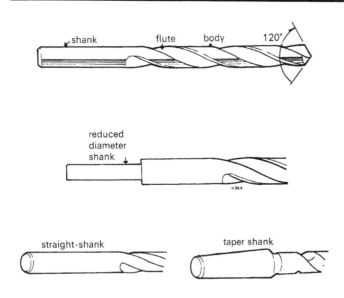

2.40 A typical drill bit (top), a reduced shank bit (center), and a tapered shank bit (bottom right)

2.41 Drill bits in the range most commonly used are available in fractional sizes (left) and number sizes (right) so almost any size hole can be drilled

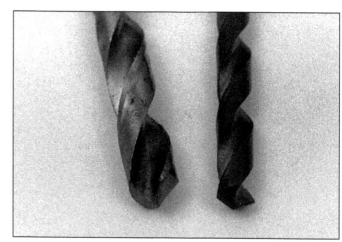

2.42 If a bit gets dull (left), discard it or re sharpen it so it looks like the bit on the right

set of sharp drill bits **(see illustration)**. Cordless drills are extremely versatile because they don't force you to work near an outlet. They're also handy to have around for a variety of non-mechanical jobs.

Twist drills

Drilling operations are done with twist drills, either in a hand drill or a drill press. Twist drills (or drill bits, as they're often called) consist of a round shank with spiral flutes formed into the upper two-thirds to clear the waste produced while drilling, keep the drill centered in the hole and finish the sides of the hole.

The lower portion of the shank is left plain and used to hold the drill in the chuck. In this section, we will discuss only normal parallel shank drills **(see illustration)**. There is another type of bit with the plain end formed into a special size taper designed to fit directly into a corresponding socket in a heavy-duty drill press. These drills are known as Morse Taper drills and are used primarily in machine shops.

At the cutting end of the drill, two edges are ground to form a conical point. They're generally angled at about 60-degrees from the drill axis, but they can be reground to other angles for specific applications. For general use the standard angle is correct - this is how the drills are supplied.

When buying drills, purchase a good-quality set (sizes 1/16 to 3/8-inch). Make sure the drills are marked "High Speed Steel" or "HSS". This indicates they're hard enough to withstand continual use in metal; many cheaper, unmarked drills are suitable only for use in wood or other soft materials. Buying a set ensures the right size bit will be available when it's needed.

Twist drill sizes

Twist drills are available in a vast array of sizes, most of which you'll never need. There are three basic drill sizing systems: Fractional, number and letter **(see illustration)** (we won't get involved with the fourth system, which is metric sizes).

Fractional sizes start at 1/64-inch and increase in increments of 1/64-inch. Number drills range in descending order from 80 (0.0135-inch), the smallest, to 1 (0.2280-inch), the largest. Letter sizes start with A (0.234-inch), the smallest, and go through Z (0.413-inch), the largest.

This bewildering range of sizes means it's possible to drill an accurate hole of almost any size within reason. In practice, you'll be limited by the size of chuck on your drill (normally 3/8 or 1/2-inch). In addition, very few stores stock the entire range of possible sizes, so you'll have to shop around for the nearest available size to the one you require.

Sharpening twist drills

Like any tool with a cutting edge, twist drills will eventually get dull **(see illustration)**. How often they'll need sharpening depends to some extent on whether they're used correctly. A dull twist drill will soon make itself known. A good indication of the condition of the cutting edges is to watch the waste emerging from the hole being drilled. If the tip is in good condition, two even spirals of waste metal will be produced; if this fails to happen or the tip gets hot, it's safe to assume that sharpening is required.

With smaller size drills - under about 1/8-inch - it's easier and more economical to throw the worn drill away and buy another one. With larger (more expensive) sizes, sharpening is a better bet. When sharpening twist drills, the included angle of

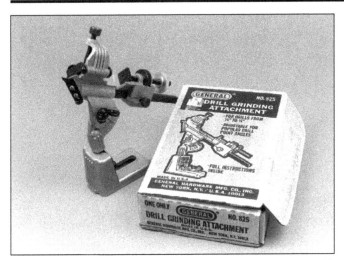

2.43 Inexpensive drill bit sharpening jigs designed to be used with a bench grinder are widely available - even if you only use it to re sharpen drill bits, it'll pay for itself quickly

2.44 Before you drill a hole, use a center punch to make an indentation for the drill bit so it won't wander

the cutting edge must be maintained at the original 120-degrees and the small chisel edge at the tip must be retained. With some practice, sharpening can be done freehand on a bench grinder, but it should be noted that it's very easy to make mistakes. For most home mechanics, a sharpening jig that mounts next to the grinding wheel should be used so the drill is clamped at the correct angle **(see illustration)**.

Drilling equipment

Tools to hold and turn drill bits range from simple, inexpensive hand-operated or electric drills to sophisticated and expensive drill presses. Ideally, all drilling should be done on a drill press with the workpiece clamped solidly in a vise. These machines are expensive and take up a lot of bench or floor space, so they're out of the question for many do-it-yourselfers. An additional problem is the fact that many of the drilling jobs you end up doing will be on the engine itself or the equipment it's mounted on, in which case the tool has to be taken to the work.

The best tool for the home shop is an electric drill with a 3/8-inch chuck. Both cordless and AC drills (that run off household current) are available. If you're purchasing one for the first time, look for a well-known, reputable brand name and variable speed as minimum requirements. A 1/4-inch chuck, single-speed drill will work, but it's worth paying a little more for the larger, variable speed type.

Most drill motors require a key to lock the bit in the chuck, but some have keyless chucks which are tightened by hand. When removing or installing a bit, make sure the cord is unplugged to avoid accidents. Initially, tighten the chuck by hand, checking to see if the bit is centered correctly. This is especially important when using small drill bits which can get caught between the jaws. Once the chuck is hand tight, use the key to tighten it securely - remember to remove the key afterwards!

Preparation for drilling

If possible, make sure the part you intend to drill in is securely clamped in a vise. If it's impossible to get the work to a vise, make sure it's stable and secure. Twist drills often dig in during drilling - this can be dangerous, particularly if the work suddenly starts spinning on the end of the drill.

Start by locating the center of the hole you're drilling. Use

a center punch to make an indentation for the drill bit so it won't wander. If you're drilling out a broken-off bolt, be sure to position the punch in the exact center of the bolt **(see illustration)**.

If you're drilling a large hole (above 1/4-inch), you may want to make a pilot hole first. As the name suggests, it will guide the larger drill bit and minimize drill bit wandering. Before actually drilling a hole, make sure the area immediately behind the bit is clear of anything you don't want drilled.

Drilling

When drilling steel, especially with smaller bits, no lubrication is needed. If a large bit is involved, oil can be used to ensure a clean cut and prevent overheating of the drill tip. When drilling aluminum, which tends to smear around the cutting edges and clog the drill bit flutes, use kerosene as a lubricant.

Wear safety goggles or a face shield and assume a comfortable, stable stance so you can control the pressure on the drill easily. Position the drill tip in the punch mark and make sure, if you're drilling by hand, the bit is perpendicular to the surface of the workpiece. Start drilling without applying much pressure until you're sure the hole is positioned correctly. If the hole starts off center, it can be very difficult to correct. You can try angling the bit slightly so the hole center moves in the opposite direction, but this must be done before the flutes of the bit have entered the hole. It's at the starting point that a variable-speed drill is invaluable; the low speed allows fine adjustments to be made before it's too late. Continue drilling until the desired hole depth is reached or until the drill tip emerges at the other side of the workpiece.

Cutting speed and pressure are important - as a general rule, the larger the diameter of the drill bit, the slower the drilling speed should be. Also, the harder the material is, the slower the drilling speed should be. With a single-speed drill, there's little that can be done to control it, but two-speed or variable speed drills can be controlled. If the drilling speed is too high, the cutting edges of the bit will tend to overheat and dull. Pressure should be varied during drilling. Start with light pressure until the drill tip has located properly in the work. Gradually increase pressure so the bit cuts evenly. If the tip is sharp and the pressure correct, two distinct spirals of metal will emerge from the bit flutes. If the pressure is too light, the bit won't cut properly, while excessive pressure will overheat the tip.

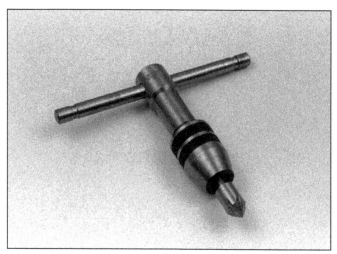

2.45 Use a large drill bit or a countersink mounted in a tap wrench to remove burrs from a hole after drilling or enlarging it

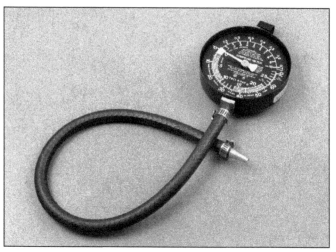

2.46 The vacuum gauge indicates intake manifold vacuum, in inches of mercury (in-Hg)

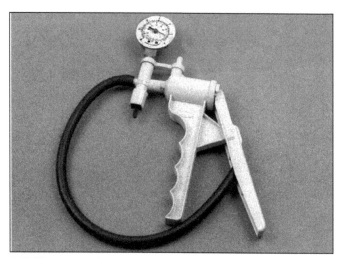

2.47 The vacuum/pressure pump can create a vacuum in a circuit, or pressurize it, to simulate the actual operating conditions

Decrease pressure as the bit breaks through the workpiece. If this isn't done, the bit may jam in the hole; if you're using a hand-held drill, it could be jerked out of your hands, especially when using larger size bits.

Once a pilot hole has been made, install the larger bit in the chuck and enlarge the hole. The second bit will follow the pilot hole - there's no need to attempt to guide it (if you do, the bit may break off). It is important, however, to hold the drill at the correct angle.

After the hole has been drilled to the correct size, remove the burrs left around the edges of the hole. This can be done with a small round file, or by chamfering the opening with a larger bit or a countersink **(see illustration)**. Use a drill bit that's several sizes larger than the hole and simply twist it around each opening by hand until any rough edges are removed.

Enlarging and reshaping holes

The biggest practical size for bits used in a hand drill is about 1/2-inch. This is partly determined by the capacity of the chuck (although it's possible to buy larger drills with stepped shanks). The real limit is the difficulty of controlling large bits by hand; drills over 1/2-inch tend to be too much to handle in anything other than a drill press. If you have to make a larger hole,

or if a shape other than round is involved, different techniques are required.

If a hole simply must be enlarged slightly, a round file is probably the best tool to use. If the hole must be very large, a hole saw will be needed, but they can only be used in sheet metal.

Large or irregular-shaped holes can also be made in sheet metal and other thin materials by drilling a series of small holes very close together. In this case the desired hole size and shape must be marked with a scribe. The next step depends on the size bit to be used; the idea is to drill a series of almost touching holes just inside the outline of the large hole. Center punch each location, then drill the small holes. A cold chisel can then be used to knock out the waste material at the center of the hole, which can then be filed to size. This is a time consuming process, but it's the only practical approach for the home shop. Success is dependent on accuracy when marking the hole shape and using the center punch.

Troubleshooting tools

Vacuum gauge

The vacuum gauge **(see illustration)** indicates ported or intake manifold vacuum, in inches of mercury (in-Hg).

Vacuum/pressure pump

The hand-operated vacuum/pressure pump **(see illustration)** can create a vacuum, or build up pressure, in a circuit to check components that are vacuum or pressure operated. It can also be used as a vacuum gauge.

Safety items that should be in every shop
Fire extinguishers

Buy at least one fire extinguisher **(see illustration)** before doing any maintenance or repair procedures. Make sure it's rated for flammable liquid fires. Familiarize yourself with its use as soon as you buy it - don't wait until you need it to figure out how to use it. And be sure to have it checked and recharged at regular intervals. Refer to the safety tips at the end of this chapter for more information about the hazards of gasoline and other flammable liquids.

Gloves

If you're handling hot parts or metal parts with sharp

2.48 Buy at least one fire extinguisher before you begin work - make sure it's rated for flammable liquid fires and **KNOW HOW TO USE IT!**

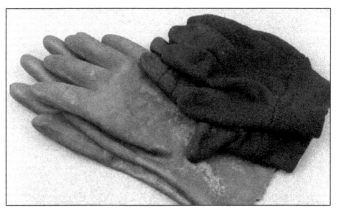

2.49 Get a pair of heavy work gloves for handling hot or sharp-edged objects and a pair of rubber gloves for washing parts with solvent or carburetor cleaner

2.50 One of the most important items you'll need in the shop is a face shield or safety goggles, especially when you're hitting metal parts with a hammer, washing parts in solvent or grinding something on the bench grinder

edges, wear a pair of industrial work gloves to protect yourself from burns, cuts and splinters **(see illustration)**. Wear a pair of heavy duty rubber gloves (to protect your hands when you wash parts in solvent.

Safety glasses or goggles

Never use a drill, hammer and chisel or grinder without safety glasses **(see illustration)**. Don't take a chance on getting a metal sliver in your eye. It's also a good idea to wear safety glasses when you're washing parts in solvent or carburetor cleaner.

Storage and care of tools

Good tools are expensive, so treat them well. After you're through with your tools, wipe off any dirt, grease or metal chips and put them away. Don't leave tools lying around in the work area. General purpose hand tools - screwdrivers, pliers, wrenches and sockets - can be hung on a wall panel or stored in a tool box. Store precision measuring instruments, gauges, meters, etc. in a tool box to protect them from dust, dirt, metal chips and humidity.

Maintenance and repair techniques

There are a number of techniques involved in maintenance and repair that will be referred to throughout this manual. Application of these techniques will enable the home mechanic to be more efficient, better organized and capable of performing the various tasks properly, which will ensure that the repair job is thorough and complete.

Fasteners

Fasteners - nuts, bolts, studs and screws - hold parts together. Keep the following things in mind when working with fasteners: All threaded fasteners should be clean and straight, with good threads and unrounded corners on the hex head (where the wrench fits). Make it a habit to replace all damaged nuts and bolts with new ones. Almost all fasteners have a locking device of some type, either a lockwasher, locknut, locking tab or thread adhesive. Don't reuse special locknuts with nylon

or fiber inserts. Once they're removed, they lose their locking ability. Install new locknuts.

Flat washers and lockwashers, when removed from an assembly, should always be replaced exactly as removed. Replace any damaged washers with new ones. Never use a lockwasher on any soft metal surface (such as aluminum), thin sheet metal or plastic.

Apply penetrant to rusted nuts and bolts to loosen them up and prevent breakage. Some mechanics use turpentine in a spout-type oil can, which works quite well. After applying the rust penetrant, let it work for a few minutes before trying to loosen the nut or bolt. Badly rusted fasteners may have to be chiseled or sawed off or removed with a special nut breaker, available at tool stores.

If a bolt or stud breaks off in an assembly, it can be drilled and removed with a special tool commonly available for this purpose. Most automotive machine shops can perform this task, as well as other repair procedures, such as the repair of threaded holes that have been stripped out.

Tightening sequences and procedures

First, install the bolts or nuts finger-tight. Then tighten them one full turn each, in a criss-cross or diagonal pattern. Then return to the first one and, following the same pattern, tighten them all one-half turn. Finally, tighten each of them one-quarter turn at a time until each fastener has been tightened to the proper torque. To loosen and remove the fasteners, reverse this procedure.

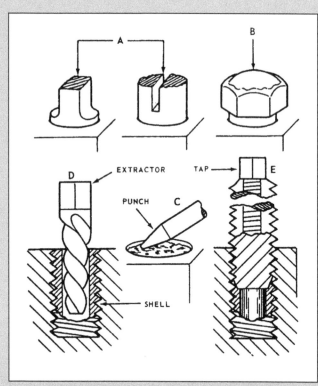

2.51 There are several ways to remove a broken fastener

A   File it flat or slot it
B   Weld on a nut
C   Use a punch to unscrew it
D   Use a screw extractor (like an E-Z-Out)
E   Use a tap to remove the shell

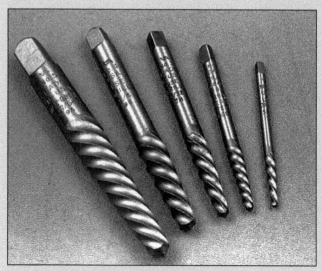

2.52 Typical assortment of E-Z-Out extractors

How to remove broken fasteners

Sooner or later, you're going to break off a bolt or screw inside its threaded hole. There are several ways to remove it. Before you buy an expensive extractor set, try some of the following cheaper methods first.

First, regardless of which of the following methods you use, be sure to use penetrating oil. Penetrating oil is a special light oil with excellent penetrating power for freeing dirty and rusty fasteners. But it also works well on tightly torqued broken fasteners.

If enough of the fastener protrudes from its hole - and if it isn't torqued down too tightly - you can often remove it with vise-grips or a small pipe wrench. If that doesn't work, or if the fastener doesn't provide sufficient purchase for pliers or a wrench, try filing it down to take a wrench, or cut a slot in it to accept a screwdriver (**see illustration**). If you still can't get it off - and you know how to weld - try welding a flat piece of steel, or a nut, to the top of the broken fastener. If the fastener is broken off flush with - or below - the top of its hole, try tapping it out with a small, sharp punch. If that doesn't work, try drilling out the broken fastener with a bit only slightly smaller than the inside diameter of the hole. For example, if the

hole is 1/2-inch in diameter, use a 15/32-inch drill bit. This leaves a shell which you can pick out with a sharp chisel.

If that doesn't work, you'll have to resort to some form of screw extractor, such as E-Z-Out (**see illustration**). Screw extractors are sold in sets which can remove anything from 1/4-inch to 1-inch bolts or studs. Most extractors are fluted and tapered high-grade steel. To use a screw extractor, drill a hole slightly smaller than the O.D. of the extractor you're going to use (Extractor sets include the manufacturer's recommendations for what size drill bit to use with each extractor size). Then screw in the extractor, making sure it's centered, and back it - and the broken fastener - out. Extractors are reverse-threaded, so they won't unscrew when you back them out.

A word to the wise: Even though an E-Z-Out will usually work, it can cause even more grief if you're careless or sloppy. Drilling the hole for the extractor off-center, using too small or too big a bit for the size of the fastener you're removing, or breaking off an E-Z Out will only make things worse. So be careful!

How to repair broken threads

Sometimes, the internal threads of a nut or bolt hole can become stripped, usually from over tightening. Stripping threads is an all-too-common occurrence, especially when working with aluminum parts such as intake manifolds or other soft metals used in carburetor manufacturing.

Usually, external or internal threads are only partially stripped. After they've been cleaned up with a tap or die, they'll still work. Sometimes, however, threads are badly damaged. When this happens, you've got three choices:

1)   Drill and tap the hole to the next suitable oversize and install a larger diameter bolt, screw or stud.

2)   Drill and tap the hole to accept a threaded plug,

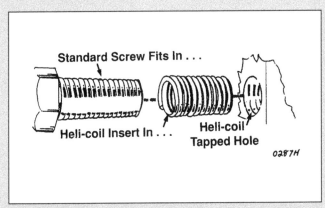

Standard Screw Fits In . . .

Heli-coil Insert In . . .

Heli-coil Tapped Hole

0287H

2.53 To install a Heli-Coil, drill out the hole, tap it with the special included tap and screw in the Heli-Coil

then drill and tap the plug to the original screw size. You can also buy a plug already threaded to the original size. Then you simply drill a hole to the specified size, then run the threaded plug into the hole with a bolt and jam nut. Once the plug is fully seated, remove the jam nut and bolt.

3)   The third method uses a patented thread repair kit like Heli-Coil or Slimsert. These easy-to-use kits are designed to repair damaged threads in spark plug holes, straight-through holes and blind holes. Both are available as kits which can handle a variety of sizes and thread patterns. Drill the hole, then tap it with the special included tap. Install the Heli-Coil (**see illustration**) and the hole is back to its original diameter and thread pitch.

Regardless of which method you use, be sure to proceed calmly and carefully. A little impatience or carelessness during one of these relatively simple procedures can ruin your whole day's work and cost you a bundle if you wreck an expensive component.

Component disassembly

Disassemble components carefully to help ensure that the parts go back together properly. Note the sequence in which parts are removed. Make note of special characteristics or marks on parts that can be installed more than one way, such as a grooved thrust washer on a shaft. It's a good idea to lay the disassembled parts out on a clean surface in the order in which you removed them. It may also be helpful to make sketches or take instant photos of components before removal.

When you remove fasteners from a component, keep track of their locations. Thread a bolt back into a part, or put the washers and nut back on a stud, to prevent mix-ups later. If that isn't practical, put fasteners in a fishing tackle box or a series of small boxes. A cupcake or muffin tin, or an egg crate, is ideal for this purpose - each cavity can hold the bolts and nuts from a particular area (i.e. carburetor mounting studs and nuts, fuel bowl components, etc.). A pan of this type is helpful when working on assemblies with very small parts, such as the carburetor or valve train. Mark each cavity with paint or tape to identify the contents.

When you unplug the connector(s) between two wire harnesses, or vacuum line connections, it's a good idea to identify the two halves with numbered pieces of masking tape - or a pair of matching pieces of colored electrical tape - so they can be easily reconnected.

Gasket sealing surfaces

Gaskets seal the mating surfaces between two parts to prevent lubricants, fluids, vacuum or pressure from leaking out between them. Age, heat and pressure can cause the two parts to stick together so tightly that they're difficult to separate. Often, you can loosen the assembly by striking it with a soft-face hammer near the mating surfaces. When a part refuses to come off, look for a fastener that you forgot to remove.

Don't use a screwdriver or pry bar to pry apart an assembly. It can easily damage the gasket sealing surfaces of the parts, which must be smooth to seal properly. If prying is absolutely necessary, use an old broom handle or a section of hard wood dowel.

Once the parts are separated, carefully scrape off the old

gasket and clean the gasket surface. You can also remove some gaskets with a wire brush. If some gasket material refused to come off, soak it with rust penetrant or treat it with a special chemical to soften it, then scrape it off. You can fashion a scraper from a piece of copper tubing by flattening and sharpening one end. Copper is usually softer than the surface being scraped, which reduces the likelihood of gouging the part. The mating surfaces must be clean and smooth when you're done. If the gasket surface is gouged, use a gasket sealer thick enough to fill the scratches when you reassemble the components. For most applications, use a non-drying (or semi-drying) gasket sealer.

Hose removal tips

Warning: *If the vehicle is equipped with air conditioning, do not disconnect any of the A/C hoses without first having the system depressurized by a dealer service department or a service station (see the Haynes Automotive Heating and Air Conditioning Manual).*

The same precautions that apply to gasket removal also apply to hoses. Avoid scratching or gouging the surface against which the hose mates, or the connection may leak. Take, for example, radiator hoses. Because of various chemical reactions, the rubber in radiator hoses can bond itself to the metal spigot over which the hose fits. To remove a hose, first loosen the hose clamps that secure it to the spigot. Then, with slip-joint pliers, grab the hose at the clamp and rotate it around the spigot. Work it back and forth until it is completely free, then pull it off. Silicone or other lubricants will ease removal if they can be applied between the hose and the outside of the spigot. Apply the same lubricant to the inside of the hose and the outside of the spigot to simplify installation. Snap-On and Mac Tools sell hose removal tools - they look like bent ice picks - which can be inserted between the spigot and the radiator hose to break the seal between rubber and metal.

As a last resort - or if you're planning to replace the hose anyway - slit the rubber with a knife and peel the hose from the spigot. Make sure you don't damage the metal connection.

If a hose clamp is broken or damaged, don't reuse it. Wire-type clamps usually weaken with age, so it's a good idea to replace them with screw-type clamps whenever a hose is removed.

3 Carburetor fundamentals

Introduction

Let's face it: Carbureted fuel systems have become rather complicated. When you open the hood of a modern vehicle, you're confronted by a nightmarish labyrinth of hoses, lines, linkages, rods, springs, tubes, valves and wires, most of them attached directly to the carburetor. Yet, for all its seeming complexity, the carburetor itself needn't be that difficult to understand. Like any mechanical component that looks formidable when viewed as a single assembly, the carburetor is simpler if you break it down into its individual parts and subsystems. Each of these parts and subsystems is, by itself, pretty elementary.

You probably already know something about carburetors. At the very least, you know the carburetor is a mechanical device that mixes gasoline with air to form a combustible mixture. And you may even be familiar with some of the bits and pieces - floats, jets, etc. - that constitute the modern carburetor. But the fact that you're reading this manual means you probably want to know more.

How does a carburetor really work? This question has stumped weekend do-it-yourselfers and professional mechanics alike at one time or another. Everyone seems to have some working knowledge about carburetors. But as with so many other automotive components, a little knowledge about carbs can be a dangerous thing. So this Chapter goes back to square one and starts from scratch. That way - perhaps even at the risk of insulting your intelligence - we can eliminate any misconceptions you may have picked up somewhere. Once you fully understand the information in this Chapter, you'll have the kind of theoretical background it takes to make sound judgments regarding adjustments, rebuilds and modifications to any carburetor in this manual.

What is a carburetor?

Basically, the carburetor is a device that "senses" the amount and speed of air flowing through it, then meters the proper ratio of "atomized" (spray of fine droplets) fuel into the airstream. And it must do all this while constantly varying the fuel ratio to match changes in engine speed and load determined by the demands of the driver and the engine's operating conditions.

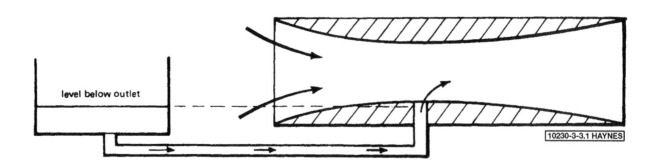

3.0 Pressure differential - the tendency of a higher pressure to push a fluid toward a lower pressure area - is one of the three principles on which carburetors are based; in this illustration, the float bowl is vented to atmospheric pressure (14.7 psi at sea level), so when the engine is running, the fuel in the bowl is pushed toward the relatively lower pressure areas at the throttle plate or the venturi, depending on driver demand

The demands of the driver:
> *Idle*
> *Acceleration*
> *Cruising*
> *High speed*
> *Deceleration*

The engine's operating conditions:
> *Cold or warm start-up*
> *Cold or warm operation*
> *High or low engine load*
> *High or low crankshaft speed*
> *High or low manifold vacuum*
> *High or low venturi vacuum*

Finally, it must do all the above quickly and accurately enough to ensure good driveability, mileage and performance, all the while providing an air/fuel ratio that produces tailpipe emissions low enough to comply with state and Federal law.

Obviously, carburetion on a modern automobile is a much more complicated proposition than it was 50, or even 25, years ago. Does this mean you can no longer service your own carburetor? Absolutely not! Once you understand the basic operating principles of carburetors and the circuits they use to meet the above demands, you can rebuild and adjust any Holley carburetor in this manual. So first, let's review those operating principles, then we'll examine the seven circuits you'll find on most of the carburetors covered in this book. Finally, we'll discuss some of the more typical features found on modern Holleys.

Basic Operating Principles

Pressure differential

Many people think that an engine sucks air into its com-

bustion chambers and the carburetor simply dumps some fuel into this airstream as it passes through the carb throat. In a sense, this is true. But ask them **why** - or how - it does this, and you're likely to get some vague answer about main jets and throttle valves and venturis.

What is really happening inside a carburetor during its operation? Quite a lot, actually. Most carburetors have seven different circuits to deal with every conceivable combination of driver demand and engine operating condition. But before we look at those circuits, let's review the three important factors that determine which circuit operates - and when. The first of these is **pressure differential**.

A gas or a liquid acted upon by two different pressures is pushed by the higher pressure toward the lower pressure area. In other words, a pressure differential causes the gas or liquid to move toward the lower pressure area **(see illustration)**.

Let's look at an everyday example of how a pressure differential works. When you sip a drink through a straw, you're sucking the liquid up the straw, right? True, but what are you really doing when you suck on the straw? You're creating a pressure differential. The atmospheric pressure bearing down on the surface of the liquid in the glass is a constant 14.7 pounds per square inch (psi) at sea level. When you suck on the straw, you lower the atmospheric pressure in your mouth, creating a less-than-atmospheric (relatively low) pressure inside your mouth. In other words, you're creating a partial vacuum. So the liquid in the glass is forced up through the straw and into your mouth.

What's pressure differential got to do with carburetors? Everything. It's pressure differential that the idle and main circuits use to push fuel out of the float bowl (which is vented to atmospheric pressure), up through a tube or passage to the low-pressure throat of the carburetor where it's dispersed into the airstream moving through the venturi. We'll get to float bowls and main circuits in a moment. Now let's look at the second operating principle of the carburetor, **manifold vacuum**.

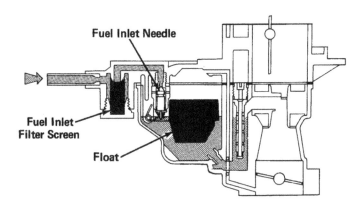

Fuel Inlet Needle

Fuel Inlet Filter Screen

Float

3.1 Carburetors store their fuel in a reservoir known as the float bowl or fuel bowl; the float bowl shown here is for a Model 5200

Manifold vacuum

What is vacuum? In science, it's the total absence of air. In automotive circles, vacuum is the term used to describe the condition of lower-than-atmospheric pressure. Vacuum is usually expressed in inches of mercury (in-Hg). The numerically higher the figure for in-Hg, the higher the vacuum.

The term "manifold vacuum" refers to the condition of relative vacuum that exists in the intake manifold when the engine is running. When they're in operation, the intake and exhaust valves and the pistons work together to create an air pump. As the intake valve opens and the piston moves down on its intake stroke, a relative vacuum is created inside the volume displaced in the cylinder. Why? Because, the air at atmospheric pressure in the intake system upstream from the carburetor can't fill the cylinder as quickly as the piston can move downward toward the bottom of its stroke. So the vacuum inside the cylinder is fairly strong, not quite so strong in the intake manifold and weaker still as we get farther away from the empty cylinder and closer to the mouth of the intake system, where the air is at atmospheric pressure. However, there's always a significant difference in pressure between the outside air and the air in the intake manifold. And if the choke plate is closed during engine warm-up, or if the carburetor throttle valve is closed, which occurs at idle and during deceleration, the pressure difference is even more pronounced. Idle circuits in the carburetor use the high manifold vacuum condition to draw fuel from the float bowl and into the airstream during idle conditions.

The Venturi effect

Another way a pressure difference is created is called the **venturi effect**. Imagine a tube with air flowing through it. If the walls of the tube are straight, i.e. the same diameter, from one end to the other, the air molecules flow through the tube at the same speed. But what if we pinch our tube in the middle somewhere? Now the air molecules must crowd together to get through the bottleneck. This bottleneck restriction is known as a *venturi* . When the air molecules arrive at the venturi, they not only crowd together, they also speed up. Once they get through the venturi, they slow down again and resume their original speed.

In physics, the venturi effect, which is named after G.B. Venturi, an Italian physicist (1746 - 1822) is expressed like this: "If gas velocity increases, gas pressure decreases." The molecules in any material, air included, are always moving around a lot. The molecules of air moving through our hypothetical tube are bouncing off the walls of the tube and off of each other, which is what creates the pressure. As they speed up through the venturi, they become more directional, so they have less time to bounce off the walls or each other, so the pressure decreases.

Carburetor Circuits

Introduction

A carburetor is basically a collection of cleverly designed circuits. These circuits provide various ways by which fuel can be atomized and mixed with incoming air for each specific operating condition. Some circuits, like the float circuit, work all the time; others, like the main or high-speed circuit, work only during certain operating conditions.

Most of the circuits listed below are found in every Holley carburetor in this manual (and in most other carburetors as well):

- The fuel inlet system or the float circuit
- The idle circuit
- The transfer or off-idle circuit
- The high-speed or main metering circuit
- The power circuit
- The accelerator pump circuit
- The choke circuit

Fuel inlet system or float circuit

The float bowl

The carburetor stores its fuel in a reservoir known as the **float bowl** or **fuel bowl (see illustration)**. Staged two- and four-barrel Holley carburetors have primary barrels that open

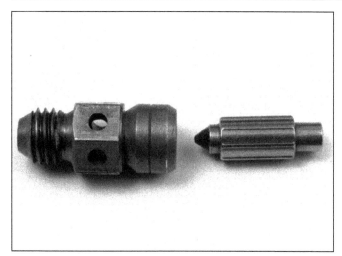

3.2 A typical inlet valve assembly: The replaceable screw-in seat is on the left, the tapered needle on the right. Note the *viton* tip on the needle; viton tips conform more snugly to the seat even with low closing pressures

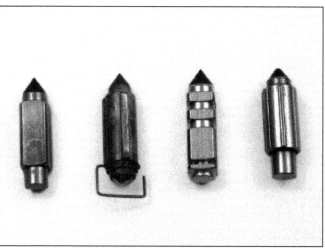

3.3 Needles come in a variety of shapes and sizes. Some needles have a wire clip, like the second needle from the left; this clip hooks around the tang on the end of the float lever

during normal operation, and secondary barrels that open whenever extra power is needed. On most Holleys, the primary and secondary barrels have separate float bowls (the 4360 is a notable exception).

Fuel must be maintained at a certain level in the bowl to assure correct fuel metering under all operating conditions. The fuel circuits depend on the float system to maintain this specified level; they're calibrated to deliver the correct mixture only when the float level is adjusted correctly. The carburetor's ability to withstand sudden changes in attitude - turning, acceleration and deceleration - also depends on correct float height.

Let's look at how a typical float circuit works. Pressurized fuel enters the float bowl through an inlet valve or float valve. The float bowl can be an integral part of the carburetor casting, or it can be a separate casting that bolts onto the carb body. Both designs are used by Holley. The inlet valve is a replaceable screw-in unit consisting of a tapered **needle** and a **seat (see illustrations)**. As the fuel fills the bowl, it lifts the float. The float is attached to a hinged lever arm that pushes on the needle as the float rises. When the fuel rises to a preset level in the bowl, the needle seals against the seat, and no more fuel can enter the float bowl. But as engine operation drains the bowl, the float drops again, pulling the needle off its seat, and allowing more fuel into the bowl. When the fuel level rises to its maximum level, the float closes the needle against its seat again. And so on.

Thus, the fuel level in the float bowl is maintained at or very near the preset level by the continually rising and falling float. Maintaining the fuel in the float bowl at or near its maximum level is critical to the proper operation of the other circuits. In fact, the float level is the overall determinant of the air/fuel mixture.

The bowl vent

The float circuit also acts as a vapor separator. An air passage or **bowl vent** connects the float bowl to the carburetor's inlet air horn. Venting the float bowl to outside air (or the charcoal canister on vehicles with emissions control systems) depressurizes the fuel being pumped into the float bowl by the

fuel pump. Once inside the bowl, fuel is no longer pressurized; it's at vented pressure, which is about the same as the outside air. Vapors trapped in the fuel as it's pumped from the tank escape through the bowl vent so pressure doesn't accumulate inside the bowl. The vent is usually positioned as close to the center of the fuel bowl as possible, and it's high enough to prevent fuel from sloshing into the air horn during hard stops or cornering.

There's also another vent connecting the float bowl to the evaporative (charcoal) canister is used on 1970 and later Holleys. When heat from the engine warms up the fuel in the bowl, the vapors are vented to the canister, then drawn into the intake manifold when the engine is running.

Most Holley carburetors are equipped with large fuel bowls. After a hot soak (engine idling or stopped after it's been heated up), the fuel pump may deliver spurts of liquid fuel and intermittent pockets of fuel vapor. The float bowl capacity must be generous enough to supply fuel to all circuits until the pump is purged of vapor.

The float

On some older carburetors, the float is made of thin brass stampings soldered together. The trouble with brass floats is that they can develop leaks along those soldered seams, fill with gasoline and sink. A sunken float opens the fuel inlet valve, overfills the float bowl and causes an excessively rich air/fuel mixture. So newer floats are made from *nitrophyl* , a closed-cell material that never leaks (though it can slowly absorb gasoline and lose its buoyancy) and is impervious to gasoline and fuel additives **(see illustrations)**. Since 1987, Holley has also offered a molded hollow plastic float. Whatever material is used, the float must be buoyant enough to shut the inlet valve when the bowl is full. The lever to which the float is attached is usually designed to provide a mechanical advantage to the buoyancy of the float itself: The longer the lever length, the greater the leverage of the float.

Float vibration is usually caused by engine or vehicle vibration or bouncing. A vibrating float can allow the inlet valve to admit fuel into the float bowl even when the bowl is full,

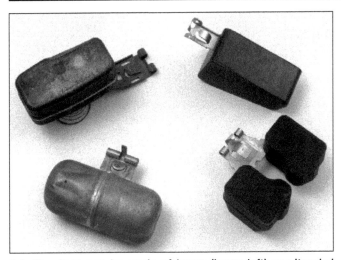

3.4 The float can be made of brass (lower left) or nitrophyl (a cellular plastic material) and it can be side-hinged or center-hinged

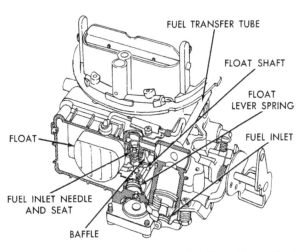

3.5 This late-model four-barrel has a side-hinged brass float and uses an internal fuel inlet valve

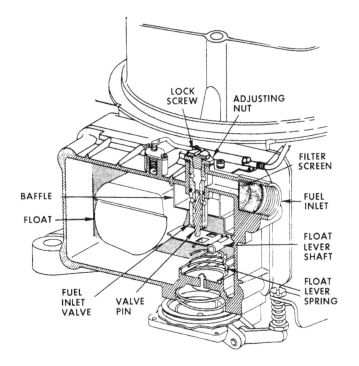

3.6 This late-model four-barrel has a side-hinged nitrophyl float and an externally adjustable fuel inlet valve

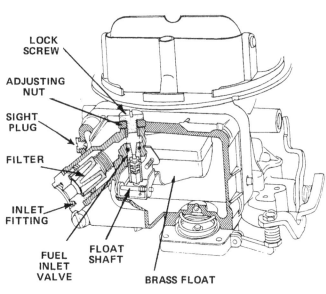

3.7 Here's a high-performance four-barrel with a center-hinged brass float and an externally adjustable fuel inlet valve

which can cause wide variations in the amount of fuel available to the other circuits. On some models, a float spring is added under the float or on a tang to minimize float vibration **(see illustration)**. This device is known as a **float bumper spring**. The float spring is especially helpful in preventing *float drop*, which sometimes opens the inlet valve during radical maneuvers such as accelerating or decelerating through a sharp turn. Some high performance carbs also use a small spring inside the inlet valve itself instead of - or in addition to - the float bumper spring.

Every Holley carburetor has a specified **float height** or float setting. This is the float location that closes the inlet valve when the required fuel level is attained. How does the manufacturer decide how high to set the float? It must be high enough to provide fuel to all the circuits, even during fast starts and stops and high cornering speeds, but not so high that it will spill fuel when driven or parked on an 18-degree slope (a 32-percent grade) On most Holleys, the float height is set by measuring float position in relation to a gasket surface while the carburetor is disassembled. On some Holleys, you can adjust a threaded inlet valve assembly to set the fuel level without disassembling the carburetor.

**3.8 On some models, a float bumper spring minimizes float vibra-
tion (some high performance carbs also use a smaller spring in
addition to, or instead of, the float bumper spring, inside the inlet
valve assembly)**

A center-pivot float (pivot axis *parallel* with the vehicle
axles) is the best design for high-speed cornering because the
float isn't affected much by centrifugal force. Side-hung floats
(pivot axis *perpendicular* to the axles) work slightly better than
center-pivoted designs during heavy acceleration and heavy
braking conditions. Some carbs have a rear bowl fuel level
that's 1/16- to 1/8-inch lower than the primary bowl level. This
forces the fuel to rise higher before it can slosh over through
the main discharge nozzle or vents during hard stops.

The rounded end of the needle rests against the float lever
arm, or is attached to the arm by a small *pull clip*, or hook. As
the float rises, the tapered end of the needle closes against the
inlet-valve seat. Some inlet needles are hollow; inside, a small
damper spring and pin cushion the needle against road shock
and vehicle vibration. Most Holley inlet-valve needles are steel
with *viton* tips. Viton-tipped needles resist dirt and conform
well against the seat even with low closing pressures.

Inlet valves are supplied with various seat openings. Each
inlet valve assembly and valve seat (if they're integral parts of a
single assembly) should be stamped (not all of them are) with a
number which indicates seat opening in thousandths of an
inch. For example, a valve assembly marked 130 has a .130-
inch diameter seat opening. A viton-tipped needle and seat
should have the letter "S" after the three-digit number.

Seat diameter and orifice length are the determinants of
fuel flow at a certain pressure: A larger opening flows more
fuel; a smaller opening flows less. Seat size must be generous
enough to fill the bowl quickly for quick acceleration after a
hot-soak period (sitting parked with a hot engine) and for extended periods of high fuel demand, such as wide open throttle at high rpm. Smaller needle seats are better than larger
seats at controlling hot fuel in the lines because vapor pressure
in the fuel line has less area to push against to force the needle
off its seat. So always use the smallest possible inlet valve that
will work. And remember: You must readjust the float level anytime you change the needle and seat, because you're changing the fuel-pressure forces against which the float must be
balanced.

The choke system

When it's cold, an engine doesn't start easily, nor does it
run right while it's warming up. Gasoline doesn't atomize well
when it's cold, especially when it's flowing through cold metal
passageways in the intake manifold. And the starter spins the
engine at such a low speed (50 to 75 rpm) that there's little
manifold vacuum available to pull fuel from the float bowl until
the engine catches and goes to fast idle. Obviously, cold starts
pose a formidable problem to the carburetor. Without a good
strong vacuum signal, little fuel will be drawn from the float
bowl, and what little fuel does dribble into the carburetor throat
will likely condense back to a liquid state as soon as it contacts
the cold metal walls of the intake manifold runners and the
cylinders.

The choke plate

When the throttle plate on a carburetor is almost closed,
there's a high vacuum below the plate. This high vacuum
draws fuel from the idle discharge port below the plate, as well
as pulling air past the partially open throttle plate. The available
air/fuel mixture is adequate for warm engine idling. But the idle
circuits alone cannot provide a sufficiently rich mixture during
cold starts. So there's another plate in the upper end of the
carburetor bore, above the main nozzle. When it's closed, this
plate, which is known as the **choke plate** or **choke valve** partially blocks off the airhorn. The vacuum that occurs beneath
the choke plate, although not as strong as the vacuum below
the throttle plate, is sufficient to pull fuel from the main metering circuit (and sometimes even from the air bleed holes!). This
extra fuel, added to the fuel being drawn out of the idle circuits
just below the throttle plate, provides an extremely rich mixture
(about five parts air to one part fuel). Why is such a rich mixture
necessary? First, there's very little manifold vacuum to help vaporize the fuel. Second, the manifold is cold, so most fuel immediately recondenses and puddles onto the manifold surfaces. Third, the fuel itself is cold, so it's not very volatile.
Fourth, the fuel from the main metering system is more liquid
than vapor because there's little air velocity available to assist
in atomizing it. Liquid fuel can't be evenly distributed to the
cylinders. When it arrives there, it will not burn well. During
starting, only a small portion of the fuel actually reaches the
cylinders as a well-vaporized mixture.

The vacuum break diaphragm

While the engine is starting, a thermostatic bimetal spring
holds the choke plate tightly closed, to provide the rich mixture
needed for starting, but as soon as the engine starts, the choke
plate must be opened enough to let in some air so the engine
can run on a slightly leaner mixture. This is accomplished in
two ways: First, the off-center installation of the choke plate
shaft on most models allows airflow to open the choke slightly
to start leaning out the mixture as soon as the engine catches
and starts. Second, the choke plate is pulled down to a preset
position by a vacuum-actuated piston or a **vacuum break diaphragm (see illustrations)**, also known as a **pull-off diaphragm** or **qualifying diaphragm,** as soon as vacuum is
available (immediately after the engine starts). On some models, a temperature-modulated diaphragm varies this preset
opening with ambient temperature. When the choke assumes
this partially-open position, it's still providing a 20 to 50 percent richer-than-normal mixture as the engine warms up and
the choke "comes off" (opens all the way).

On many older and some newer automatic choke designs,
the choke plate is pulled down to its preset partial opening by
a piston connected to engine vacuum **(see illustrations)**. The
piston pulls the choke plate to this partially open position as

3.9a A disassembled **Model 5200** automatic choke assembly: Note the vacuum break diaphragm housing (arrow); the diaphragm inside pulls the choke plate slightly open as soon as the engine starts

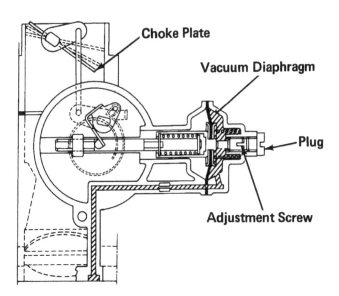

3.9b A cutaway of a typical modern Holley automatic choke mechanism: Note the adjustment screw in the vacuum break diaphragm housing; this screw lets you adjust the amount of partial choke opening

3.10a On some automatic chokes, such as this unit on a Holley four-barrel, the choke plate is pulled down by a piston (arrow) connected to engine vacuum; the piston pulls the choke plate to a partially open position as soon as the engine starts, applying vacuum to the piston; note the bimetal assembly in the choke cover

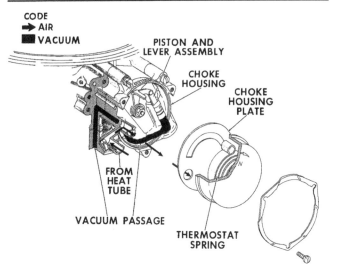

3.10b A cutaway of a typical Holley choke assembly with a vacuum piston

soon as the engine starts and vacuum is applied to the piston. In some older carburetors, this vacuum piston was internal, part of the choke housing or carburetor casting. The vacuum piston design was a source of several problems at one time. The piston gummed up and stuck; or the housing distorted because of engine heat, causing the piston to hang up; or the plug in the end of the piston bore fell out. When these problems occurred, the carburetor had to be removed, the housing torn down and the piston freed. If the housing was warped, the piston bore had to be reamed so the piston could operate freely again. Many newer carburetors have abandoned the piston setup for an external bolt-on vacuum diaphragm unit that connects to the intake manifold by a hose. The diaphragm might leak or rupture, but not as frequently as the piston used to stick. And replacing a diaphragm is a lot easier than freeing a stuck piston. Now back to starting the engine!

The fast idle cam

Depending on the ambient temperature, a richer mixture alone may be insufficient to keep the engine running during en-

gine warm-up. Sometimes, a higher idle speed is needed to overcome the friction of all those cold parts rubbing together, or the engine will stall. So the choke plate is connected to a **fast idle cam** by a linkage rod **(see illustration)**. Before a starting a cold engine, press the throttle all the way to the floor and release it. This allows the choke to close and positions the fast idle cam so the throttle plate is held partially open. While the fast idle cam is set, the engine idle is high (800 to 1100 rpm, or even higher, on a cold engine).

What if you absent-mindedly pump the throttle several times - without pushing it all the way to the floor - while trying to start a cold engine? The engine floods. And if you continue to crank the flooded engine while pumping the accelerator pedal without completely opening the throttle, the situation worsens. With the choke still closed on a cold engine, your chances of starting the engine aren't good. But if you push the

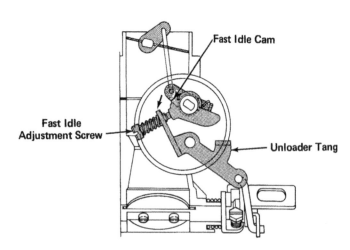

3.11 The choke plate is connected to a fast-idle cam to run the engine at a fast idle while it's warming up; if you flood the engine during starting, push the throttle down to the floor and the unloader tang pushes the choke open so some extra air can lean out the mixture

throttle all the way to the floor, a little tang on the throttle lever called the **unloader** will contact the fast idle cam and push the choke open far enough to admit some additional air into the engine to lean out the mixture. An unloader can be bent to the desired adjustment so that it opens the choke the right amount.

Unloaders are also helpful when starting problems arise because the engine is still hot but the choke thermostat coil is cold. When the engine is turned off, the choke coil and its housing cool off more quickly than the engine itself. After an hour or so, this difference in temperature is substantial. If - when you try to start the still-warm engine - the choke coil has cooled off to 75-degrees or less, it closes the choke all the way, sending a super rich air/fuel mixture into the engine. Sometimes, this mixture is just too rich for the engine to fire. Here's where the unloader comes to the rescue. If you simply apply full throttle, the unloader linkage opens the choke plate and admits fresh air into the carburetor. The leaner mixture is combustible, so the engine starts.

As the engine warms up, the choke is gradually opened by the bimetal spring in an automatic choke, or by you if the choke is a manual unit. This action also reduces the idle speed to a normal "curb idle."

Automatic chokes

On high performance carburetors and on older street carbs with a manual choke, you must learn how much to choke the engine to get it started, and you must also learn how to not choke it so much that you flood it. You must also remember to turn off the choke as the engine warms up, or the fuel mileage will be so bad you'll be stopping at every other gas station. Modern street carburetor chokes do all this automatically. There are two basic types of automatic choke: *integral* and *divorced*.

Integral chokes

Integral chokes use a metal tube to route heated air from a stove around the exhaust manifold or the exhaust-heat crossover to a thermostatic bimetal spring inside a housing on

the side of the carburetor. Some integral chokes use electricity or engine coolant to heat the bimetal spring.

An integral choke heated by hot air sometimes closes at the wrong time. A choke is unnecessary during hot starts. But because there's no flow of heated air past the bimetal spring while the engine is off - and because the thermostatic spring cools off significantly faster than the engine - the choke can close again, even though the engine could very well be warm enough for a choke-less start. Integral chokes heated by engine coolant - known as *hot water chokes* - can suffer from the same problem. Another problem with integral chokes heated by hot air is deterioration with age. Carbon build-up in the hot air tube reduces or blocks the flow of heated air to the spring. When this happens, the choke either opens very slowly or doesn't open at all.

Integral electric chokes are usually found on aftermarket replacement carburetors which must fit a wide variety of intake manifolds. The spring housing contains a heating element that is connected to a 12-volt ignition-switch-controlled power source. When the engine is started, electrical current is applied to the heating unit and the spring undergoes the same type of bimetal reaction created by exhaust heat. Electric chokes allow for a great deal of versatility on custom carb/manifold installations; they're easy to hook up without plumbing or other attachments to the engine. They generally need only one wire, usually from the ignition switch. (Don't use the coil or other ignition components as an electrical power source; the current drain could adversely affect the operation of the ignition system.) But there are several disadvantages to electric chokes:

• Current is drained from the battery when power requirements are already high (during starting).

• An electric choke resets itself when the engine is turned off, even if it's already warmed up, so the choke comes back on when the engine is restarted.

• If the engine isn't started right after the ignition is turned on, the choke may open even though the engine is still cold (Powering the choke from the alternator circuit - so it only gets power when the engine is actually operating - is a quick fix for this problem).

• The choke can come off while the engine is still cold.

Choke assemblies are adjusted and set at the factory for the make and model vehicle on which they're installed. The phenolic housings on integral automatic chokes have index marks to help you adjust the choke plate setting if the need arises. The factory setting closes the choke on a new carburetor when the choke bimetal spring is about 70-degrees F. If you want to readjust the choke angle setting at this temperature, tap the carb slightly to overcome any shaft friction and allow the choke plate to rotate to the angle established by the bimetal spring. If you want less or more choke, move the choke index one mark, then try it. Don't make big choke adjustments all at one time. The arrows on the housing tell you which way to move the index to change the choke operating characteristic (either LEAN or RICH). Rotating the housing in the rich direction increases spring preload, causing the choke plate to open more slowly; turning it in the lean direction has the opposite effect. You will seldom need to move a choke more than one index mark from the factory setting. Finally, be forewarned! The choke covers on many emissions-era carburetors are riveted to prevent tampering. If the choke must be adjusted, you'll have to drill out the rivets and replace them with sheet metal screws.

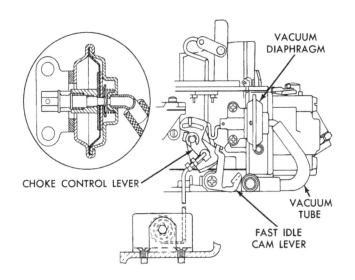

3.12 A typical divorced choke with its thermostatic element mounted on a pad on the intake manifold

Divorced or remote-type chokes

Divorced or **remote-type** chokes **(see illustrations)** use a thermostatic bimetal spring mounted right on the intake manifold or in a well in the exhaust-heat passage of the intake manifold. A mechanical linkage rod from the bimetal spring operates the choke lever on the carburetor. Divorced chokes respond accurately to actual engine operating conditions because the choke only operates when the engine is cold. The choke is gradually closed by a thermostatic spring as the engine warms up. Engine vacuum is used to draw heated air from a *stove* on the exhaust manifold into the choke housing, causing the spring to relax and allowing the choke to open. Most choke plates have an offset shaft, so the choke will open by its own weight if there is no spring pressure holding it closed. Divorced chokes are usually non-adjustable. Some early bimetal units had a provision for adjustment similar to that described above for integral chokes, but it was eliminated on later units.

Electrically-assisted chokes

Limiting the duration of choke on-time to the absolute minimum is critical because of hydrocarbon and carbon monoxide emission levels. At first, engine heat was used to control the choke plate. Some earlier designs used a water-heated cap choke design in which a heater hose was routed directly against the choke cap. Others ran the heated coolant directly through the cap. Newer designs have exchanged the water-heated choke for one with an electric assist. Cap-type chokes use a ceramic heating element hooked to the choke coil and to a bimetallic temperature switch which draws power from the alternator on a constant basis. At a temperature of less than 60-degrees F. (all temperature specs are approximate), the switch stays open, preventing current from reaching the ceramic heating element, and allowing normal choking action to take place using the thermostatic coil. But as the temperature climbs above this cutoff point, the sensor switch closes to let current pass to the heating element, causing the choke coil spring to pull the choke plates open within 90 seconds.

Divorced, or remote, well-type chokes with an electric assist operate in a similar manner, but the heating element is separate from the choke and is wired to an external control switch, which is connected to the ignition switch **(see illustration)**. Turning the key to the "On" position activates the sensor, which passes current at about 60-degrees F., but cuts it off at 110-degrees F.

The idle and low-speed circuit

The engine requires a richer mixture at idle than it does during part-throttle operation. Unless the idle mixture is rich enough, slow and irregular combustion will occur, mainly because of the dilution of the intake charge by residual exhaust gases that occurs during idle. It's the idle circuit's job to provide this rich mixture. The idle system must also keep the engine running at the specified idle speed even when accessories such as the alternator, air-conditioning compressor or power-steering pump are dragging down rpm. And, on vehicles with an automatic transmission, the idle system has to compensate for the load on the engine imposed by the transmission.

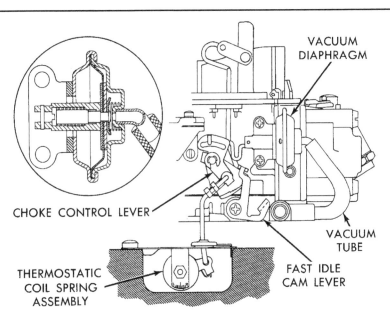

3.13 A typical divorced choke with its thermostatic element mounted in a well in the exhaust heat passage of the intake manifold

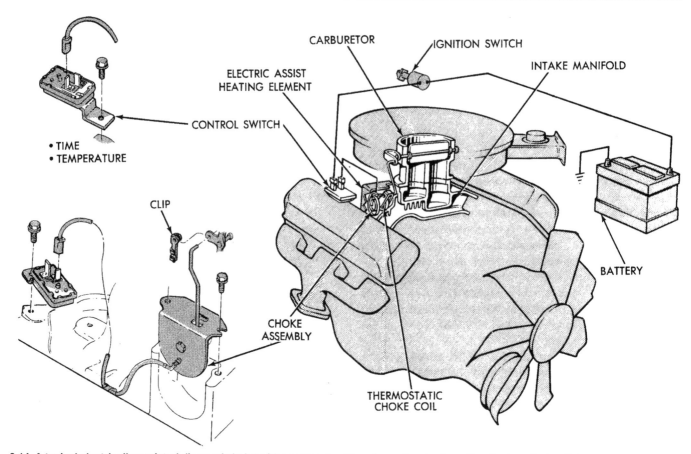

3.14 A typical electrically-assisted divorced choke with separate heating element and externally-wired control switch

Recall that the carburetor delivers fuel into the incoming airstream because of the difference in pressure between the air in the float bowl, which is vented to atmosphere, and the air below the throttle plate, which is under a relative vacuum the strength of which is contingent on the angle of the throttle plate. The idle circuit is a good example of how this works. At idle and low speeds, insufficient air is drawn through the venturi to operate the main metering system. However, intake-manifold vacuum is high on the engine-side of the throttle plate because the nearly closed plate restricts airflow through the carburetor. This high vacuum creates the pressure differential needed to operate the idle system.

The idle circuit (**see illustrations**) begins at the main jet, through which fuel can enter the **idle well**. The fuel level in the main well is established by the float system. Anytime the engine is turned off, or the throttle plate is open too much to create a low-pressure area at the curb-idle port, the fuel in the main well remains at this level. But when the engine is at idle, *i.e.* the throttle plate is closed or almost closed, fuel is drawn up through an **idle tube** from the idle well. The lower end of this tube may have a calibrated hole in it. This hole is known as the **idle feed restriction**, **idle orifice** or **idle jet**. The idle jet usually picks up fuel after it has already passed through the main jet, but the main jet is considerably larger than the idle jet, so it has no effect on the idle circuit. Once the fuel reaches the top of the idle tube, it flows through a crossover passageway which is higher than the level of fuel in the bowl. At the point where it turns the corner and heads back down toward its final destination, outside air is introduced into the idle circuit

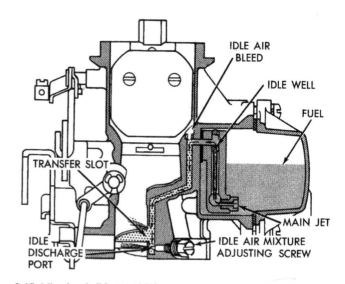

3.15 Idle circuit (Model 1920)

through an **idle air bleed**. From here, the air/fuel *emulsion* (the fuel is no longer in liquid form; it's *trimmed*, or leaned out, into an emulsion of globules of fuel droplets and air bubbles) heads down, through another passageway, toward the vacuum side of the throttle plate. On some carburetors, a calibrated **channel restriction** is located just below the idle air bleed, instead of or in addition to the one at the lower end of the idle tube.

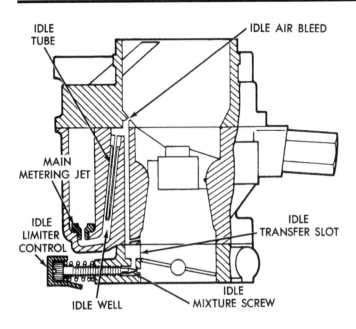

3.16 Idle circuit (Model 1940/1945)

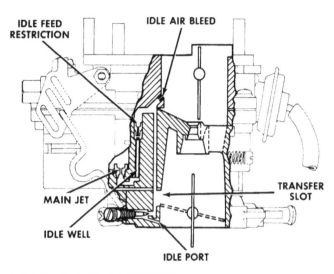

3.17 Idle circuit (Model 2210/2245)

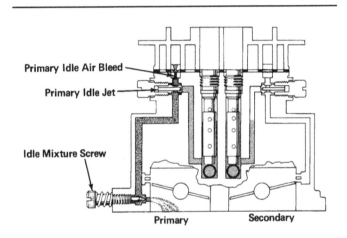

3.18 Idle circuit (Model 5200/5210)

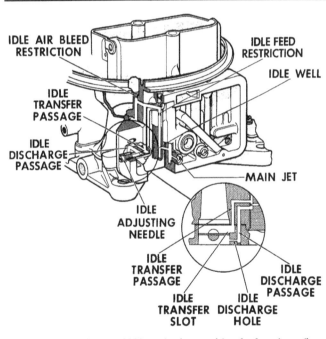

3.19 Idle circuit (Model 2300, and primary side of a four-barrel)

Increasing the size of the idle air bleed orifice reduces the pressure differential across the bleed, which decreases the amount of fuel pulled over from the idle well. Increasing idle air bleed size leans out the idle mixture, even if the idle-feed restriction is unaltered. Conversely, decreasing idle air bleed size increases pressure drop in the system and richens the idle mixture.

It should be noted that the actual routing of idle fuel and air/fuel emulsion differs between models. On conventional Holleys, as on typical Carters, Rochesters, etc., the idle circuit is in the main body casting. On modular Holleys, such as 4150s and 4160s, the routing is inside the metering block. But conventional or modular, idle circuits follow the same general route: The idle passages direct the fuel up toward the top of the carburetor, where bleed air is admitted, before bringing it back down a vertical feed passage to the curb-idle and off-idle discharge ports. Why? Because the fuel in the idle circuit must be raised to a level higher than the level of the fuel in the bowl. Otherwise, the discharge port would just drain the float bowl!

The curb-idle port and the mixture adjusting screw

The idle mixture enters the airstream flowing past the throttle plate at the **idle discharge hole** or **curb-idle port (see illustration).** The amount of mixture that can flow through the idle port is determined by the strength of the pressure differential at the mouth of the discharge hole and the area of the port exposed by the throttle plate, and by an adjustable needle, known as the **mixture adjusting screw** or **idle mixture needle**, situated right at the discharge hole. The needle tip of the screw protrudes into the curb-idle port. The mixture screw simply controls the amount of fuel emulsion allowed to flow through the curb-idle port. If the screw is screwed in (clockwise), it allows less mixture to pass; if it's screwed out, it allows more mixture to pass. When the air/fuel emulsion boils out into the air flowing past the partially open throttle plate, it atomizes

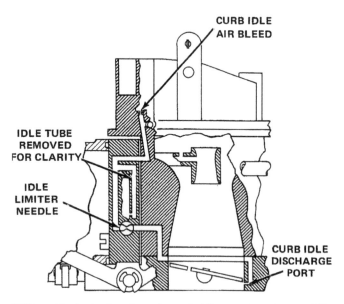

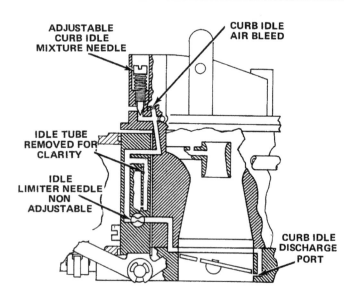

3.20   Typical four-barrel curb-idle and primary idle-transfer circuits

3.21   Typical four-barrel curb-idle circuit with adjustable curb-idle mixture needle and non-adjustable idle limiter needle

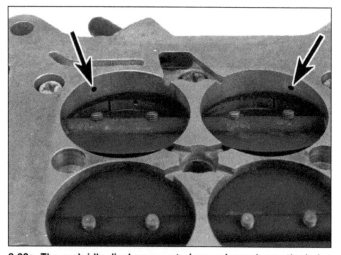

3.22a   The curb idle discharge ports (arrows) are always the holes closest to the flange

into the actual mixture that allows the engine to idle. There's a mixture screw for each primary bore in the carburetor. On most models, turning the mixture screws in leans the mixture; turning them out richens it. **Note:** *The mixture screws on some carburetors work exactly opposite - turning the screws out will lean the idle mixture and turning them in will richen it. On these models this will usually be indicated by a sticker with a counterclockwise arrow and the word LEAN on it.* Mixture screws have very little effect on engine idle speed, which is controlled by the angle of the throttle plate opening.

Conventional designs locate the idle mixture adjusting screws so that the needle tip protrudes into the discharge port. But they're located in the metering block on Holley modular carbs such as the 4150 and 4160, some distance from the actual port. However, both setups are functionally identical. The mixture adjusting screw simply controls the amount of fuel emulsion allowed to reach the discharge port. Where you put the screw itself is irrelevant.

The angle of the throttle plate opening is set by a **throttle stop screw,** not by the throttle bore itself. This prevents the throttle from sticking in the bore. It also makes the idle system less sensitive to mixture adjustments. The idle airflow specification of each Holley carburetor is set at the factory. It should not be readjusted to seat the throttle in the bore (especially if it has a diaphragm-operated secondary).

By the late Seventies, idle mixture adjustment had been virtually eliminated. The factory idle adjustment was protected by an **idle limiter cap,** or the idle adjustment needle was hidden beneath a plug. If you're servicing an emissions-era Holley, you'll find an idle limiter cap on the end of the idle mixture screw. This cap, which is installed at the factory after the idle mixture has been set, prevents tampering with the idle mixture adjustment by limiting the idle mixture screw adjustment to about 3/4-turn. Once the cap is installed, it can't be removed without destroying it, so if it's missing, you know the idle mixture has been fiddled with.

The idle transfer port

At this point, the engine is idling smoothly on air entering around the throttle plate, combined with an emulsified mixture of air and fuel entering each primary throttle bore through an idle port with an adjustable mixture screw. Now the problem is how to obtain a smooth transition from curb idle to cruising speed. As the throttle opens, vacuum diminishes at the idle port, so less fuel will be drawn from this port. So a second hole, known as the **idle transfer port** or the **idle transfer slot** is drilled into the vertical passage feeding the curb-idle discharge port, only a little higher up, closer to the venturi, right above the edge of the throttle plate at its idle position **(see illustrations).** When the idle circuit is metering fuel through the curb-idle port, the idle transfer slot serves as a lower air bleed for the idle system. This extra bleed leans the air/fuel mixture further, which improves its ability to atomize when it's discharged through the curb-idle port.

When the engine is idling, i.e. the throttle plates are closed, the idle transfer slot is *above* the vacuum that occurs just below the throttle plate at idle. But as the throttle plate is

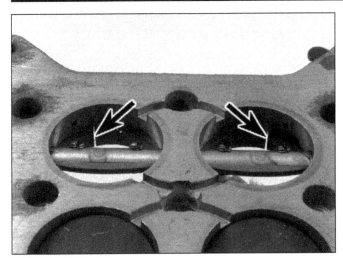

3.22b The transfer holes (arrows) are usually slots (they're easier to machine), but some models use a series of holes instead

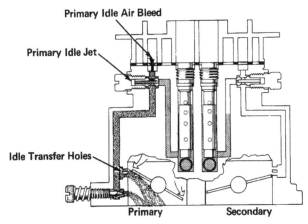

IDLE TRANSFER

3.22c A cutaway of a typical idle transfer circuit (Model 5200/5210)

opened, the transfer slot is progressively uncovered. The vacuum just below the edge of the plate moves up, growing progressively weaker at the curb-idle discharge port and stronger in the area of the transfer slot. The transfer port ceases its function as the lower air bleed for the curb-idle port and begins to discharge air/fuel emulsion. As the throttle plate opens wider, vacuum at the curb-idle port grows weaker and weaker. Less and less fuel is drawn out of the curb-idle port, while more and more fuel comes out of the transfer port, because it's closer to the area of stronger vacuum right below the plate. At some point, the throttle is open far enough - and there's enough airflow through the venturi - so that fuel begins to flow out of the main nozzle. On a well-designed carburetor, the transition from the idle and idle transfer ports to the main circuit is smooth and unnoticeable. **Note:** *Except on older Holleys (such as the 2110), the transfer port is always a slot, not a hole.*

The cruising or main-metering or high-speed circuit

Introduction

Once the engine speed or airflow increases to the point at which the idle and accelerating pump systems are no longer needed, the **main-metering** system **(see illustrations)** takes over. This system supplies the correct air/fuel mixture to the engine during cruising-and-higher speeds.

The throttle plate

When you step on the gas pedal, you open the **throttle valve** or **throttle plate** in the base of the carburetor. Interestingly, the throttle shaft is slightly offset in the throttle bore (about 0.020-inch on primaries and about 0.060-inch on secondaries). In other words, the throttle plate area on one side of the throttle shaft is larger than the area on the other side of the shaft. Why? Because, believe it or not, the throttle plates are actually *self-closing*! There are two reasons why they're set up this way. First, because of the sizable closing force generated when manifold vacuum is high - which it is at idle - throttle plates with offset shafts return to idle more consistently. Second, offset shafts are a safety measure. In the event you should accidentally fail to reconnect the linkage or throttle return spring, they prevent over-revving the engine because airflow past the throttle plates tends to close them.

3.23 Looking through the top of a big Holley (Model 4160 shown); note the venturis and venturi boosters for each primary and secondary barrel (main metering circuit nozzles face down and discharge into these boosters)

One popular misconception about the throttle is that it controls the *volume* of air/fuel mixture that goes into the engine. Actually, engine displacement never changes, so the volume of air pulled into the engine is constant for any given speed. What the throttle controls is the *density* or *mass flow* of the air pumped into the engine by piston action: The charge density is lowest (the air/fuel mixture is thinner) at idle, and highest (air/fuel mixture is thicker) at WOT. Put another way, the cylinder can fill its volume more completely when the throttle plate is open. A dense charge has more air mass, so higher compression and burning pressures - and higher power output - can be achieved. The throttle controls engine speed and power by varying the density of the charge - not the volume of air - supplied to the engine.

The venturi

The partial vacuum created by the downward stroke of the pistons draws air into the air cleaner and through the carburetor where the air picks up fuel, and goes into the engine. As the air passes through the bore of the carb, it goes through a

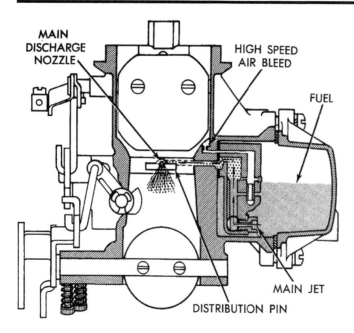

3.24a Main-metering circuit (Model 1920)

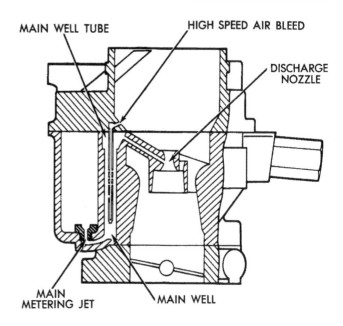

3.24b Main-metering circuit (Model 1940/1945)

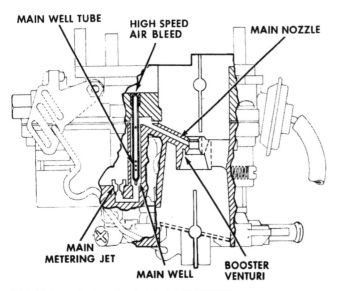

3.25 Main-metering circuit (Model 2210/2245)

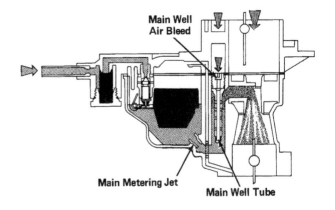

3.26 Main-metering circuit (Model 5200/5210)

smooth-surfaced restriction - the **venturi** - that's slightly narrower than the rest of the carb bore. The venturi pinches down this flowing column of air, then allows it to widen back to the diameter of the throttle bore. This inrushing air column has a certain pressure. To get through the constricted area, it must speed up. Remember that in a venturi, there's a pressure drop, or relative vacuum, which increases in proportion to the speed of the air flowing through it (which in turn is determined by engine speed and throttle plate position). A gentle diverging section - which starts at the smallest area of the venturi and continues to the lower edge of the tapered section - recovers most of the pressure.

If you take a hollow tube and run it in a straight line from inside the fuel bowl to this low-pressure area, the suction of the partial vacuum will draw fuel out of the bowl. The venturi

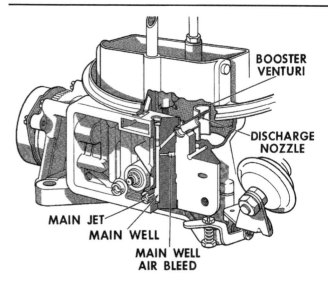

3.27a Typical four-barrel main-metering circuit

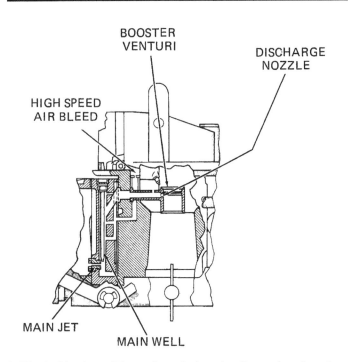

BOOSTER
VENTURI

DISCHARGE
NOZZLE

HIGH SPEED
AIR BLEED

MAIN JET

MAIN WELL

3.27b A side view of the main-metering circuit on a four-barrel, showing the idle tube in the main well

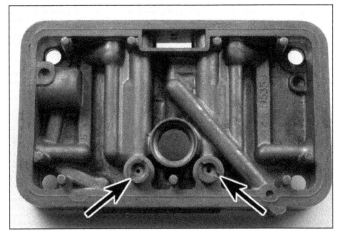

3.28a On high performance Holleys such as this Model 4160, the main jets (arrows) are screwed into the bottom of the metering block

3.28b Typical replaceable main metering jets; note the jet size (arrows) engraved on the side

supplies fuel in proportion to the mass of air moving through it. The size of the venturi determines the pressure difference available for the main metering system: The smaller the venturi, the greater the pressure difference, the sooner the main system is activated and the better the mixture of fuel and air. The engine speed at which the main system begins to feed fuel into the venturi depends on the displacement of the engine that's pulling air through the carburetor.

For a given engine size, a bigger carb needs a higher engine speed to activate the main system, a smaller carburetor, a lower rpm. Venturi size also determines the maximum amount of air available at wide-open throttle. If the venturi is too small, top-end horsepower is reduced, even though the carburetor may provide a good fuel/air mixture at cruising speeds. So manufacturers compromise. They use carbs with smaller-than-optimal airflow capacities to get good fuel atomization, vaporization and good distribution, which enhances economy for the stop-and-go and cruising speeds at which most of us drive most of the time. High performance carburetors have a small *primary* venturi with a bigger *secondary* venturi for better top-end flow.

Although it might seem reasonable to assume that the theoretical low-pressure point and the point of highest velocity would be at the venturi's minimum diameter, fluid friction moves this point about 0.030-inch below the point of smallest diameter. This low-pressure, high velocity point of maximum suction is known as the *vena contracta*. Most modern carbs have double, even triple, venturis known as venturi *boosters* (which we'll get to in a minute) to strengthen the suction at the vena contracta The center of the discharge nozzle or the trailing edge of the boost venturi is always placed at this point because it supplies the strongest vacuum *signal* for the main metering system. This signal travels to the main and power systems via an *asperator channel*. The asperator channel is always plumbed to the point of greatest "depression" *(i.e. the*

lowest vacuum) in the booster venturi on an upward angle to the vertical fuel pickup channel. This passage, in turn, delivers fuel from the main and power circuits to the booster venturi.

The main jet and the main mixing well

Fuel for the main-metering circuit is stored in a special chamber known as the **main mixing well**, or simply the **main well**, a vertical passage cored into the metering block (on all 2300, 4150/4160, 4165, 4175 and 4500 models) or into the main casting on other models. The amount of fuel that enters the main mixing well from the float bowl is determined by the **main jet (see illustrations)**, a calibrated orifice screwed into the bottom of the well, through which fuel must travel from the float bowl to the main well. The size *and* the shape of this hole determines how much fuel can pass through the jet for a given pressure difference (for detailed information on jet sizes and how they affect fuel metering, refer to Chapter 8). At a low pressure difference, little fuel flows through the main jet. And since the jet provides very little restriction to low flow, little turbulence is created as the fuel comes out of the jet into the main mixing well. But as the throttle is opened further, higher airflow through the venturi creates a greater pressure difference in the main metering system. As the flow through the main jet increases, more fuel must go through the small hole in the main jet, creating greater turbulence. Interestingly, as the pressure difference increases across the main jet, the proportion of fuel flowing through it for a given pressure difference is less. So the air/fuel mixture at higher speeds becomes leaner. Up to a point, this increased leanness improves fuel economy. Of course, if you open the throttle still further, the fuel supplied by the main-metering circuit may be insufficient; under these conditions, the power circuit or the accelerator pump circuit, or both, may have to be activated. We'll get to those two circuits in a moment.

With low pressure in the venturi acting in the same manner as a suction pump, there is no reason why fuel should exit the discharge nozzle in anything other than a liquid state. Remember our pressure difference analogy of sucking fluid through a straw? Suction alone initiates flow, but it doesn't provide any means of converting liquid gasoline to a spray mist. Some atomization occurs when a stream of liquid is introduced into a column of fast-moving air. But many of the gasoline droplets are still too large to be atomized thoroughly enough for complete combustion. What's needed is a complete liquid-to-mist breakdown between the fuel bowl and the discharge nozzle. This process is known as **pre-atomization** or **emulsification**. An emulsion is a light, frothy mixture of fuel and air.

The main well air bleed and the emulsion tube

The **main air bleed**, also known as a **high speed bleed**, which is located on the flat surface surrounding the inlet horn, makes this emulsification possible. An air bleed "senses" total air pressure; it's unaffected by airflow variations. With an air passage leading from the air bleed to the emulsion tube, the same suction (low pressure) force that causes fuel flow now draws *air* into the delivery tube as well as fuel. The effect is similar to what happens when you try to suck up the last little bit of liquid in the bottom of an iced drink with a straw: Much of the liquid that surrounds the ice cube is drawn up the straw with some amount of air. In some instances, the liquid breaks down into smaller droplets as it travels up the straw.

However, the air bleed's function isn't just emulsifying the fuel. It also exerts control over fuel flow by "bleeding off" some of the suction force or "signal" which exists at the discharge nozzle. By varying the size of the bleed, the amount of suction (vacuum) required to initiate fuel flow can be specified. As bleed size is enlarged, vacuum necessary to initiate fuel flow is increased. Conversely, a reduction in bleed size reduces vacuum requirements. The air bleeds contained in production carburetors are precisely measured restrictions, sized such that fuel flow is started at specified airflow rates.

The main well air bleed is the part of the main metering circuit that affects the air/fuel ratio. Because of the pressure difference created by the venturi (and the booster venturi, which we'll get to in a minute), this small precisely machined hole draws air into the main well. It gets there either through an external passage in the metering plate that leads to the **mixing tube** or **emulsion tube** located in the center of the well, or through a passage to the external channel in the metering block and then directly into the main well via a series of holes connecting the channel to the well. If the pressure difference is strong enough, the level of fuel in the well drops below the level of fuel in the float bowl and air is added to the fuel via the emulsion tube. When this happens, the final mixture is leaned out. As air volume through the carburetor increases, so does the leaned-out mixture ratio. If the vehicle is pushed to a very

high speed, the main metering circuit alone will deliver a very lean mixture.

The main well air bleeds contained in production carburetors are precisely-machined, fixed-dimension restrictions, sized so that fuel flow is initiated at specific vacuum levels (some *feedback* or "closed-loop" carbs have variable, electronically-controlled air bleeds - more on those when we get to feedback carburetors). Holley engineers advise against modifications to bleed size. Without the sophisticated equipment needed to analyze fuel flow, it's impossible to accurately assess the result of such alterations.

The emulsion tube is a small brass tube with holes in it; these tiny holes allow air from the air bleed to break up the fuel for better atomization. About 2/3 of the way up the well, the air from the airbleed/emulsion tube is allowed to mix with the fuel in the well. Here's how it works: As fuel flow through the main system increases, the main jet begins to restrict fuel flow, and the fuel level in the main well drops below the fuel level in the float bowl. Air from the main well airbleed flows through the perforations in the mixing tube, lowering the level of fuel inside the tube as well. The air in the emulsion tube bubbles out through the tiny holes in the tube and mixes with the fuel in the well, forming an "emulsion" of fuel and air bubbles that's ready for atomization.

By now, it should be obvious that *the level of fuel in the float bowl - and, therefore, the main mixing well - is critical to the main-metering circuit*. If it's correct, there will be fuel available just below the tip of the main nozzle, ready to be sucked into the venturi. But if the fuel level in the bowl is low, the fuel level inside the main nozzle will also be low, i.e. too far below the tip of the main nozzle. This condition will delay the introduction of fuel from the nozzle into the venturi for a brief instant when you step on the throttle, and the engine will run lean or have a *flat spot*. If the fuel level in the bowl is too high, fuel will actually drip from the end of the nozzle, whether the engine is running or shut off. This condition can seriously affect mileage when the engine is running, and allow gasoline to drip into the intake manifold when it isn't. In other words, the float level determines the distance the fuel must be lifted to flow out the main delivery tube and into the venturi area. The lower the float level, the greater the pressure difference necessary to move fuel up into the main delivery tube. And a higher float level lets more fuel flow for a given airflow through the carburetor, which makes the mixture richer; a lower float level setting causes the mixture to be leaner. Obviously, the float level affects not only the idle and transfer circuits, as discussed previously, but it also has a marked effect on the main-metering circuit as well.

The main nozzle

The tube from the main mixing well to the venturi is called the **main delivery tube** or **main nozzle**. When the air velocity in the venturi reaches the point at which it produces a vacuum signal strong enough to create a difference in pressure between the nozzle tip and the (vented-to-atmosphere) fuel in the mixing well, the fuel in the delivery tube begins traveling from the well to the venturi.

The booster venturi

If the discharge nozzle in the venturi were nothing more than a tube, an unacceptably high airflow would be needed to develop a signal strong enough to pull fuel from the main well. Unless it's necked down to a configuration that would strangle the engine at high speed, the venturi alone cannot provide a pressure difference great enough to pull the emulsified air/fuel mix out of the tube at low air velocities. The pressure difference must be amplified somehow. And that's what the **boost** or **booster venturi** does - it's a signal amplifier. By amplifying the signal available for main system operation, the boost venturi enables the main circuit to function well even at lower speeds and, therefore, lower airflows. This ability is even more advantageous in high-performance carburetors because a boost venturi doesn't seriously impede the carburetor's airflow capacity.

Booster venturis have a number of advantages, the most important of which are the following. First, the trailing edge of the booster venturi discharges at the low-pressure point (vena contracta) in the main venturi. So boost venturi airflow is always accelerated to a higher velocity because it "senses" a greater pressure differential than does the main venturi. The air/fuel mixture which emerges from the boost venturi is traveling faster than the surrounding air, so there's a "shearing" effect between the two airstreams which enhances fuel atomization.

Booster venturis also enhance cylinder-to-cylinder fuel distribution. The "ring" of air flowing between the booster and the main venturis channels the charge toward the center of the airstream, which means less of it ends up clinging to the carb wall below the venturi, and more of it reaches the hot intake manifold where it can be further vaporized before being sucked through each individual intake runner into a cylinder. Bottom line: More air/fuel mixture actually *gets* to the cylinders. The intensity and direction of the air/fuel mixture can be tailored to each specific carb/intake manifold/engine combination in the dyno room by varying the shape of the booster and by adding bars, cut-outs, tabs and/or wings to the basic booster shape.

Some of these boosters are less-than-ideal at distributing fuel evenly around the throttle bore. Poor manifold designs on some vehicle have exacerbated this problem, so Holley engineers either added a tab or milled off a portion of the booster's trailing edge. These modifications "shape" the low-pressure area so that the fuel is pulled into a more effective distribution pattern as it exits the discharge nozzle.

Boosters also offer advantages to the manufacturer. Holley can use one casting for many different engine applications by simply altering the shape and size of the booster venturi for each specific application. Holley's ability to utilize several basic booster designs without changing the basic carburetor casting is one reason why its carburetors are so versatile and relatively inexpensive. The cost of swapping a booster is far less than the cost of recasting an intricate carburetor body.

Finally, it's difficult to build a carburetor with a correctly *shaped* main venturi that will still fit under the hood of a typical modern automobile. Ideally, manufacturers want a 20-degree entry angle and a 7 to 11-degree diverging angle below the vena contracta. However, it's usually impossible to make a carburetor tall enough to permit such a configuration. Booster venturis allow a much *shorter* main venturi so the carb clears the hood. Carburetor designers achieve essentially the same results with one or two booster venturis stacked in the main venturi as they can achieve with a single longer venturi of the "optimal" length.

Vane type boosters

When government regulations began to mandate lower exhaust emissions during the Seventies, Holley responded with the Economaster line of carbs. One of these units, the two-barrel 2300, uses a **vane-type booster**. Eight small

spokes (the vanes) connect an inner circle to the outer circumference of the booster, creating eight corresponding low-pressure areas on their undersides. This not only enhances fuel distribution, it also improves atomization, because the localized low-pressure area beneath each vane helps break up the fuel into smaller droplets.

Annular-discharge boosters

However, distribution and atomization aside, vane-type boosters are very restrictive to airflow, so their application is limited to the street, where good fuel economy is a priority. For high airflow *and* good atomization and distribution, racers and high performance enthusiasts turn to another Holley booster design that departs from conventional wisdom: the **annular-discharge booster**, which was introduced in 1980. In this design, fuel is channeled through the body of the booster and enters the airstream through eight tiny holes located around the inner circumference of the booster. Annular-discharge boosters produce a much stronger vacuum signal than conventional booster venturis; they're the booster of choice among racers, because high performance engines often use long duration camshafts that reduce manifold vacuum at lower engine speeds.

Every booster design creates a vacuum signal of a different intensity, so you can see that fuel metering requirements and air bleed dimensions must also vary from one carburetor to the next, even though they may have identical throttle bore and venturi diameters. By juggling these variables, Holley engineers can create a wide array of fuel delivery characteristics.

But that doesn't mean *you* should try to fiddle with air bleed dimensions or booster venturis. Without the proper tools and diagnostic equipment, altering either bleeds or boosters is virtually impossible. Bleeds aren't designed to be modified. Period. And boosters aren't replaceable without special tools (and even if they were, you'd be upsetting the balance of a very carefully designed system). Reckless, misguided enthusiasm can ruin a perfectly functioning carburetor. If you want to fiddle with the main metering circuit, experiment with replacing the main jets (see Chapter 8). Swapping jets is the only way a do-it-yourselfer can alter fuel flow characteristics in the main metering circuit.

The power circuit

When you want extra power from your engine for acceleration or passing, the power circuit **(see illustrations)** provides the richer mixture that's needed to do the job. Here's how it works. Manifold vacuum is a very good indicator of engine load. Vacuum is usually stronger at idle, weaker at open throttle. Think of the power valve as a "switch" operated by manifold vacuum. As the load increases - you come to a hill, for example, or pull out to pass another car - you must open the throttle valve more to maintain a certain speed or to increase your speed. But as you open the throttle plate, manifold vacuum drops because, with the throttle plates open, there's less restriction to the air flowing through the carburetor. In fact, if it weren't for the power circuit, this condition would normally result in a *leaner* air/fuel mixture just when you need a richer one.

A vacuum passage in the carburetor directs manifold vacuum to the power valve, which usually consists of a spring-loaded piston or a spring-loaded diaphragm. At idle and normal cruising speeds, manifold vacuum holds the diaphragm or piston closed against spring pressure. But when the throttle plate is opened, manifold vacuum drops. This drop in manifold vacuum is the signal to the power circuit to open up. Below a

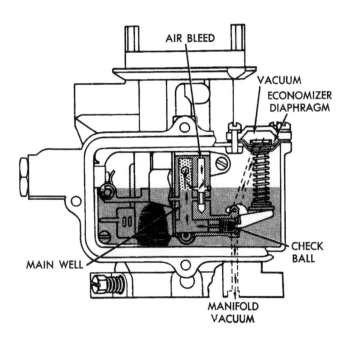

3.29 Power circuit (Model 1920)

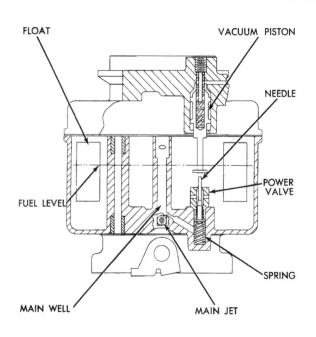

3.30 The Model 1940 and 1945 both have similar power circuits, but the 1945 circuit is known as a gradient fuel enrichment system because it uses a valve with a tapered or staged valve stem; here's a typical Federal Model 1945 with a gradient power enrichment system

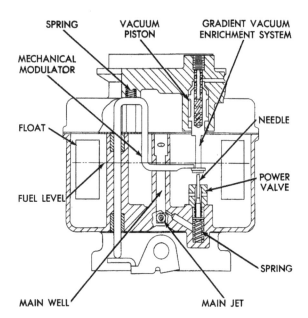

3.31 Typical California Model 1945 gradient power enrichment system

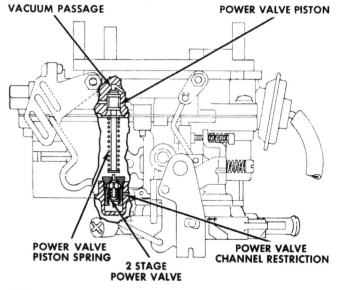

3.32 Power circuit (Model 2210)

preset point, usually about six inches of mercury (in-Hg), there's insufficient vacuum to keep the piston or diaphragm closed. The spring overcomes manifold vacuum and opens the power valve. Fuel flows through the power valve and through a pair of power-valve channel restrictions (PVCRs) **(see illustration)** to merge with the fuel already flowing into the main well from the main jet. The result is a richer mixture. How much richer depends on the size of the PVCRs, tiny holes located in the shoulder adjacent to the threaded bore for the power valve

in the metering plate. PVCRs vary in size from .030 to .120-inch diameters, depending on the metering plate application. The important point here is that it's the PVCR diameter, not the openings in the power valve, that regulates the amount of fuel admitted through the circuit (except for those applications which use a two-stage power valve - see below).

When you no longer need the extra power, you lift the accelerator pedal, the throttle plates close, manifold vacuum rises and the vacuum signal acts on the diaphragm or piston to overcome power-valve-spring tension, closing the power valve and shutting off the added fuel supply.

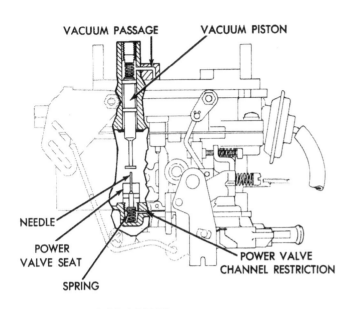

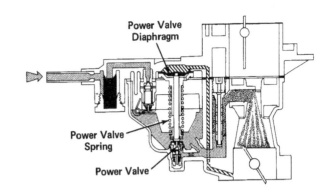

POWER FUEL SYSTEM (PRIMARY)

3.34 Power circuit (Model 5200/5210)

3.33 Power circuit (Model 2245)

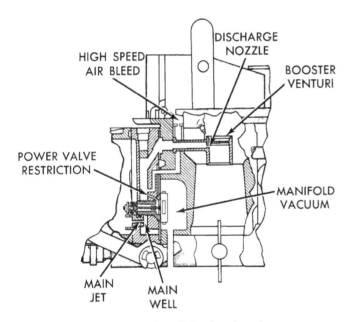

3.35a Power circuit on a typical Holley four-barrel

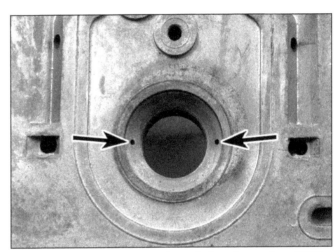

3.35b The power valve channel restrictions (arrows), or PVCRs, determine the amount of additional fuel that the power circuit can flow into the main well

3.35c A typical primary side metering block for a Model 4160; this block contains the drilled passages for the idle, main and power circuits between the float bowl and their corresponding channels in the carburetor body

High-performance Holleys (such as the 2300, 4150, 4160, 4165, 4175 and 4500) use dual or single-stage **diaphragm-type power valves** which screw into an interchangeable metering block or plate **(see illustrations)**. You can select from a wide variety of valves with different flow areas and opening points. The opening-point sizes range from 2.5 to 10.5 in-Hg. The last digits in the part number of a particular single-stage power valve indicate the opening point in inches of mercury (in-Hg) for that valve when a decimal point is placed ahead of the last digit. This number is also stamped on the valve. For example, 125-105 opens at 10.5 in-Hg; 125-25 opens at 2.5 in-Hg.; etc. The power valve opening point is another variable Holley uses to find the best trade-off between economy, exhaust emissions and driveability. The opening point can occur

fairly late on OEM replacement carburetors in order to meet state or Federal emission standards and still provide good economy.

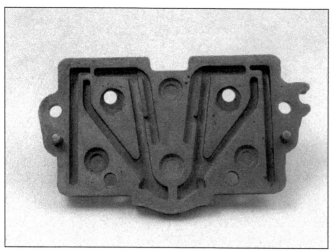

3.35d A typical secondary side metering plate, again for a Model 4160; note the absence of any provision for a power valve (some high performance Holleys - such as those used for drag racing - use metering blocks instead of plates on the secondary side so they can use secondary power valves)

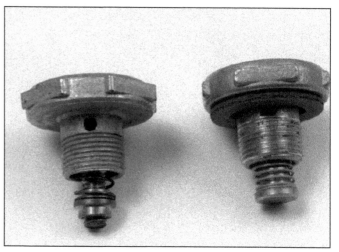

3.35e Typical diaphragm-type power valves: The older one on the left uses drilled holes; the one on the right uses "picture window" slots. Drilled-hole types are used only on two-stage power valves; all modern diaphragm-type Holley power valves use picture windows because they flow more fuel

Manifold vacuum fluctuates significantly at idle and low speed on some high-performance engines in response to *valve timing* as well as throttle position or engine load. A power valve must not open and close in response to these variations. When you select a power valve for racing applications, select it very carefully (refer to Chapter 8).

Holley manufactured earlier power valves with four or six drilled holes; newer power valves have a pair of "picture-window" slots **(see illustration)**. Power valves with drilled holes are no longer available as replacement parts for performance carbs; and they don't flow as much fuel as picture-window designs. However, they are still used on two-stage power valves (see below) and you may run one in some factory-installed applications. But all modern Holley high-performance two and four-barrel carburetors use picture-window power valves with enough flow capacity for two 0.095-inch PVCRs. If you're rebuilding an older carb with the four or six-hole variety of power valve, you can retrofit one of the newer picture-window valves. But read Chapter 8 before messing with power valve or channel restriction modifications. Altering the diameter of the power valve channel restrictions must be done with restraint. If you drill them out too much, you may have to replace the metering plate to restore good performance.

Holley carburetors for the street - especially those manufactured for 1973 and later engines with EGR systems - use two-stage power valves **(see illustration)**. Staged power valves open partially at one vacuum level, then open fully when manifold vacuum falls to a lower level. The first stage in these valves opens a metering orifice smaller than the PVCR. Two-stage power valves are used on some replacement carburetors to help certain engines pass emission tests. They're also available as replacement parts for vehicles with a relatively low power-to-weight ratio, such as recreational vehicles and station wagons. Don't use a two-stage power valve before reading Chapter 8.

Piston-type power valves are used in 1940, 1945, 1946, 2210, 2245, 2280, 2360, 4360 and 5200 models. Spring force is set by the factory to open at a specified manifold vacuum. This setting can be changed by adding or deleting shims from the spring at the foot of the piston stem. The greater the num-

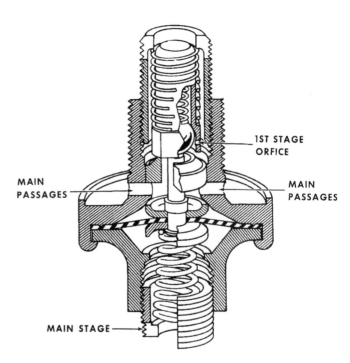

3.36 Cutaway of a two-stage power valve

ber of shims, the higher the manifold vacuum at which the valve opens. If the spring is shortened or the shims removed, the valve opens at a lower manifold vacuum.

Accelerator pump circuit

Fuel is quite heavy compared to air. Remember that the only thing that keeps it well vaporized at idle is the high manifold vacuum produced by the closed throttle plate. But when you open the throttle abruptly to accelerate from idle, manifold vacuum drops to zero. Also recall that the only thing that keeps fuel well vaporized at cruising speeds is the high venturi vacuum produced by the speed of the air rushing through the car-

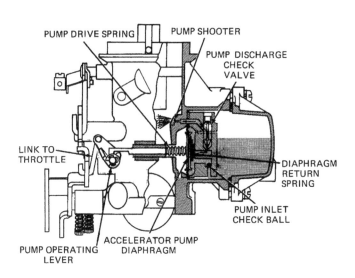

3.37 Accelerator pump circuit (Model 1920)

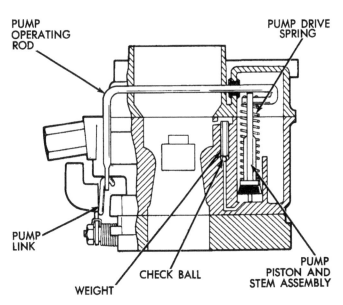

3.38 Accelerator pump circuit (Model 1940/1945)

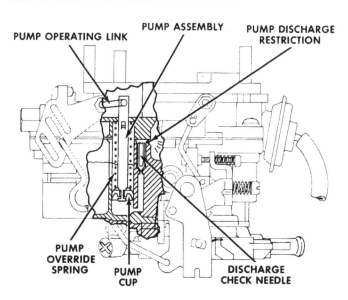

3.39 Accelerator pump circuit (Model 2210/2245)

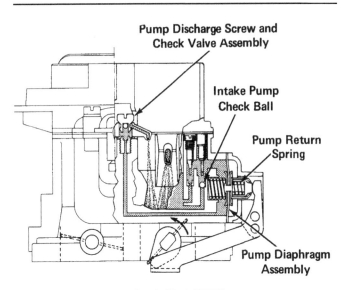

3.40 Accelerator pump circuit (Model 5200)

buretor. But when you open the throttle quickly to accelerate from cruising speeds, airspeed through the carb - and therefore venturi vacuum - lags behind engine speed momentarily. In either case, some of the bigger fuel droplets in the vaporized air/fuel mixture condense back into liquid form (an especially thorny problem on big-port and large-plenum manifolds since there's more surface area for fuel to condense onto). A throttle bore that's suddenly opened needs a significant increase in fuel to replace the fuel which falls out of suspension. But the throttle plate is already too far past the idle and idle-transfer ports for them to help out. And, without a strong vacuum signal in the venturi area, the main circuit can't flow a drop. The result? Air rushes through the carburetor, but fuel flow is momentarily halted, so the mixture sucked into the cylinders is ultra-lean. A moment later, some pressure difference returns to activate the main circuit, but by then it's too late - the engine hesitates or stumbles on the lean mixture.

In other words, when the throttle is moved quickly, there's

a time lag between demand (throttle opening) and main system activation. Part of the reason for this phenomenon is engine load. Put the transmission in neutral, rev up the engine and what happens? The engine accelerates smoothly. Why? Because when the throttle is opened in the absence of a load, engine speed can build smoothly and quickly. So manifold vacuum returns to its normal 14 to 17 in-Hg rather quickly. But put the transmission in gear and do the same thing. Now the engine may stumble badly if the accelerator pump is dysfunctional or missing. Why? Because when the engine is put under a load, airflow and vacuum stay low for a lot longer than when the engine is unloaded. During this period, the carburetor flows little fuel because without airflow and vacuum, the regular circuits can't do much.

The **accelerator pump** system **(see illustrations)** closes this hole in the carburetor's fuel delivery envelope by squirting a pressurized burst of raw fuel into the carburetor right above

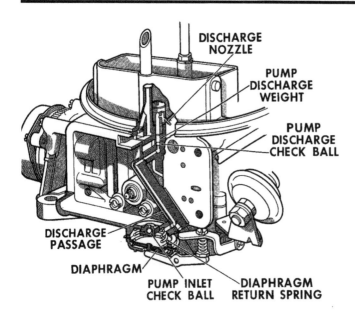

3.41b A typical diaphragm type pump is a simple affair: Pump lever, diaphragm and spring - that's all there is to it!

3.41a Four-barrel accelerator pump circuit with a diaphragm-type pump and pump inlet check ball

the venturi. The mechanically-actuated accelerator pump is linked to the throttle to give one squirt of fuel each time the throttle is opened quickly. When you open the throttle, the pump linkage operates a **diaphragm,** or a **plunger** or **piston** inside a pump **chamber** or **well.** This raises the fuel pressure in the pump well, which forces the **pump-inlet ball** or **check valve** onto its seat so fuel won't escape from the pump well. And the pressure also raises the **discharge needle** or **discharge check ball** off its seat, allowing the fuel to be injected or discharged through a **discharge nozzle** or **"shooter"** into the venturi. Some shooters are aimed at the throttle plate or against the bore. But on carbs with venturi boosters, shooters are usually aimed right at the booster. The shot is pulled toward the trailing edge of the booster by air streaming into the carb, breaking up the fuel for better vaporization.

As the throttle moves toward its closed position, the throttle-to-pump linkage returns to its original position, pulling the piston or diaphragm back to its "at-rest" position and creating a vacuum in the pump cavity, which pulls the pump-inlet ball or check valve off its seat to allow the pump well to refill from the float bowl. This vacuum in the pump well also pulls the discharge needle or check ball onto its seat, so the next shot of fuel pulled from the float bowl stays in the pump well until the next squirt. The discharge needle or ball is really a check valve. It is fit into the pump discharge passage to allow fuel from the pump to be discharged - but not to allow air to enter the passage when the pump piston is returning to its "up" position. If there were no check valve, air would flow through the pump discharge jet, down the passage and into the space under the pump plunger as it rose and no fuel would flow into the pump cylinder from the float bowl. The weight of a discharge check needle is sufficient to keep it closed against the vacuum signal created by the air rushing by the shooter nozzles, so fuel won't be pulled out of the pump circuit. However, the discharge check ball used on some models is too light to prevent fuel from dribbling out of the shooters. On these models, an **anti-pullover** discharge nozzle prevents the shooters from pulling fuel out of the pump passage. But because the ball is so light,

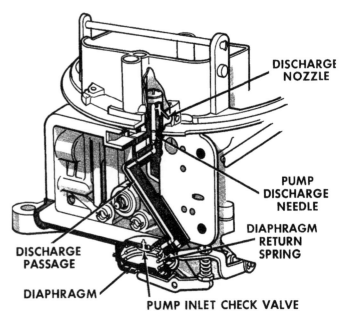

3.42 Four-barrel accelerator pump circuit with "mushroom" or "umbrella" type plastic inlet check valve

excess vapors in the pump well can escape past it and out into the carburetor through the shooters. This feature is handy on diaphragm-type pumps with rubber inlet valves because vapors can't escape past these inlets back to the float bowl. They can only be purged from the pump when it's activated or when the pressure goes up enough to raise the discharge ball or needle off its seat.

Holley street carbs and OEM replacement units use **piston-type** or **diaphragm-type** accelerator pumps. Piston pumps don't offer quite as positive a seal as diaphragm-type pumps. And when they're activated, some amount of fuel always leaks out of the chamber between the piston cup and the chamber cylinder wall. Eventually, the plunger (cup) material shrinks and hardens, and the pump leaks more fuel than it discharges at the nozzle.

All Holley high performance carbs (in fact, all four-barrels except the 4360) use diaphragm-type units **(see illustrations)**

3.43a Typical steel discharge check ball valve on a diaphragm-type accelerator pump: Although it doesn't respond as quickly as the newer "mushroom" or "umbrella" type plastic inlet check valve, a steel ball is held only .012-inch off its seat by its wire retainer, so response time is still fast

because they respond more quickly than piston types and because leakage of fuel from the pump chamber is virtually nonexistent. Regardless of the type of pump inlet used - steel check ball or rubber "mushroom" or "umbrella" check-valve - neither type leaks measurably. The "umbrella" type valve, a rather recent addition to the Holley high-performance line, offers one theoretical advantage over the check ball: It's closed when at rest, so fuel is discharged from the pump chamber the moment that the throttle is yanked open. The steel check ball, on the other hand, is normally in its open position and must be seated (closed) before fuel moves to the discharge nozzle. However, if you invert the bowl and measure the clearance between the check ball and its retainer, you'll find that the clearance is only about 0.012-inch (*i.e.* the retainer holds it a mere 0.012-inch off its seat), so you'd need a pretty sensitive throttle foot to tell the difference in pump response time!

Diaphragm-type pumps have capacities of 30 cc or 50 cc, depending on the application. (The pumps don't actually squirt 30cc or 50cc at one time. This is the amount they will discharge over ten full strokes of the throttle - three or five cc's per stroke.) You can determine the capacity of the pump on your carb by the size of the pump cover: The smaller pump has a thinner (about 1/4-inch) cover; the larger pump has a thicker (about 1/2-inch) cover. "Double-pumpers" have two accelerator pumps - a 30cc unit for the primary and a 50cc unit for the secondary.

Pump discharge volume and delivery rate are determined by the pump cam and linkage adjustments. Generally speaking, the volume of fuel discharged through the nozzle is determined by the number of degrees that the throttle shaft is rotated and the rapidity with which that rotation occurs. But this relationship between linkage motion and volume of fuel discharged is adjustable. A nylon **cam** on the throttle lever raises the pump link lever, which moves the other end of the lever downward, forcing the override spring/nut and bolt assembly against the pump operating lever. The opposite end of the operating lever therefore pivots upward, compressing the spring-loaded pump diaphragm and forcing fuel out of the discharge nozzle. When the throttle is returned to its idle position, spring pressure returns the diaphragm to its "at-rest" position, which allows the pump chamber to refill. Pump cam "lift" (the shape

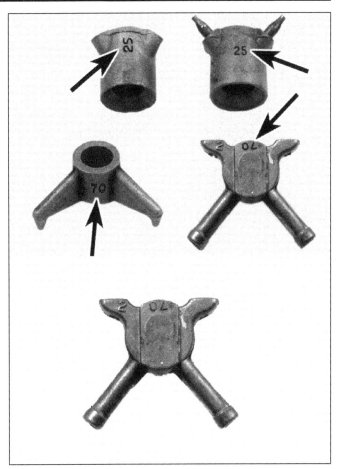

3.43b Typical replaceable shooters; the size is stamped on the body

of the cam) and the relationship of the cam to the throttle lever (it has two different mounting positions) are what determine the characteristics of pump action. The total lift of the cam affects the length of the pump stroke and therefore the capacity available from the pump. Cam profile or shape controls the "curve" of the pump system, i.e. how the pump "comes on" (quickly or slowly) For example, a cam with a sharp nose effects quick and strong pump action to produce a quicker pressure rise, a cam with a gentler rise produces just the opposite effect, etc. Cams are color-coded (black, white, red, blue, orange, green or pink) to identify each profile. High performance Holleys are usually equipped with a white cam. Besides changing the cam, you can also alter the adjustment of the pump linkage itself. But before you tinker with the accelerator pump, refer to Chapter 8.

Another way to adjust discharge volume is by changing the shooters, which, like pump cams, are replaceable on Holley high performance carbs. The rate of discharge is governed by shooter hole size **(see illustration)**. For example, a larger hole allows fuel to be discharged more quickly and with less pressure than a smaller hole. Shooters are available with hole diameters ranging from 0.021 to 0.052 inch. The size of the discharge hole is stamped on the shooter. Again, before you start switching shooters, read Chapter 8.

Secondary systems

In the early days, all carburetors had only one throat. When large six- and eight-cylinder engines came into the pic-

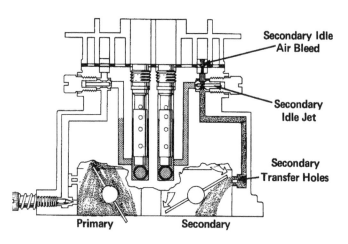

3.44 Secondary idle and transfer circuit on a Model 5200

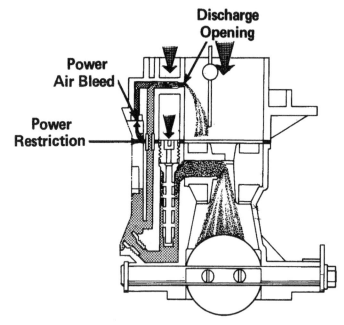

3.45 Secondary power circuit on a Model 5200

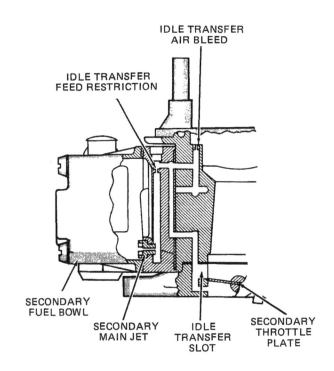

3.46 Typical four-barrel secondary idle and transfer circuit

ture, someone discovered that the fuel was distributed more evenly if the carburetor was manufactured with two throats, or barrels. The parts in the second throat were exactly the same as in the first. Each throat had its own main jet and main nozzle, and each had its own idle system. One float system supplied both barrels with fuel. In other words, the two-barrel was simply a single casting with two one-barrels side-by-side. One barrel served each level of the traditional dual-level, cross-H manifold. For a period of many years, American sixes and V8s were equipped with these two-barrel carburetors. But they were still **single-stage** carburetors. In other words, both throttle plates opened simultaneously and both barrels worked like two identical one-barrels metering fuel at the same time.

Then in the late Forties, manufacturers began to emphasize performance. More airflow meant more power, so these single-stage two-barrels got bigger, but not necessarily better, because even though they contributed to higher performance, they created a host of driveability problem associated with those big barrels because their metering range was simply too narrow to cope with real-world conditions. Large venturis demanded higher engine speeds to initiate fuel flow from main systems. Some of these carbs had cleverly designed idle systems which could hide a late-starting main circuit. But then, as now, idle circuits were controlled by manifold vacuum so there wasn't much they could do to hide this hole in the fuel delivery envelope during the low-vacuum, low-airflow conditions that occurred, for example, during a high-load, low-rpm situation. Of course, these low-vacuum, low-airflow conditions also adversely affected fuel mixing and vaporization. Uneven cylinder-to-cylinder filling and rough-running engines during low-airflow conditions were all-too-common. Something had to be done. So in the early Fifties, carburetor engineers turned to **staged** carburetors for V8s. But it would be nearly 20 more years before mid-size American sixes got staged carbs - the first staged two-barrel was installed on the 1970 Ford Pinto. By this time, staged carbs had been in use for years on European and Japanese vehicles because the conflicting demands of performance and economy outweighed higher production costs. By the mid-Eighties, staged two-barrels were common on small displacement/low-horsepower engines in US-built vehicles.

The primary and secondary main metering circuits in staged Holleys are virtually identical - except for the 4160 and 4175, which use a metering plate instead of a metering block with removable jets. The differences in their design are determined by what they're used for and when they're used. In effect, the secondary side is simply another carburetor that opens later than the primary side **(see illustrations)**. It has its own idle and main metering systems and (on some models built since 1967) its own accelerator pump. Some secondaries even have their own power systems. The differences from one design to another are determined by what they're used for and

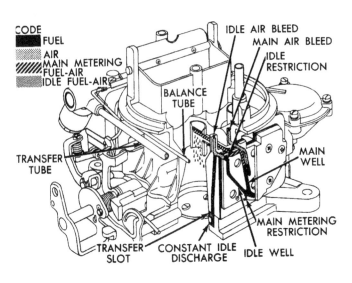

CODE
- FUEL
- AIR
- MAIN METERING FUEL-AIR
- IDLE FUEL-AIR

IDLE AIR BLEED
MAIN AIR BLEED
IDLE RESTRICTION
BALANCE TUBE
MAIN WELL
TRANSFER TUBE
MAIN METERING RESTRICTION
TRANSFER SLOT
CONSTANT IDLE DISCHARGE
IDLE WELL

3.47 Secondary circuit (Model 4160)

when they're used. Think of the secondary system in a staged carburetor as another carburetor in parallel with the primary system. This "parallel-carburetor" strategy expands the metering range of the carburetor significantly.

Staged carburetors use smaller venturis to get the main metering system going, and to provide good vaporization and air/fuel mixing during cruising and light load, *i.e.* lower rpm, conditions. Then they use larger venturis for top-end power at higher engine speeds. The primary side of a staged carb is similar to a single-stage carbs. The slightly smaller venturi size is the only obvious visual difference. The two secondary barrels (or single secondary barrel on a staged two-barrel) are operated only when maximum airflow is needed for more power.

During starting, a secondary vacuum break diaphragm operates from a thermal vacuum valve which senses the temperature of the heated carburetor air. During cold weather starts, the choke opens fully before the thermal valve reaches its operating temperature, so the secondary vacuum break has no effect on choke action. But for warm weather starts, the thermal valve opens to actuate the secondary break, which then opens the choke a bit more. This secondary breaking action is delayed a few seconds by a built-in bleed, giving the engine sufficient time (usually between 8 and 20 seconds) to stabilize its operation before the choke opens a second time.

A secondary idle circuit offers several advantages. It distributes the mixture better at low engine speeds and stabilizes engine idle, so leaner idle settings can be used, which cleans up the exhaust emissions at idle. A secondary idle system also keeps the fuel level from flooding the secondary float bowl in the event that the needle and seat should develop a slight leak, or open as a result of float bounce. Finally, because it's always pulling some fuel from the secondary bowl during idle conditions, a secondary idle system prevents fuel from sitting in the secondary bowl long enough to turn stale (don't laugh - it can happen in congested urban driving environments where the secondary main metering circuit is seldom employed, particularly if the driver is conservative). A constantly replenished secondary float bowl also prevents the inlet needle valve from becoming "varnished" closed. However, since the secondary idle circuit isn't really essential to engine operation, there's usually no provision for mixture adjustment.

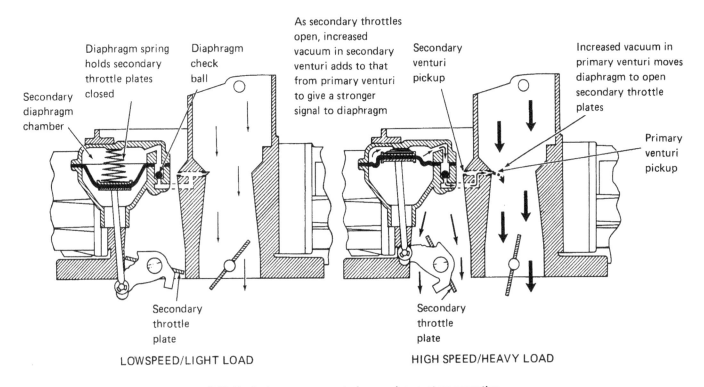

Secondary diaphragm chamber

Diaphragm spring holds secondary throttle plates closed

Diaphragm check ball

As secondary throttles open, increased vacuum in secondary venturi adds to that from primary venturi to give a stronger signal to diaphragm

Secondary venturi pickup

Increased vacuum in primary venturi moves diaphragm to open secondary throttle plates

Primary venturi pickup

Secondary throttle plate

Secondary throttle plate

LOWSPEED/LIGHT LOAD

HIGH SPEED/HEAVY LOAD

3.48 Typical vacuum-operated secondary system operation

Holley secondary systems are activated by mechanical linkage or by a vacuum-actuated diaphragm. Holleys with mechanical secondaries are generally favored by high performance enthusiasts and hot rodders with manual transmissions in their vehicles and by most drag racers regardless of what transmission they use. Mechanically-actuated secondary throttle plates are opened by linkage connected to the primary throttles. This linkage is usually "**progressive**," meaning it doesn't begin to open the secondaries until the primaries have reached about 40 to 50-percent (40 to 45-degrees) of their maximum opening angle. This setup provides pretty good low-speed driveability and fuel economy. Because the primaries are already open 40-degrees or more by the time secondary throttle opening commences, the opening rate of the secondary throttles has to be quicker so that both the primaries and the secondaries will reach wide-open-throttle simultaneously. Direct, or 1:1, linkage is usually used only on racing carburetors. The secondary throttles are closed by a return spring, by airflow (the throttles are offset on their shafts) and by a return link from the primary throttles.

Rapid rotation of the secondary throttle plates from closed to wide open can often result in a serious hiccup if no additional fuel is immediately available. In other words, this is the same situation that occurs on the primary side, and Holley solves it the same way - with an accelerator pump. Primary and secondary accelerator pump circuits are virtually identical except that secondaries use a larger 50 cc pump. Virtually all Holley carbs with a mechanical secondary have a second accelerator pump. Only the 4360 has mechanical secondaries and no secondary pump.

The problem with mechanical secondaries is that - even though they deliver higher performance - they still waste gas. Most of us can live with a little less performance and a little better gas mileage. That's why diaphragm-actuated secondaries **(see illustration)** are popular for the street, particularly on vehicles with automatic transmissions. Vacuum-activated secondaries are controlled by a diaphragm which reacts to airflow - first through the primary venturis, then through the secondary venturis as well - to operate the secondary throttles. The idea is to use only enough airflow to fulfill engine requirements. Initially, the diaphragm responds to the vacuum created in a pickup tube in one of the primary venturis. Some of the vacuum signal goes to the upper chamber in the diaphragm, pulling the secondary throttle plates open. As engine speed - and therefore airflow - picks up, so does vacuum at the venturi, raising the diaphragm even further. Once a reasonable airflow is established in the secondaries, another vacuum signal - captured by another pickup tube in one of the secondary venturis - is applied to the diaphragm, which then has enough vacuum power to open the throttle plates are far as necessary to satisfy engine requirements. At lower engine speeds, this pickup tube in the secondary also serves as a bleed to prevent the secondary throttles from being actuated at the wrong time.

In other words, how far the secondary throttle plates open depends, at first, on how much airflow there is through the primary barrels. As the secondary throttle plates open further and airflow is established in the secondary barrels, the vacuum signal from the primary venturis to the diaphragm is augmented by vacuum from one of the secondary barrels through the bleed opening. On most models, the secondary throttles don't open all the way until and if you put your foot to the floor and hold it there long enough to rev out the engine to its maximum speed.

If a carburetor with diaphragm-actuated secondaries is too large for the engine, the diaphragm automatically limits the "size" of the carburetor because the diaphragm only opens the secondary throttles to supply the amount of mixture needed. The secondaries also remain closed when the primary throttles are opened wide at low rpm. This eliminates small bogs and flat spots. And it means the same carburetor can be used with a wide range of engine displacements, gear ratios, vehicle weights, etc.

The bleed works in conjunction with a check ball (in the vacuum passage to the diaphragm) to provide smooth, steady secondary opening. The check ball allows vacuum to get through the passage at a controlled rate through the bleed (a groove in the check-ball seat). When rpm is reduced, the secondaries aren't needed, so the check ball raises, releasing vacuum from the diaphragm, allowing air to flow freely back into the top side of the diaphragm chamber. The diaphragm spring, which is fully compressed when the secondary throttle plates are wide open, pushes the diaphragm back down, rotating the secondary throttle shaft to its closed position (for information on the relationship between this spring and throttle opening rates - and how to alter that relationship - refer to Chapter 8). Airflow acting against the offset plates also helps to close the throttle plates. And a mechanical link positively closes the secondaries when the primaries are closed. Diaphragm-operated secondaries have a low opening effort or pedal "feel" because the throttle linkage only has to open the primaries and actuate one accelerator pump.

Since secondary throttles are only used for full power and high-load conditions, Holley engineers have dispensed with power valves on many carburetors and compensated by installing relatively rich secondary jetting instead. Those models which do have secondary power valves usually have a problem with fuel slosh and spill-over during deceleration because their secondary main jets are too big. Adding a power valve allows the use of a smaller secondary main jet size. Some racing carbs are also fitted with secondary power valves because fuel flow through the largest available jets is insufficient. But the need for more fuel usually occurs only on large displacement engines fitted with a single 4500. On 4160s and 4175s, there's no provision for a secondary power valve because they use a metering plate instead of a block. The main circuit metering restrictions in the plate are drilled on a vertical - not a horizontal - plane, which minimizes fuel slosh through the restriction during deceleration.

Carburetor controls

At this point, we have a carburetor that will do the job. But modern carburetors must not only do their job - they must do it cleanly enough to satisfy state and Federal emissions regulations. And they must offer good driveability. To accomplish these often conflicting goals, a number of gadgets have been hung on Holley carburetors. The most important of these devices are described below.

Anti-dieseling solenoids

Have you ever turned the ignition off, only to discover that the engine kept running in a kind of jerky, jumpy way? Then you're already acquainted with run-on, or *dieseling*. Dieseling problems were virtually unheard of before the smog-reduction era began. Nowadays, with the lean mixtures and high idle speeds needed to keep engines from puffing out smog at idle, it's not uncommon to turn the key off and have the engine keep running, especially when it's really hot. Why does the engine do this? Because the throttle is opened so far. To fix it, Holley engineers designed a gadget which closes the throttle all the way when you turn off the ignition. It's called an **anti-dieseling solenoid**.

When the ignition is on, the solenoid is energized and the plunger protruding from the end of the solenoid holds the throttle open at its normal curb idle. Idle speed is adjusted either by screwing the solenoid in or out of its bracket or by adjusting the threaded plunger. While the plunger is energized, the idle-speed screw is held off its seat. Turning the ignition off de-energizes the solenoid, which retracts the plunger and lets the throttle drop back so that the idle-speed screw goes against its seat on the carburetor. So there are two idle adjustments on solenoid-equipped carburetors: The curb idle is set at the solenoid; the key-off idle is set with the usual idle-speed screw. To operate properly, the solenoid must be wired so that it goes off with the ignition switch. If it is mistakenly wired so that it is hot all the time, then it won't drop back when the key is turned off, and the engine will diesel.

Anti-stall dashpot

Let's say your vehicle has an automatic transmission, and you accelerate away from a light by stomping authoritatively on the throttle, then suddenly change your mind as you realize that you're on a collision course with an 18-wheeler which also wants to occupy the same intersection at the same time as you. You have no choice but to lift your foot off the accelerator pedal and stomp on the brakes. And your engine, of course, dies - unless you have an **anti-stall dashpot** fitted to your carburetor. On a vehicle with a manual transmission, this situation would never occur, because you'd quickly de-clutch and the engine would simply rev up a little. But an automatic doesn't disengage when you lift your foot off the gas; it immediately pulls the engine down to a normal idle. This kills the engine because a big glob of air/fuel mixture is still on the way to the cylinders. In effect, the combustion chambers are flooded by this mixture. The anti-stall dashpot solves this problem by letting the engine return to its normal idle *slowly*, giving the combustion chambers time to burn off the extra-rich mix.

Dashpots used on Holley carburetors are mounted on the carburetor. A dashpot is easy to identify: It consists of a chamber, diaphragm and spring. A plunger protrudes from the chamber and just touches the throttle linkage. When the throttle pushes against the plunger, the diaphragm tries to force air out of the chamber, but the only exit is through a small hole, which lets the plunger in slowly. The spring returns the plunger for the next stroke.

Dashpots are adjustable, usually by varying the length of the plunger with a screw. The dashpot should let the throttle return slowly enough so the engine doesn't die, but not so slowly that the vehicle tries to keep going through an intersection!

Bowl vents

The **external bowl vent** is sometimes erroneously referred to as an anti-percolation vent. But it's not the same thing. Anti-percolation vents were once used to vent the main well on many carburetors, but they're no longer used on modern Holleys. An external bowl vent allows hot fuel vapors to escape from the float bowl through an external tube which routes the fumes to the evaporative emissions control system (the charcoal canister). If these vapors were not vented from the bowl, they could get through the internal vent, which could create too rich a mixture during idling and make hot starts difficult. Bowl vents are opened at idle by mechanical linkage. Once the throttle is above idle, the bowl vent closes and the internal bowl vent takes over.

An **internal bowl vent** is a tube that looks something like a main nozzle, except that it is larger and is located up high near the choke plate. Any fuel vapors that escape from the float bowl during acceleration and cruising conditions are allowed to escape right into the carburetor, where they're sucked into the intake manifold along with the air/fuel mixture and burned. The internal bowl vent is placed in the airstream through the carburetor at an angle so that air is forced into the float bowl above the fuel level. This slightly pressurizes the fuel in the bowl and helps to push the fuel out the main nozzle.

Deceleration controls

During deceleration, particularly when you snap the throttle plate shut very quickly, a high vacuum is created in the intake manifold. This high vacuum is caused by the engine turning over at higher-than-idle speeds with the throttle plate closed, which pulls extra fuel out of the idle circuit, creating a very rich mixture, so rich that the engine misfires. And misfiring increases emissions.

One way to reduce this high emission-causing manifold vacuum is to have the throttle held open slightly. Another solution is to provide a special passage to let air bypass the throttle valve. Air allowed past the throttle valve reduces the manifold vacuum. Then the mixture will lean out. The device on the carburetor that holds the throttle open slightly during deceleration is known as the **dashpot** or throttle controller.

Hot idle compensators

A cold engine needs a very rich mixture because the gasoline doesn't vaporize well when surrounded by all that cold metal. In order to get enough gasoline vaporized so a cold engine will run, we choke the carburetor to provide a rich mixture. Logically, you might think that if a cold engine needs more fuel, then an extremely hot engine needs less fuel. This is true. When an engine is idling in traffic on a hot day, underhood temperatures are hot enough to cook on. Engines start to run rough at idle when they get this hot because the increased temperature can improve fuel vaporization so well that the mixture is just rich enough to upset the idle.

The **hot idle compensator** is a little air valve that allows fresh air to enter the manifold and lean the mixture when the engine is hot. Various types of hot idle devices are used on Holley carburetors. Some models (like the 5200 and others) use a bimetal thermostatically controlled air bleed valve which admits additional air into the idle system and relieves over-rich idle conditions during prolonged hot idling (and subsequent high engine temperatures). On 5200 models, you'll find a small bimetal spring housed in a little recess in the carb casting right above the throttle linkage. When the engine is hot, this spring opens the small check ball built into the end of the spring, which opens an air passage in the carb casting. The hot idle compensator valve can be replaced by removing the access plate and cork gasket. Although the location of the hot idle compensator valve may vary from model to model, they all look and work pretty much like the device on the 5200.

Idle solenoids

Because of the extremely lean idle mixtures mandated by emissions regulations, it's become necessary to increase idle speeds. Higher idle speeds reduce misfiring caused by lean mixtures and they reduce emissions. Many engines use an electric **idle solenoid** to control idle speed at higher rpm. This solenoid is attached to the carburetor by a bracket. A solenoid plunger contacts a tab on the throttle lever and is energized when the ignition switch is turned to On. The plunger contact is adjusted by moving the solenoid, or by adjusting the plunger or an adjustment screw on the throttle lever.

When the ignition is turned off, the solenoid is de-energized and the idle speed setting is controlled by the low-idle screw. The resulting lower idle speed minimizes "dieseling" or "run on." Before turning off the ignition, the engine should be allowed to return to idle for several seconds. The solenoid doesn't have enough energy to extend the plunger from its "off" position. When you start the engine, you must open the throttle lever so the solenoid plunger can be extended.

Feedback carburetors

Internal combustion engines which use gasoline for fuel produce three toxic emissions - hydrocarbons (HC), carbon monoxide (CO) and oxides of nitrogen (NOx). How much HC, CO and NOx a vehicle is allowed to emit is determined by Federal and state laws which began in the late Sixties and which have since then become progressively stricter. Carburetor development evolved accordingly.

In 1975, the catalytic converter was introduced as a means of reducing HC and CO. The first generation of converters were known as oxidation catalysts because they converted HC and CO emissions into harmless byproducts by adding oxygen to them. Carbon monoxide was converted to carbon dioxide (CO_2) (which is a "greenhouse" gas, though it's otherwise harmless) and hydrocarbons were converted into water (H_2O) and carbon dioxide. Oxides of nitrogen were reduced by lowering compression ratios and adding an exhaust gas recirculation (EGR) valve that lowered combustion temperatures by diluting the fresh intake charge of air/fuel mixture with a small portion of exhaust gases.

These modest measures worked for a few years. Then, in the late Seventies, the Environmental Protection Agency (EPA) mandated much stricter regulations regarding exhaust emissions. These stringent new regulations left carburetor manufacturers like Holley in a quandary. Should they scrap their highly-refined line of carburetors, developed over a 75-year period, and make the technically and financially costly jump to electronic fuel injection? Or should they develop more accurate fuel metering systems for their existing product lines? Most carburetor manufacturers, including Holley, must have decided that the technology, the parts suppliers and public acceptance for electronically-controlled fuel injection were still a few years off. Holley elected to stick with carburetors for a few more years. The result was the **feedback carburetor system**, which is worth looking at not just because it's the state-of-the-art in current carburetor design - and probably the final stage in the long evolution of the carburetor - but because it's what you probably have on the family car if it was manufactured in the late Seventies or the early to mid-Eighties.

To achieve the further reductions in NOx mandated by the late-Seventies EPA Regulations, auto manufacturers turned to a more sophisticated type of catalytic converter known as the *three-way* catalyst. A three-way cat not only oxidizes HC and CO, it *removes* oxygen from NOx emissions. This portion of the TWC, which is located in the front portion of the cat, is known as a *reduction* catalyst. Here's how it works: Oxygen released from NOx combines with HC and CO. NOx becomes just plain nitrogen (N2), which is an inert gas, and the HC and CO become water and carbon dioxide, just like an oxidation catalyst. The only catch was, for a TWC to work right, the engine had to emit the right *balance* of HC, CO and NOx. This balance occurs when the air/fuel mixture ratio is close to the **stoichiometric** ratio of 14.7:1. To achieve this goal, Holley engineers turned to the **closed-loop** or **feedback** system of engine information sensors supplying data to a microprocessor (computer) that manipulates a mixture adjuster in the carburetor.

An **exhaust gas oxygen sensor** (EGO sensor or O2 sensor), which is installed in the exhaust manifold or the exhaust pipe (upstream to the catalyst, is the most important of these information sensors because the amount of oxygen in the exhaust varies in proportion to the presence of the three principal pollutants in a very predictable manner. Which means the oxygen content of the exhaust gas can be used to adjust the air/fuel mixture. The O2 sensor compares the oxygen content in the exhaust with the oxygen content in the outside air. The difference between the two is expressed as a low-voltage *analog* signal (a variable voltage output). This analog signal is sent to an electronic control unit (ECU) that converts it to a *digital* signal and interprets it by comparing it to a *map* (program) stored inside its memory. If the oxygen content increases - exhaust gas is lean - the voltage output decreases; if the oxygen content decreases - exhaust gas is rich - the voltage output increases. The ECU looks at this data and alters the air/fuel mixture accordingly by adjusting a **solenoid-controlled mixture control valve,** sometimes called a **duty-cycle solenoid.**

As long as the oxygen sensor/computer/duty-cycle solenoid keep the air/fuel ratio close to stoichiometric, the three-way cat can do its job of removing harmful pollutants from the exhaust. If the mixture gets too lean, NOx goes up; if the mixture gets too rich, HC and CO go up. It's a neat system, but there's one catch. The oxygen sensor must be heated up - usually to around 500-degrees F. - before it will work correctly. So the closed-loop system is worthless during cold starts. Under these conditions, the system must revert to **open-loop control**. Mixture control is set at a fixed value by the program in the ECU memory until the engine is sufficiently warmed up and the oxygen sensor is ready to go to work. When that point is reached, the system goes into closed-loop control.

The system also goes into open-loop mode during wide-open throttle (WOT) conditions. Whenever the **manifold pressure sensor** and the **throttle position sensor** tell the ECU that the engine is being operated at WOT, the ECU puts the system into open loop and supplies a rich mixture based on instructions stored in the program rather than oxygen sensor feedback. When the heavy throttle condition ends, the ECU returns the system to closed-loop mode.

On some carburetors, the duty-cycle solenoid only operates on the main system by opening and closing a restriction to supplement the main jet. Other carbs have valves that also open and close an auxiliary air bleed affecting idle mixture as well. Mixture is controlled by the length of time the valve is held open or closed. The ECU's output to the valve is constantly altered as it monitors the different drivetrain sensors, oxygen, throttle position, manifold pressure, coolant temperature, etc.

The good news is that most feedback carburetor designs are quite similar to conventional carburetors. The big difference is usually in the main metering circuit. A duty-cycle solenoid operates a fuel metering jet. The duty-cycle solenoid fuel passage is parallel to the main jet fuel passage.

When the duty-cycle solenoid is closed, fuel delivered to the venturi discharge nozzle must come through the main jet. This is a rather lean mixture. When the solenoid is energized, fuel through it and the main jet is delivered to the venturi discharge nozzle. This is a rather rich mixture.

The duty-cycle solenoid is energized by the computer control system at a frequency of about 10 times a second. The longer the solenoid is held open each time it's opened, the richer the mixture. The computer determines the open-to-closed time ratio by inputs from sensors in the vehicle. The exhaust gas oxygen sensor is the main input, but engine temperature, manifold vacuum, atmospheric pressure and other factors may be considered in the computer program.

If the exhaust gas sensor senses that there is little or no oxygen present in the exhaust gas flow, the computer will increase the amount of time the solenoid remains closed, leaning out the mixture. Less gasoline in the cylinder during combustion will leave more oxygen in the exhaust gas.

This information is then sent to the computer to affect the next input to the duty-cycle solenoid. The overall effect is that the air/fuel mixture will stay very close to the stoichiometric ratio of 14.7:1. Very few harmful exhaust emissions will be produced, and fuel mileage will be better.

There are several styles of feedback carburetors. They usually use a duty-cycle solenoid to control the air/fuel ratio in the main metering circuit. They may, however, control airflow to the air bleed of the mixing well or fuel to the mixing well in some other way. The end result is the same. The air/fuel ratio is carefully controlled using a computer, sensors and a carburetor with a duty-cycle solenoid.

Carburetor types

One-barrel carbs

One-barrel carbs have one venturi, one throttle plate and one set of metering circuits to meet all air/fuel requirements. They usually have an airflow capacity of about 150 to 300 cubic feet per minute (cfm) and are used primarily on smaller four and six-cylinder engines.

Single-stage two-barrel carbs

Single-stage two-barrel carbs are basically two one-barrel carbs with a common main body. They have one float bowl, float, choke and accelerator pump. The metering systems for both barrels work simultaneously, but the idle mixtures for each barrel are adjusted separately. Single-stage two-barrels have two throttle valves that open and close together on a common shaft. These carbs have airflow capacities of about 200 to 550 cfm and are used on many six and eight-cylinder engines.

Two-stage two-barrel carbs

Two-stage two-barrel carbs were introduced in the early Seventies to improve fuel economy and performance and to lower emissions on four and small six-cylinder engines. They use a single float and float bowl but have two barrels and two throttles that operate semi-independently of one another. The primary barrel is smaller than the secondary barrel. It has idle and transfer circuits and supplies the air/fuel mixture at low and moderate loads and speeds. The primary barrel also usually includes the main metering and power systems and the accelerator pump.

The throttle of the secondary barrel opens at higher speeds and loads when the primary throttle is about half-open. The secondary barrel supplies additional fuel and air to the engine. It includes transfer, main metering and power systems. The secondary throttle can be operated by vacuum or mechanical linkage. Some of these carburetors have a choke only for the primary barrel; others have chokes for both barrels. Most two-stage two-barrel carbs have airflow capacities of about 150 to 300 cfm (they're usually used on small engines).

Four-barrel carbs

Four-barrel carbs are used mostly on V-8 engines. They have airflow capacities of about 400 to 800 cfm (and some high-performance models can flow over 1000 cfm!). Four-barrels have two primary and two secondary barrels in a single body. At low and moderate loads and speeds, the two primary barrels work together like a single-stage two-barrel carb. At higher speeds and loads, the secondary barrels work in a fashion similar the secondary barrels on a two-stage two-barrel.

The primary barrels have idle, low-speed, high-speed and power systems. They also contain the choke and accelerator pump. The secondary barrels have separate high-speed and power systems and may even have a separate idle system and accelerator pump. One model, the 4360, has a single float bowl (though it still has two floats) for all four barrels. All other models have separate bowls for the primary and secondary barrels.

4 Carburetor identification

Holley carburetors are identified by the stamped numbers usually located on the side or the air horn of each carburetor. There are two different numbers that identify a carburetor; a model number, and a list number (sometimes referred to as the part number). It is very important to know that most often the list number will designate the exact model. Included in this chapter is a list of some of the most common Holley carburetor part numbers *(see the tables at the end of this Chapter)* along with the model numbers.

The model number describes the class, type and general features of the products with that group. For example, the 4160 describes one model of the four-barrel carburetor. The first digit of the number denotes the number of barrels (throttle bores). The model 2210, 2300 and 2360 all begin with the number two and all are two barrels. The 3150 and 3160 are three - barrel carbs. The last three positions describe differences in the construction of the carburetor, compared to others in the same family.

However, this doesn't hold true in every case. Not all two-barrel models start with the number two - model 5200 is a two-barrel. Carburetors for use with computer control systems usually begin with the number six. Also, on later model carburetor, some of the numbers listed with the part number are strictly for the date of production.

The list number is assigned to a component as it is put into circulation (this includes fuel pumps, ignition parts and other Holley products). Some list numbers include letter prefixes.

The sales number is for identifying popular replacement parts for a particular carburetor, fuel pump, etc. Sales numbers include a prefix number, a dash number and a suffix number and are used primarily for ordering and stocking products. You won't find this number on the carburetor (or other Holley product) - it'll be on the box or the price sheet. Because of this, you really don't need to be concerned with sales numbers.

Identifying your particular Holley carburetor may be very difficult because of the many variations and updates. The automotive parts person is best suited to detect the model and kit number from their detailed automotive parts counter catalogs. If there is any problem identifying the carburetor, have the carburetor in front of you when speaking with the auto parts person and he/she will see to it that you get the correct parts.

Whenever installing any carburetor that was not originally intended for use on your vehicle, or whenever any modification is to be made (changing float bowls from side-hung to center-hung, converting a manually operated choke to an electrically operated one, etc.) be sure to consult the Holley Standard Replacement Catalog for specifics.

Now for some descriptions of the carburetors covered by this manual.

One-barrel carburetors

Models 1904/1908/1920/1960   Used on Chrysler slant-six engines, farm and industrial applications.

Models 1940/1945/1946/1949   1974 introduction Ford, Chevrolet and Chrysler - car, truck and industrial use. May be
6145/6146/6149 (feedback versions)   marked "Autolite."

Two-barrel carburetors

Models 2210/2211/2245   Introduced in late 60's. GM, Chrysler application.

Models 2280/6280 (feedback version) 1964 through 1973 GM.
                                     1978 through 1987 Chrysler.

Models 2300/2305/2380   Introduced in late 60's.
                                     Single and multiple carb applications.

Models 2360/6360 (feedback version) Introduced in 1983. GM Rochester Dual-Jet replacement.

Models 5200, 5210/5220   Introduced in early 70's. Ford, AMC, GM, VW/Audi four-cylinder engines. Many other
6500/6510/6520 (feedback versions)   applications with the use of an adapter plate.

Models 2100/2110   Originally installed on mid 30's to late 50's Ford cars and trucks. Also popular as an
                                     aftermarket carburetor for high-performance VW applications.

Four- (and three-) barrel carburetors

Models 4150/4160/4180/4190   Introduced in 1957 on some Ford engines. Used on Chevrolet and Chrysler high-
3150/3160 (three-barrel versions)   performance engines later. Most popular carburetor for high-performance aftermarket
                                     applications.

Models 4165/4175   Introduced in 1971 as a bolt-on replacement for spread-bore carburetors (GM,
                                     Chrysler, Ford). The 4165 has mechanically operated secondaries; the 4175 has
                                     vacuum-operated secondaries and is available in a feedback version.

Models 4010/4011   Introduced in the late 80's. High performance replacement for original equipment
                                     applications. Constructed of aluminum. 4010 has standard Holley square flange; 4011
                                     based on Rochester Quadrajet spread-bore configuration.

Model 4360   Introduced in 1976. Replacement for the Rochester Quadrajet on '65 - '82 GM
                                     vehicles. Also replaces the Thermoquad on '75 and later Chrysler products.

Model 4500   Introduced in 1970 for non-street use only. Two flow rates: 1050 and 1150 cfm.

4.1 Model 2300/2305/2380 (arrow points to list number)

Models 2300/2305/2380

The model 2300 (see illustration) has been used in factory-installed single and triple-carburetor applications and has also enjoyed much success on the racetrack, in classes where only single two-barrel carburetors are allowed. Depending on the model, airflow ratings vary from approximately 200 to 600 cfm. It is very important to know that the 2300 is constructed exactly like the front half of a 4150 or 4160. All overhaul steps are exactly the same except for the addition of the back-half on the other models.

The idle adjustment on some 2300's is reversed. The screw turns IN to richen and OUT to make it lean, exactly the opposite of normal adjustment. There is a sticker or stamp on the carburetor that identifies these models.

On triple carburetor set-ups, the outboard carburetors don't have chokes or accelerator pumps, but use metering plates. The center carburetor has a choke, a diaphragm-type accelerator pump and a metering block with traditional screw-in main jets and a power valve. The throttle valves on the outboard carburetors are vacuum operated.

4.2 Model 4160 (arrow points to list number)

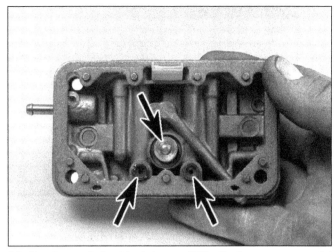

4.3 Here's a metering block from a 4150 - the lower arrows point to the main jets; the upper arrow points to the power valve

4.4 The 4160 doesn't use a metering block on the secondary side - it has a metering plate, with fixed-size orifices instead of screw-in jets and no power valve

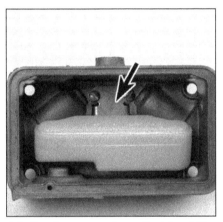

4.5 Here's a center-hung float bowl (the arrow points to the float pivot) - these are generally the most desirable type of float bowls for oval track or road racing use, since fuel sloshing side-to-side will have little effect on the float level

4.6 A side-hung float bowl is good for drag racing because fuel sloshing to the rear will have little effect on the float level (this holds true for a conventionally mounted carburetor, with the float bowls situated in line with the vehicle. If the carburetor is mounted sideways, like in some dual-carb setups, you'd want to run center-hung floats).

Warning: *The throttle bores on model 2300 carburetors are offset towards the rear of the carburetor. Be very careful when installing the base gasket onto the manifold - if it isn't in perfect alignment it could cause the throttle plates to hang up in the open position. Be sure to check the throttle action by hand before starting the engine.*

In 1986 Holley introduced a high-performance carburetor for four-cylinder engines, called the model 2305. It's capable of either 350 or 500 cfm and uses many model 2300 parts. However, unlike the model 2300, the throttle plates on the 2305 operate independently of each other The secondary throttle plate is opened by a mechanical linkage, but doesn't start opening until the primary throttle plate has opened 40-degrees.

Models 4150/4160/4180/4190/3150/3160

The carburetors in this family are all quite similar to one another. These are the carburetors that most likely come to mind when people think of aftermarket four-barrel carburetors **(see illustration)**. The 4150 made its debut in 1957 on some Ford models, and later gained popularity among other manu-

facturers. Flow ratings vary from 390 to 850 cfm. The 3150 and 3160 models are three-barrel versions of the 4150 and 4160. Besides that, all other component descriptions and service procedures are the same as the 4150 and 4160.

The 4150 and 4160 are very similar to each other. The difference lies in the secondary metering block - the 4160 doesn't have one. Instead, it has a thin metering plate with fixed-size orifices to regulate fuel flow instead of screw-in jets and a power valve like the 4150 **(see illustrations)**. The 4160 is basically just a lower cost version of the 4150. It can easily be turned into a 4150 by purchasing a conversion kit made by Holley.

4150's also are equipped with two accelerator pumps - one on the primary side and one on the secondary side. Most of these have mechanically operated secondaries, while all versions of the 4160, except one, have vacuum operated secondaries (the large vacuum diaphragm housing on the side of the carburetor is the quickest way to distinguish the 4160).

Two different style float bowls have been used on these carburetors - center-hung and side-hung **(see illustrations)**.

4.7 Model 4010 (Model 4011 similar) - arrow indicates location of list number

Center-hung floats are generally more desirable for oval track and road racing applications. For drag racing the side-hung style is preferred. Side-hung bowls can be replaced with center-hung bowls, and vice-versa. The float level on either type of bowl is externally adjustable (except for on one certain style of side-hung bowl).

The 4180 is a modified 4160, designed to meet strict emissions control regulations yet still offer enhanced performance. The metering block has been redesigned and the idle mixture adjustment screws have been eliminated. They are located in the throttle body, are factory set and sealed at the factory to prevent tampering. The discharge boosters are also different.

The 4190 is similar to the 4180. Its usage is limited almost exclusively to Ford heavy-duty truck applications.

Models 4165/4175

These models were introduced in 1971 as a direct replacement for Rochester Quadrajet and Chrysler Thermoquad carburetors. The 4165 and 4175 were big favorites for replacing the Rochester Quadrajet and Carter Thermo-Quad carburetors. They are actual "bolt-on" items - even the stock air cleaner and throttle linkage can be used!

Although these carburetors resemble the 4150 and 4160 models, they really aren't the same at all. Most of the parts are completely different. The most obvious difference is the spread-bore throttle body configuration. The mounting holes are spaced at 4-1/4 X 5-5/8 inches, and the secondary throttle plates are much larger than the primaries.

The 4165 can be considered as the 4150 counterpart. It has mechanically actuated secondaries. It is available in two versions: standard and high-performance. The standard version is equipped with side-hung float bowls and ports for all the necessary vacuum and emissions control hook-ups. The high-performance models are similar, but they feature center-hung float bowls, and may also necessitate the purchase of an aftermarket air cleaner and fuel line to complete the installation, which sort of takes them out of the "bolt-on" replacement category.

The 4175 is similar to the 4165, but has vacuum-actuated secondaries and no accelerator pump on the secondary side. It is intended for use on trucks, motorhomes and other vehicles where mechanical secondaries would cause the engine to bog

under low speed, high throttle angle operating conditions. It also works well on many GM and Chrysler passenger cars. It's available as a replacement for Quadrajets with mixture control solenoids (feedback systems) and works with the existing ECM.

Model 4500

Also known as the "Dominator," this carburetor is intended for the race track or extreme high-performance use only - it is not legal for street use. It's available in two flow rates - 1050 and 1150 cfm. They are easily identifiable by their extremely large (two-inch) throttle bores and the lack of a choke tower (all 4500's since 1983 have been manufactured without choke towers, since everyone who uses these carburetors removes the towers anyway). The main body and the throttle body are cast as one unit. The linkage between the primary and secondary throttle bores is concealed on the underside of the throttle body. All models in this family are "double pumpers."

Operation and overhaul procedures are similar to that of the 4150.

Models 4010/4011

In the late 1980's Holley realized the need for a new line of performance carburetors and came up with Models 4010 and 4011 **(see illustration)**. While there was nothing actually wrong with their existing carburetors, they did evolve from original equipment units. The 4010 and 4011 were designed from scratch, not being based on any existing carburetor.

The 4010 is based on the standard Holley square bore throttle body, while the 4011 is based on the Rochester Quadrajet spread bore configuration. Each carburetor is available in two versions - mechanically actuated secondaries or vacuum actuated secondaries. The models with mechanically actuated secondaries are equipped with two accelerator pumps; one on the primary side and one on the secondary side (these are called double-pumpers).

Each unit is available in two different airflow ratings. The 4010 is rated at 600 or 750 cfm; The 4011 is available in either a 650 or 800 cfm version.

These carburetors are made out of aluminum and weigh about half that of a 4160, or approximately six pounds. The throttle body is an integral part of the main body. The float bowls are also integral with the main body - they're not separable like on many other Holleys. The floats are of the center-hung type and the float levels are externally adjustable. The airhorn forms the entire top of the carburetor and serves as the cover for the float bowls. Marine versions are available, too.

Other features include hookups for emission control hoses, a manual choke linkage (on most models) and many plastic parts. Because of the latter feature, extreme care must be taken when overhauling one of these carburetors - you don't want to immerse any of these plastic components in carburetor cleaner, since most of these solutions will attack plastic. **Note:** *These carburetors use Torx head screws, so you'll have to buy a No. T-20 and a No. T-25 Torx driver if you plan to work on them.*

Model 4360/2360/6360

The model 4360 **(see illustration)** was first introduced in 1976 as a direct bolt-on replacement for the Rochester Quadrajet. It is compatible with most 1965 through 1982 GM vehicles that use the Q-jet. It can also replace the Thermo-Quad carburetor on 1975 and later Chrysler products. **Note:** *A*

4.8 Model 4360 - if you think this carburetor looks kind of like a Rochester Quadrajet, that's because it is a direct bolt on replacement for the Q-Jet! It also replaces the Thermoquad carburetor on Chrysler products (arrow indicates location of list number)

4.10 Model 2280

feedback version of the 4360 is not available. 1982 and later GM models with computer-controlled carburetors must use Holley's model 4175.

The 4360 is built on a spread bore platform, just like the Quadrajet. It uses mechanically actuated secondaries, which begin to open when the primary throttle plates reach an angle of 40-degrees. The carburetor is made up of three aluminum castings; the air horn, main body and throttle body.

The Model 2360 is basically the same as the primary side of the 4360. It was designed as a direct bolt-on replacement for the Rochester Dual-Jet carburetor.

Model 6360 is a feedback version of the 2360. This carburetor is for use on 1981 and later GM vehicles with computer-controlled carburetors.

Models 5200/5210/5220/6500/6510/6520

These carburetors were designed for small (mainly four-cylinder) engines, and are sometimes referred to as the "Holley-Weber" **(see illustration)**. They are staged, or progressive carburetors. The secondary throttle plate is mechanically actuated. They have been used on many foreign and domestic compact cars and small trucks.

Models 6500, 6510 and 6520 use duty-cycle solenoids (mixture control solenoids) and are for use with computer-controlled (closed loop) systems.

4.9 Model 5200, also known as the Holley-Weber (arrows indicate location of the list number)

These carburetors are most easily identifiable by the square choke towers and (on some models) the four air cleaner mounting studs protruding from the airhorn. All models have automatic chokes. Some are electric and others are thermostatically operated by the engine coolant.

Models 2280/6280

Introduced in the '70's, the Model 2280 **(see illustration)** was used mainly on original equipment installations, although it's also available as an aftermarket item. From 1978 to the late 1980's it was installed on Chrysler product 318 cubic inch V8's in passenger cars and trucks. It can also be used as a replacement for the Carter BBD on earlier Chrysler products, as well as the Rochester 2GC on many GM models.

The 2280 is made up of three aluminum castings - the air horn, the main body and the throttle body. Most models are equipped with a divorced choke system while a few are equipped with a manual choke. The accelerator pump isn't like the pumps on most Holleys. It's a piston type, not a diaphragm type.

The Model 6280 is the same as the 2280, except it has a duty cycle (mixture control) solenoid for use on vehicles with closed loop (computer controlled) systems.

Models 2210/2211/2245

The model 2210 **(see illustration)** carburetor was originally designed for the Chrysler 360 and 383 cubic inch engines. Aftermarket carburetors are also offered for other Chevrolet and Chrysler models.

Model 2245 was introduced on 1974 Chrysler products. It's very similar to the 2210, but the air horn is different (it'll accept a stock GM air cleaner) and the power valve is different. The 2211 is also quite similar, but the base is wider to replace a Rochester 2GC carburetor.

These carburetors feature three-piece construction - an aluminum throttle body, a zinc air horn and zinc main body. A piston type power valve and a side hung float are used. Airflow is rated at 250 cfm for the 2211 and 380 cfm for the 2210 and 2245.

Models 2100/2110

These very straightforward two-barrel carburetors **(see illustration)** were originally installed on 1930's-era Ford cars and trucks. The early versions were equipped with mechanical

4.11 Model 2210 (arrow points to list number)

4.12 Model 2110 (2100 similar) - arrow points to list number

4.13 Model 1940 - this is a Holley carburetor, even though it says Autolite on the side (arrow points to list number)

4.14 Model 1920 - arrow points to location of list number

chokes. Later (mid-50's) versions were equipped with electric chokes.

These carbs are made of three separate castings; a zinc air horn, a zinc main body and a cast iron throttle body. The boosters are also cast from zinc. A positive drive, plunger-style accelerator pump is used. The main difference between the 2100 and the 2110 is in the booster design, but the air horn and main body castings are slightly different, too.

Although these carburetors were pretty much retired from original equipment installations in the late 50's, they regained popularity in the 70's amongst Volkswagen enthusiasts looking for an affordable high-performance carburetor for modified VW engines. This resurrected carburetor was nick-named the "Bug Spray."

These carburetors are available in 200 or 300 cfm versions. The 200 cfm model is recommended for most applications and will give the most satisfactory performance for all-around driveability. The 300 cfm model is recommended for highly-modified engines (large displacement, flowed dual-port heads, etc.).

These carburetors aren't equipped with vacuum ports, so if you plan to install one on your VW you'll have to purchase an aftermarket, centrifugal-advance-only distributor. An aftermarket intake manifold will also be required, since the stock manifold won't accept the larger carburetor, and for the sake of performance you'd want to do this anyway.

Models 1940, 1945, 1946, 1949, 6145, 6146 and 6149

This group of carburetors were installed mostly on Ford in-

line six-cylinder engines used in industrial applications (tractors, machinery etc.), cars and trucks. The carburetor is usually marked "Autolite" on the side. These models **(see illustration)** are also available as Holley aftermarket replacements.

These are downdraft, one-barrel carburetors. They are all fairly similar to each other. The differences lie mainly in the design of the choke, power valve, throttle body and cfm rating.

They are constructed of three castings; the air horn and main body are made of zinc, the throttle body is aluminum. The accelerator pump is a spring-driven plunger type.

Models 6145, 6146 and 6149 are for use on vehicles with computer-controlled (or "closed loop") fuel systems. They are equipped with a duty cycle solenoid (sometimes called a mixture control solenoid) to regulate fuel flow.

Model 1920, 1904, 1908 and 1960

The Model 1920 **(see illustration)** first made its appearance on Chrysler 170 and 225 CID slant-six engines. Models 1904, 1908 and 1960 are very similar and were used on Ford cars and trucks. It is also commonly installed on farm equipment and other industrial engines.

The 1920 is basically a one-barrel downdraft carburetor consisting of three major castings. The throttle body is aluminum. The metering body and the fuel bowl are zinc. Some early models used a glass fuel bowl.

A metering block is installed between the float bowl and main body. It contains the main jet and accelerator pump check valves. The fuel bowl contains a side-hung, closed-cell (nitrophyl) float. The accelerator pump is either a spring-driven or cam-driven style. The choke may be manual or automatic (divorced or integral).

0-1850 (4160)
Universal Application, vacuum secondaries, non-emission, man. choke, 600 c.f.m.

0-2818 (4150)
Recommended for the off-road use on Chevrolet 1964-69 327-350 C.I.D. engs.
Vacuum secondaries, automatic choke, 600 c.f.m., non-emission carb. side inlet, manual trans. recommended.

0-3310 (4160)
Recommended for 350-455 C.I.D. engs., non-emission, vacuum secondaries, man. choke, 750 c.f.m.

0-3418-1 (4150)
Designed for 1966-67 corvette, 427 eng. (425 H.P.). Vacuum secondaries, square bore, single pump, secondary metering block, 855 c.f.m., non-emission.

0-4118 (4150)
Recommended for single quad high-rise manifolds. Designed & recommended only for off/road performance use, or for street use where local emission control legislation permits, non-emission carb.
Vacuum secondaries, automatic choke, 725 c.f.m.

0-4412 (2300)
Designed & recommended only for off/road performance use or street use where local emission control legislation permits. Manual choke, 50cc pump. 500 c.f.m., non-emission

0-4452-1 (4160)
1968-70 390, 428, 429 C.I.D. engs.
Vacuum secondaries, auto choke, 600 c.f.m. emission carb. Ford kick-down lever.

0-4548 (4160)
Designed for Ford/Mercury 1961-67; 289, 332, 352, 390 and 428 engines. Vaccum secondaries, square bore, single pump, 450 c.f.m.

0-4742 (4150)
Designed for Ford/Mercury 1966-69; 390 eng. Vacuum secondaries, square bore, single pump, secondary metering block, 600 c.f.m.

0-4749 (4160)
Designed for 1967-70 Chrysler, 383 & 440 engs. Vacuum secondaries, divorced choke, single pump, 600 c.f.m.

0-4776 (4150)
Designed & recommended only for off/road performance use or for street use where local emission control legislation permits; and recommended for small displacement engs. Mechanical secondaries, man. choke, double pump, 600 c.f.m., non-emission.

0-4777 (4150)
Designed & recommended only for off/road performance use, or for street use where local emission control legislation permits. Non-emission. Mechanical secondaries, man. choke, 650 c.f.m.

0-4778 (4150)
Designed & recommended only for off/road performance use, or for street use where local emission control legislation permits. Recommended for high output medium displacement engs. Mechanical secondaries, manual choke, double pump, 700 c.f.m., non-emission.

0-4779 (4150)
Recommended for high output medium displacement engines. Designed & recommended only for off/road performance use or for street use where local emission control legislation permits. Non-emission. Mechanical secondaries, manual choke, double pump, 750 c.f.m.

0-4780 (4150)
Recommended for large displacement engs. Designed & recommended only for off/road performance use, or for street use where local emission control legislation permits. Non-emission. Mechanical secondaries, manual choke, double pump, 800 c.f.m.

0-4781 (4150)
Recommended for high output large displacement engs. Designed & recommended only for off/road performance use, or for street use where local emission control legislation permits. Non-emission.
Mechanical secondaries, manual choke, double pump, 850 c.f.m.

0-4782 (2300)
Designed and recommended only for off/road performance use. Non-emission, center carburetor for 3 x 2 Automatic divorced choke, side pivot float. Originally designed for Dodge, Plymouth. 355 c.f.m.

0-4783 (2300)
Designed and recommended only for off/road performance use. Non-emission, outboard carburetor for 3 x 2, accelerator pump, side hung float. Originally designed for Dodge, Plymouth, 500 c.f.m.

0-6109 (4150)
Designed for off/road competition applications using 2 x 4 intake manifold. Mechanical secondaries, double pump, side inlet bowls, 750 c.f.m.

0-6210 (4165)
Direct str. replacement for all Quadra-Jet carbs on 1965-70-327, 350, 402 C.I.D. engs. Mechanical secondaries, spread bore, double pump, 650 c.f.m., emission carb.

0-6211 (4165)
Direct street replacement for all Quadra-Jet carbs on 1965-70 427, 454 C.I.D. engs. Mechanical secondaries, double pump, spread bore, 800 c.f.m., non-emission, divorced choke, side inlet.

0-6212 (4165)
Competition replacement for all Quadra-Jet carbs on 1965-70 427, 454 C.I.D. engs. Mechanical secondaries, spread bore, man. choke, double pump, 800 c.f.m., non-emission.

0-6213 (4165)
Hi-Performance direct str. replacement for all Quadra-Jet carbs on 1965-70 427, 454 C.I.D. engs. Mechanical secondaries, double pump, spread bore, 800 c.f.m., non-emission, divorced choke, center inlet.

0-6291 (4160)
Designed for 1971-70 Ford 351C. Vacuum secondaries, hot air choke, 600 c.f.m.

0-6299-1 (4160)
Designed & recommended for hi-perf. 4 cyl. Pinto-Vega applications. Manual choke, side pivot bowls, vacuum secondaries, 390 c.f.m.

0-6468 (4165)
Direct street replacement or all Quadra-Jet carbs, on 1971 350-402-454 C.I.D. engs. Mechanical secondaries, spread bore, double pump, 650 c.f.m., emission carb.

0-6497 (4165)
Direct street replacement for 1970-72, 455 C.I.D. engs. (except Toronado)
Mechanical secondaries, spread bore, double pump, 650 c.f.m., emission carb., integral choke.

0-6498 (4165)
Direct str. replacement for 1967-71 350, 400, 428, 455 C.I.D. engs. (except Ram Air)
Mechanical secondaries, spread bore, double pump, 650 c.f.m., divorced choke.

0-6499 (4165)
Performance replacement for Quadra-Jet carbs on 1971, 350, 454 C.I.D. eng. Mechanical secondaries, manual choke, spread bore, double pump, 650 c.f.m., non-emission, center inlet.

0-6512 (4165)
Direct str. replacement for all Quadra-Jet carbs on 1966-69 350, 400, 425, 455 C.I.D. engs.
Mechanical secondaries, spread bore, double pump, 650 c.f.m., emission carb., integral choke.

0-6520 (4160)
Designed for 1971 Ford 429 eng. Vacuum secondaries, hot air choke, 600 c.f.m.

0-6528 (4165)
Direct str. replacement for 1968-71 large block & 1968-70-350 C.I.D. eng. Mechanical secondaries, spread bore, double pump, 650 c.f.m., emission carb.

0-6619 (4160)
1970-72 307, 350, 402 C.I.D. engs.
Vacuum secondaries, auto choke, 600 c.f.m., emission carb.

Carburetor identification chart (1 of 6)

0-6708 (4150)
1970-72 327-400 C.I.D. engs. not for use on EGR applications. Mechanical secondaries, double pump, man. choke. 650 c.f.m. Non emission.

0-6709 (4150)
1970-72 400, 455 C.I.D. engs. not for use on EGR applications. mechanical secondaries, double pump, man. choke. 750 c.f.m. Non emission.

0-6710 (4165)
Competition replacement for all Carter Thermo-Quad carbs. on 1971-72 340, 400 C.I.D. engs. Mechanical secondaries, manual choke, spread bore, 800 c.f.m., non emission, center inlet.

0-6711 (4165)
Direct replacement for all Carter TQ carbs on 1971-72, 340, 400 C.I.D. eng. Mechanical secondaries, spread bore, double pump, 650 c.f.m., emission carb., divorced choke.

0-6772 (4165)
Direct street replacement for all Quadra-Jet carbs on 1972 350, 402, 454 C.I.D. engs. Mechanical secondaries, spread bore, double pump, 650 c.f.m., emission carb.

0-6773 (4165)
Direct str. replacement for all Quadra-Jet carbs on 1972 350, 400, 455 C.I.D. eng. (exc. Ram Air)
Mechanical secondaries, spread bore, double pump, 650 c.f.m., divorced choke, emission design.

0-6774 (4165)
Direct street replacement for 1972-350, 455 C.I.D. engs. Mechanical secondaries, spread bore, double pump, 650 c.f.m., emission carb.

0-6853 (4165)
Competition replacement for all Quadra-Jet carbs. on 1965-70 327, 350, 402 C.I.D. engs. Mechanical secondaries, spread bore, double pump, 650 c.f.m., non emission, divorced choke.

0-6895 (4150)
Universal, competition 390 c.f.m. double pump, no choke, mechanical secondaries.

0-6909 (4160)
1970-72 318-360-383-400-413 C.I.D. engs.
Vacuum secondaries, auto choke, 2 stage power valve 600 c.f.m. emission carb.

0-6910 (4165)
Competition replacement for al Quadra-Jet carbs. on 1966-73 Oldsmobile (exc. Toronado) 350-400-425-455 C.I.D. engs. & 1967-73 Pontiac 350, 400, 428, 455 C.I.D. engs. Mech. secondaries, man. choke, spread bore, 800 c.f.m., non emission, center inlet.

0-6919 (4160)
1970-72 302, 360, 390 C.I.D. engs.
Vacuum secondaries, auto choke, 2 stage power valve, 600 c.f.m., emission carb. Ford kickdown lever.

0-6946-1 (4160)
Designed for 1976 Ford Trk. 390 eng. F250, 350, hot air choke, 600 c.f.m., vacuum secondaries.

0-6947 (4160)
Designed for 1975 Ford Trk. 390 eng. F250, F350, hot air choke, 600 c.f.m., vacuum secondaries.

0-6979 (4160)
1973-74 350, 454 C.I.D. engs.
Vacuum secondaries, auto choke, 2 stage power valve, 600 c.f.m., emission carb.

0-6989 (4160)
1973-74 302 C.I.D. engs. 1973-74 304 and 1973-77 360-401 C.I.D. engs. Vacuum secondaries, auto choke, 2 stage power valve, 600 c.f.m., emission carb. Ford kickdown lever.

0-7001 (4165)
Direct str. replacement for 1973 350, 455 C.I.D. engs. (exc. Toronado)
Mechanical secondaries, spread bore, double pump, 650 c.f.m., emission carb., integral choke.

0-7002 (4175)
Direct replacement for 1973-74 350, 454 C.I.D. engs., emission carb. Vacuum secondaries, spread bore, 650 c.f.m.

0-7004 (4175)
Direct str. replacement for all Carter T.Q. carbs. on 1973-74 340-400 C.I.D. engs., emission carb., vacuum secondaries, single pump, spread bore, 650 c.f.m. Comes with divorced choke 45-84.

0-7005 (4175)
Direct str. replacement for all Carter T.Q. carbs. on 1973-74 440 (A-120) stand. engs.
Vacuum secondaries, single pump, spread bore, 650 c.f.m., emission carb. Comes with divorced choke 45-84.

0-7006 (4175)
Direct replacement for all Carter T.Q. carbs. on 1973-74 440 (A-134) perf. option engs.
Vacuum secondaries, single pump, spread bore, 650 c.f.m., emission carb. Comes with divorced choke 45-84.

0-7009 (4160)
1973-74 318-360-383-400-413 C.I.D. engs. Vacuum secondaries, auto choke, 2 stage power valve, 600 c.f.m., emission carb.

0-7010 (4160)
Universal street performance recommended for 1970-72 350-455 eng. Center pivot fuel bowls, vacuum secondaries, auto choke, 780 c.f.m.

0-7053-1 (4160)
Designed for 1974 Ford, 460 eng., hot air choke, side pivot bowls, vacuum secondaries, 600 c.f.m.

0-7054 (4165)
Direct str. replacement for all Quadra-Jet carbs. on 1973 400-455 C.I.D. engs. (exc. Ram Air & Firebird & GTO w/shaker hood)
Mechanical secondaries, spread bore, double pump, 650 c.f.m., integral choke, emission design.

0-7154 (4160)
Designed for 1972-73 Ford Cobra 351, hot air choke, vacuum secondaries, 600 c.f.m.

0-7320 (4500)
Universal, off/road, 1150 dominator for competition use only.

0-7351 (4175)
Direct replacement for all Quadra-Jet carbs. on 1974 350, 455 C.I.D. engs. (exc. Toronado)
Vacuum secondaries, single pump, spread bore, 650 c.f.m., emission carb., integral choke.

0-7397 (4175)
Direct street replacement for all 1974 350-455 eng. Vacuum secondaries, integral choke, single pump, side inlet bowls. Emissions carburetor.

0-7413 (4160)
Designed for 1973-74 Ford L.D. Trk. 460 eng., hot air choke, vacuum secondaries, 600 c.f.m.

0-7448 (2300)
Manual choke. Designed & recommended only for off/road performance use or street use where local emission control legislation permits. 350 c.f.m, non emission.

0-7454 (4360)
Direct replacement for all Quadra-Jet carbs on 1973-74 350, 400 C.I.D. engs. Mechanical secondaries, spread bore, 450 c.f.m., emission carb.

Carburetor identification chart (2 of 6)

0-7455 (4360)
Direct replacement for all Quadra-Jet carbs. on 1970-72 350-400 C.I.D. engs. Mechanical secondaries, spread bore, divorced choke, 450 c.f.m., emission carb.

0-7456 (4360)
Direct replacement for all Quadra-Jet carbs. on 1965-69 327, 350, 400 C.I.D. engs. Mechanical secondaries, spread bore, divorced choke, 450 c.f.m., emission carb.

0-7556 (4360)
Direct replacement for all Quadra-Jet equipped 1967-71 350, 400, 428, 455 C.I.D. engs.
Mechanical secondaries, divorced choke, spread bore, 450 c.f.m., emission carbs.

0-7850 (4160)
Designed for 1975 Ford 460 eng., hot air choke, vacuum secondaries, 600 c.f.m.

0-7855 (4175)
Direct replacement for all Carter Thermo-Quad carbs on 1975-76 440 (A-134) perf. option engs. 1979-84 360; 1980-83 318 Dodge Trk.
Vacuum secondaries, single pump, spread bore, 650 c.f.m., emission carb., divorced choke.

0-7955 (4360)
1970-74 302 C.I.D. engine E-100-200 series vans. Calibrated for use w/Holley Manifold #300-6
Mechanical secondaries, spread bore, integral choke w/elec. assist 450 c.f.m., emission carb. Ford kickdown lever.

0-7956 (4360)
1975-76 351 C.I.D. engs. in E-100-150-200-250 series vans. Calibrated for use w/Holley Manifold #300-6.
Mechanical secondaries, spread bore, integral choke w/elec. assist, emission carb., 450 c.f.m. Ford kickdown lever.

0-7957 (4360)
1973-76 318 C.I.D. engs. in B-100-200 series vans. Calibrated for use w/Holley Manifold #300-7, emission carb. Mechanical secondaries, spread bore, divorced choke, 450 c.f.m.

0-7958 (4360)
1973-76 360 C.I.D. engs. in B-100-200 series vans. Calibrated for use w/Holley Manifold #300-7. Mechanical secondaries, spread bore, divorced choke, 450 c.f.m., emission carb.

0-7985 (4160)
1975-76 350-400 C.I.D. engs. (Trucks & R.V.'s)
Vacuum secondaries, auto choke, 2 stage power valve, 600 c.f.m., emission carb.

0-7986 (4160)
1975-76 351W C.I.D. engs. (Trucks & R.V.'s)
Vacuum secondaries, auto choke, 2 stage power valve, 600 c.f.m., emission carb. Ford kickdown lever.

0-7987 (4160)
1975-76 318-360-400 C.I.D. engs. (Trucks & R.V.'s)
Vacuum secondaries, auto choke, 2 stage power valve, 600 c.f.m., emission carb.

0-8001 (4360)
1971-72 318 C.I.D. engine B-100-200 series vans. Calibrated for use w/Holley Manifold #300-7. Mechanical secondaries, spread bore, divorced choke, 450 c.f.m., emission carb.

0-8002 (4360)
Direct replacement for Quadra-Jet or for use w/Holley Manifold #300-19 on 1975-76 350, 400 C.I.D. engs. in G-10-20 series vans. Integral choke, 450 c.f.m., emission design.

0-8003 (4360)
On 1971-74 350 C.I.D. engs. in G-10-20 series vans.
Direct replacement for Quadra-Jet or for use w/Holley Manifold #300-19. Mechanical secondaries, spread bore, divorced choke, 450 c.f.m., emission carb.

0-8004 (4160)
1975-76 350-400 C.I.D. engs. (passenger cars)
Vacuum secondaries, auto choke, 2 stage power valve, 600 c.f.m., emission carb.

0-8005 (4160)
1975-76 302, 351W, 351M, 400 C.I.D. engs.
1975-79 304 and 1978 360 C.I.D. AMC engs.
Vacuum secondaries, auto choke, 2 stage power valve, 600 c.f.m., emission carb. Ford kickdown lever.

0-8006 (4160)
1975-76 318-360-400-440 C.I.D. engs. (passenger cars exc. lean burn)
Vacuum secondaries, auto choke, 2 stage power valve, 600 c.f.m., emission carb.

0-8007 (4160)
Designed and recommended for use on small displacement 8 cyl. & performance 6 cyl. with aftermarket intake. Vacuum secondaries, side inlet bowl. Electric choke 390 c.f.m.

0-8059 (4175)
Direct replacement for all Quadra-Jet carbs. on 1975 400, 455 C.I.D. engs., emission carb. Vacuum secondaries, spread bore, single pump, 650 c.f.m., integral choke.

0-8060 (4175)
Direct replacement for all Quadra-Jet carbs. on 1976 400, 455 C.I.D. engs.
Vacuum secondaries, single pump, spread bore, 650 c.f.m., emission carb., integral choke.

0-8082-1 (4500)
Universal off/road 1050 c.f.m. dominator for competition use only.

0-8149 (4360)
Direct replacement for stock Q-Jet on 1975-76 350-400 eng. Features integral choke, EGR, PCV, mechanical secondaries, 450 c.f.m.

0-8156 (4150)
Designed for universal off/road competition, center inlet bowls, double pumps, mechanical secondaries, no choke.

0-8158 (4360)
Direct replacement on Quadra-Jet on 1975-76 350, 400, 455 C.I.D. engs.
Mechanical secondaries, single pump, spread bore, 450 c.f.m., emission carb.

0-8162 (4150)
Designed for universal, off/road competition center inlet bowls, double pumps, mechanical secondaries. No choke.

0-8204 (4360)
Direct replacement for 1975-76 Olds 350-455 eng. Mechanical secondaries, integral choke, EGR, PCV, spread bore design, 450 c.f.m., emission design.

0-8206 (4360)
(Except '76 400 lean burn)
Direct replacement for 1975-76 Chrysler 360-440 eng. Mechanical secondaries, divorced choke, EGR, PCV, spread bore design, 450 c.f.m.

0-8207 (4160)
Designed for 1976 Ford 460 eng., hot air choke, vacuum secondaries, 600 c.f.m.

0-8276 (4175)
Direct replacement for 1974 Chevrolet L.D. Truck 350 eng., vacuum secondaries, single accelerator pump, divorced choke, spread bore design, 650 c.f.m.

0-8302 (4175)
Direct replacement for 1974 Chevrolet L.D. Truck, 454 eng., vacuum secondaries, single accelerator pump, divorced choke, spread bore design, 650 c.f.m., emission design.

Carburetor identification chart (3 of 6)

0-8516 (4360)

(Odd Fire)
1975-76 231 C.I.D. eng. V-6 (calibrated for use w/Holley Street Dominator Manifold #300-22). Mech. secondaries, spread bore, integral choke, 450 c.f.m., emission carb.

0-8517 (4360)

1975-77 262, 305 C.I.D. eng. (calibrated for use w/Holley Street Dominator Manifold #300-19). Mechanical secondaries, spread bore, integral choke, 450 c.f.m., emission carb.

0-8546 (4175)

1969-72 350 C.I.D. eng. Vacuum secondaries, single pump, auto choke, 650 c.f.m., emission carb.

0-8642 (4360)

Direct replacement for 1973-74 Buick 350–455 eng. Mechanical secondaries, integral choke, EGR, PVC, spread bore design, 450 c.f.m.

0-8677 (4360)

(Even Fire)
1977-78 C.I.D. V-6 eng. (calibrated for use w/Holley Street Dominator Manifold #300-22). Mechanical secondaries, spread bore, single pump, 450 c.f.m., emission carb., electric integral choke.

0-8679 (4175)

Vacuum secondary versions of **0-6210** 650 c.f.m., emission carb.

0-8700 (4175)

1969 396 C.I.D. eng. (Truck & Van), emission carb. Vacuum secondary, single pump, auto choke, 650 c.f.m.

0-8877 (4360)

Direct replacement for 1977 Chevy 350 eng. Mechanical secondaries, integral choke, EGR, PCV, spread bore design, 450 c.f.m.

0-8879 (4175)

Direct replacement for 1974 Chevy Trk. 350 eng. Vacuum secondaries, divorced choke, PCV, spread bore design, 650 c.f.m., emission design.

0-8896 (4500)

Universal, off/road 1050 c.f.m. dominator for competition use only.

0-8958 (4360)

Direct replacement for 1977 Chrysler 400 Lean-Burn. Mechanical secondaries, divorced choke, EGR, PCV, spread bore design, 450 c.f.m., emission design. Direct T.Q. replacement.

0-9002 (4160)

Designed for 1977-80 Ford R.V. with street dominator manifold. Vacuum secondaries, elec. choke, side inlet bowls, 600 c.f.m. Ford kickdown lever.

0-9040 (4160)

Direct replacement for 1977 Ford 460 eng. Vacuum secondaries, elec. choke side inlet bowls, 600 c.f.m. Ford kickdown lever.

0-9088 (4360)

1978 Ford passenger 302 eng. (calibrated for use w/Holley Street Dominator Manifold #300-6, 300-30Z). Mechanical secondaries, single pump, spread bore, 450 c.f.m., emission carb. Ford kickdown lever.

0-9105 (4360)

Direct replacement for 1978 Chevy 350 eng. 305 Olds eng. Mechanical secondaries, EGR, integral choke, PCV, spread bore design, 450 c.f.m., inc. Calif. emission design.

0-9112 (4360)

Designed for 1978 Chevy with street dominator manifold 300-19. Mechanical secondaries, integral choke, spread bore design, 450 c.f.m.

0-9185 (4360)

Direct replacement for 1977 Pontiac 400 eng. Mechanical secondaries, PCV, integral choke, EGR, spread bore design, 450 c.f.m., emission design.

0-9188 (4150)

Designed for 1965 Chevelle 396 eng. (425 H.P.). Vacuum secondaries single pump, square bore, single pump, secondary metering block, 780 c.f.m.

0-9192 (4360)

Direct replacement for 1978 Buick, Olds 403 eng. Mechanical secondaries, integral choke, PCV, EGR, single pump, spread bore design, 450 c.f.m., emission design.

0-9193 (4360)

Direct replacement for 1978 Olds 350 eng. Mechanical secondaries, integral choke, PCV, EGR, single pump, spread bore design, 450 c.f.m.

0-9210 (4160)

1978-79 Dodge Van 318, 360 eng. (calibrated for use w/Holley Street Dominator Manifold #300-7 or 300-29Z). Vacuum secondaries, integral choke, 2 stage power valve, 600 c.f.m., emission carb.

0-9219 (4160)

Designed for 1978 Chevy Trk. with 300-19 street dominator. Vacuum secondaries, elec. choke, 600 c.f.m., emission design.

0-9254 (4160)

1979-81 Chevrolet/GMC Truck 350 V8
1978-79 Chevrolet Passenger 350 Eng.
1978-79 Oldsmobile
1978-79 Pontiac 400 Eng.
(calibrated for use w/Holley Street Dominator Manifolds). Vacuum secondaries, integral choke, 2 stage power valve, 600 c.f.m., emission carb.

0-9375 (4500)

Universal, off/road 1050 c.f.m. dominator (Annular Discharge) competition use only.

0-9377 (4500)

Universal, off/road 1150 c.f.m. dominator (Annular Discharge) competition use only.

0-9379 (4150)

Universal, competition 750 c.f.m. (Annular Discharge) double pumper, center inlet bowls.

0-9380 (4150)

Universal, competition 850 c.f.m. (Annular Discharge) double pumper, center inlet bowls.

0-9381 (4150)

Universal, competition 830 c.f.m. (Annular Discharge) double pumper, center inlet bowls.

0-9626 (4160)

1972-79 Jeep 304, 360, 401 engs. (calibrated for use w/Holley Street Dominator Manifold #300-31Z). Vacuum secondaries, integral choke, 2 stage power valve, 600 c.f.m., emission carb.

0-9645 (4150)

Universal, competition 750 c.f.m. (methanol) double pump, center inlet bowls.

0-9646 (4150)

Universal, competition 850 c.f.m. (methanol) double pump, center inlet bowls.

0-9647 (2300)

Universal, competition 500 c.f.m. (methanol) single pump, center inlet bowls.

0-9678 (4360)

Direct replacement for 1979 Chevrolet passenger 350 eng. Mechanical secondaries, integral choke, PCV, EGR, spread bore design, 450 c.f.m.

0-9694 (4360)

1978-80 200, 229 V-6 eng. (calibrated for use w/Holley Street Dominator Manifold #300-34). Mechanical secondaries, spread bore, single pump, 450 c.f.m., emission carb., electric choke.

0-9834 (4160)

Universal application, vacuum secondaries, non emission, elec. choke, 600 c.f.m.

Carburetor identification chart (4 of 6)

0-9875 (4360)
Direct replacement for Quadra-Jet equipped 1979, 305 Chevrolet engs. Mechanical secondaries, spread bore, integral choke.

0-9895 (4175)
Direct replacement for Quadra-Jet equipped 1975-78 350 L-48, L-82 engs. Vacuum secondaries, spread bore, single pump, 650 c.f.m.

0-9923 (4175)
Direct replacement for 1975-78 Chevy Trk. (3/4 & 1 ton) 350-4 bbl., vacuum secondaries, divorced choke, spread bore design, 650 c.f.m., emission design.

0-9931 (4360)
Direct replacement for Quadra-Jet equipped 1979, 350 engs. Mechanical secondaries, spread bore, 450 c.f.m.

0-9935 (4360)
Calibrated for use w/Holley Manifold #300-28Z, 1980 267, 305 C.I.D. engs., mechanical secondaries, spread bore, 450 c.f.m.

0-9948 (4175)
Direct replacment for 1976-77 L.D. Chevrolet Trk. with 350 eng. Vacuum secondaries, integral choke, spread bore.

0-9973 (4360)
Calibrated for use w/Holley Manifold #300-37, 1977-80 2189 c.c. Toyota, mechanical secondaries, electric choke.

0-9976 (4175)
Direct replacement for 1979 Chevrolet Trk. with 350 eng. 1981 with manual trans. (over 6,000 G.V.W.) Vacuum secondaries, integral choke, single accelerator pump, emission design.

0-80073 (4175)   650 c.f.m.
Direct replacement for Quadra-Jet equipped 1983 305 Chevy/GMC truck 1982-83 Pontiac Trans-Am; Chevrolet Z-28, 305 C.I.D. vacuum secondaries. Full emission provisions and direct computer hook-up.

0-80086 (4360)
Designed for 1980 Ford Trk. L.D.; 302 Eng. with Holley Manifold #300-30Z, 300-6. Mechanical secondaries, single pump, spread bore, 450 c.f.m., emission carb. Ford kickdown lever.

0-80095 (2305)
Designed and recommended only for off/road performance use on all domestic and import 4 cyl. engs., no choke, universal staged linkage, side pivot float, 500 c.f.m.

0-80098 (4180)
Direct replacement for 1983 Ford/Mercury 302 H.O. V-8 Mustang and Capri; has full emission provisions. Vacuum secondaries, square flange. 600 c.f.m.

0-80099 (4180)
Designed for use for 1983 Ford/Mercury 302 H.O. V-8 Mustang and Capri; inc. California. Has full emission provisions. Vacuum secondaries, square flange. 600 c.f.m.

0-80111 (4180)
Direct replacement for 1979 Ford L.D. Trk., 460 C.I.D. automatic kickdown linkage. Vacuum secondaries, square flange. 600 c.f.m., full emission provisions, hot air choke.

0-80112 (4180)
Direct replacement for 1979 Ford L.D. Trk., 460 C.I.D. automatic kickdown linkage. Vacuum secondaries, square flange. 600 c.f.m., full emission provisions, electric choke.

0-80120 (2305)
Designed and recommended for off/road performance use only on all domestic and import 4 cyl. engines, manual choke, universal staged linkage, side pivot float. 350 c.f.m.

0-80128 (4175)
Designed for 1982 Chevy/GMC Trk. 350 eng., vacuum secondaries electric choke. 650 c.f.m.

0-80133 (4180)
Designed for 1981-82 Ford Trk. 460 eng., auto trans., vacuum secondaries, hot air choke. 600 c.f.m.

0-80139 (4175)
Designed for 1980 Buick, Olds, 307 eng., vacuum secondaries, divorced choke. 650 c.f.m.

0-80140 (4175)
Designed for 1981-84 Buick, Olds 307; 1981 Pontiac 307, vacuum secondaries, electric choke, 650 c.f.m.

0-80145 (4150)
Designed for all small block applications. Ford kickdown linkage, secondary metering block, race type fuel bowls, vacuum secondaries single pump. 600 c.f.m.

0-80155 (4175)
Designed for 1985 360 Dodge Trk., vacuum secondaries, electric choke. 650 c.f.m.

0-80163 (4180)
Designed for 1984-85 Ford/Mercury 302 eng., full emission provision, vacuum secondary, square flange, 600 c.f.m.

0-80164 (4180)
Designed for 1984-85 Ford Trk., 351 eng., full emission provisions, vacuum secondaries, square flange, 600 c.f.m.

0-80165 (4180)
Designed for 1981-83, 1985, Ford Trk., 460 eng., full emission provision (Incl. Calif.), vacuum secondaries, square flange, 600 c.f.m.

0-80166 (4180)
Designed for 1983-85 Ford Trk., 460 eng., full emission provision (Incl. Calif.), vacuum secondaries, square flange, 600 c.f.m.

0-80169 (4175)
Designed for 1980-85 Chevy/G.M.C. Trucks with 454 eng. Vacuum secondaries, single pump, 650 c.f.m., emission design.

0-80186 (4500)
Universal/off-road 750 C.F.M. Dominator. Double pump, annular discharge, race bowls.

0-80431 (4160)
Ford 427 2x4 w/M.T. Primary carb., 550 c.f.m., vacuum secondary, electric choke.

0-80432 (4160)
Ford 427 2x4 w/M.T. Secondary carb., 550 c.f.m., vacuum secondary, no choke.

0-80436 (4150)
Universal performance vacuum secondaries Ford kickdown, electric choke, 850 c.f.m.

0-80450 (4160)
1970-74 GM pass. cars, 307, 455 V-8; 1970-74 Chev./GMC trucks, 350-454 V-8. Calibrated for use with Holley or other aftermarket manifolds. 600 c.f.m., vacuum secondary, electric choke, emission calibrated, emission provisions.

0-80451 (4160)
1975-79 GM pass. cars, 350-455 V-8; 1979-81 Chev./GMC trucks, 350 V-8. Calibrated for use with Holley or other aftermarket manifolds. 600 c.f.m., vacuum secondary, electric choke, emission calibrated, emission provisions.

0-80452 (4160)
1975-77 Ford truck, 351W, V-8; 1977-80 Ford truck, 351M, 400 V-8; 1979-80 Ford truck, 302 V-8. Calibrated for use with Holley or other aftermarket manifolds. 600 c.f.m., vacuum secondary, electric choke, emission calibrated, emission provisions.

0-80453 (4160)
1970-74 F.M.C. pass. cars, 302 V-8; 1970-72 F.M.C. pass. cars, 390 V-8; 1977-78 F.M.C. pass. cars, 351M V-8; 1970-72 Ford truck, 360, 390 V-8; 1970-74 Ford truck, 302 V-8; 1977-78 Ford truck, 351M V-8 (F100). Calibrated for use with Holley or other aftermarket manifolds. 600 c.f.m., vacuum secondary, electric choke, emission calibrated, emission provisions.

0-80454 (4160)
1970-77 Chrysler pass. cars, all V-8; 1975-79 Dodge truck, 318, 360 V-8; 1978 Dodge truck, 400 V-8 (over 6000 GVW). Calibrated for use with Holley or other aftermarket manifolds. 600 c.f.m., vacuum secondary, electric choke, emission calibrated, emission provisions.

0-80457 (4160)
1965-69 Universal application, all V-8s. Calibrated for use with Holley or other aftermarket manifolds. 600 c.m.f., vacuum secondary, electric choke, universal throttle level w/Ford auto. trans. kickdown, non-emission.

0-80460 (4160)
1975-78 Chev./GMC trucks, 350 & 400 V-8. Calibrated for use with Holley or other aftermarket manifolds. 600 c.f.m., vacuum secondary, electric choke, emission calibrated, emission provisions.

0-82010 (2010)
Universal performance, annular booster, Ford kickdown, manual choke, 350 c.f.m.

Carburetor identification chart (5 of 6)

0-82011 (2010)
   Universal performance, annular booster, Ford kickdown, manual choke, 500 c.f.m.

0-82012 (2010)
   Universal performance, annular booster, no choke, 560 c.f.m.

0-83310 (4160)
   Universal/off-road small and big block applications. Ford kickdown linkage, vacuum secondaries, manual choke, non-emission, 750 c.f.m.

0-84010 (4010)
   Universal application, dual fuel inlet, vacuum secondaries, 600 c.f.m., manual choke, square bore flange.

0-84011 (4010)
   Universal application, dual fuel inlet, vacuum secondaries, 750 c.f.m., manual choke, square bore flange.

0-84012 (4010)
   Universal application, dual fuel inlet, mechanical secondaries manual choke, square bore flange, 600 c.f.m.

0-84013 (4010)
   Universal application, dual fuel inlet, mechanical secondaries, manual choke, square bore flange, 750 c.f.m.

0-84014 (4011)
   Universal application, dual fuel inlet, vacuum secondaries, manual choke, spread bore flange, 650 c.f.m.

0-84015 (4011)
   Universal application, dual fuel inlet, vacuum secondaries, manual choke, spread bore flange, 800 c.f.m.

0-84016 (4011)
   Universal application, dual fuel inlet, mechanical secondaries, manual choke, spread bore flange, 650 c.f.m.

0-84017 (4011)
   Universal application, dual fuel inlet, mechanical secondaries, manual choke, spread bore flange, 800 c.f.m.

0-84020 (4010)
   Universal application, square mounting flange, vacuum secondaries, Ford automatic transmission, kickdown linkage, 600 c.f.m., dual fuel inlet

0-84035 (4010)
   Universal performance, annular booster, Ford kickdown, electric choke, 600 c.f.m.

Carburetor identification chart (6 of 6)

5 Troubleshooting

General information

A malfunctioning carburetor can cause a variety of problems, ranging from obvious symptoms like a no-start condition or deiseling (run-on after the engine has been shut off) to problems that are harder to track down, such as an intermittent fuel smell, a flat spot at a certain engine rpm or decreased fuel mileage.

This Chapter provides a reference guide to the more common carburetor-related problems which may occur during the operation of your vehicle. However, many of the symptoms described could have causes related to other systems. Review the following list and make sure all of these criteria are met before attempting to diagnose carburetor problems.

To operate properly, the carburetor requires:

A constant fuel supply
All linkages and emission control systems hooked up
Good engine compression
Healthy ignition system firing voltage
Correct ignition spark timing
An airtight intake manifold
Engine at operating temperature
All carburetor adjustments performed correctly

Problems in the above areas can cause the following symptoms:

Engine cranks but won't start, or is hard to start when cold
Engine is hard to start when hot
Engine starts, then stalls
Engine idles roughly and stalls
Engine runs unevenly or surges
Engine hesitates on acceleration
Engine loses power during acceleration or at high speed
Poor fuel economy
Engine backfires

Symptom-based troubleshooting

Note: *All adjustment procedures are contained in Chapter 7.*

Engine cranks but won't start, or is hard to start when cold

- **Improper starting procedure is being used**
  Verify that the proper starting procedure - as outlined in the owner's manual - is being used.

- **There's no fuel in the gas tank**
  Add fuel. Check the fuel gauge for proper operation.

- **The choke valve is not closing sufficiently when cold**
  Adjust the choke.

- **The choke valve or linkage is binding or sticking**
  Realign the choke valve or linkage as necessary. If you find dirt or "gum," (varnish), clean the linkage with carburetor cleaner spray. **Note**: *Do not oil the choke linkage.*

- **There's no fuel in carburetor**

    1 Disconnect the primary (low voltage) wires from the coil.

    2 Disconnect the fuel line at the carburetor. Connect a hose to the fuel line and run it into an approved fuel container.

    3 Crank the engine. If there is no fuel discharge from the fuel line, check for kinked or plugged lines.

    4 Disconnect the fuel line at the tank and blow it out with compressed air, reconnect the line and check again for fuel discharge. If there's still no fuel discharge, replace the fuel pump.

    5 If the fuel supply is okay, check the following:

    a) Inspect and - if plugged, replace - the fuel filter(s).

    b) If the fuel filter is okay, remove the air horn or fuel bowl and check for a sticking float mechanism or a sticking inlet needle. If they're okay, adjust the float level.

- **The engine is flooded**

    Note: *To check for flooding, remove the air cleaner. With the engine off, look into the carburetor bores. Fuel will be dripping off the nozzles and/or the carburetor will be very wet.*

    1 Verify that you're using the correct carburetor unloading procedure. Depress the accelerator to the floor and verify that the choke valve is opening. If it isn't, adjust the throttle linkage and unloader.

    2 Dirt in the carburetor preventing the inlet needle from seating. Clean the system and replace the fuel filter(s) as necessary. If you find excessive dirt, remove the carburetor, disassemble it and clean it.

    3 Defective needle and seat. Check the needle and seat for a good seal. If the needle is defective, replace it with a Holley matched set.

    4 Check the float for fuel saturation, a bent float hanger or binding of the float arm.

    5 Adjust the float.

Engine is hard to start when hot

- **The choke valve is not opening completely**

    1 Check for a binding choke valve and/or linkage. Clean and free up or replace parts as necessary. **Note:** *Do not oil the choke linkage - use carburetor spray cleaner only.*

    2 Check and adjust the choke thermostatic coil.

    3 See if the choke thermostatic coil is binding in the well or housing.

    4 On an integral choke system, check for a vacuum leak.

- **The engine is flooded**

    Refer to the procedure under *Engine cranks but won't start.*

- **There's no fuel in the carburetor**

    1 Check the fuel pump. Run a pressure and volume test.

    2 See if the float needle is sticking to its seat, or if the float is binding or sunk.

- **The float bowl is leaking**

    Fill the bowl with fuel and check for leaks.

- **Fuel is percolating**

    Open the throttle all the way and crank the engine to relieve an over-rich condition.

The engine starts, then stalls

- **The engine does not have a fast enough idle speed when cold.**

    Adjust the fast-idle speed.

- **The choke vacuum diaphragm unit is not adjusted to specification or it's defective.**

    1 Adjust the vacuum break to specification.

    2 If it's already correctly adjusted, check the vacuum opening for proper operation as follows. **Note:** *Always check the fast idle cam adjustment before adjusting the vacuum unit.*

    a) On an externally-mounted vacuum diaphragm unit, connect a piece of hose to the fitting on the vacuum diaphragm unit and apply suction with a hand-operated vacuum pump. The plunger should move inward and hold vacuum. If it doesn't, replace the unit.

    b) On an integral vacuum piston unit, remove the cover and visually inspect the piston and vacuum channel. If the piston is corroded or sticking, replace the assembly.

- **The choke coil rod is out of adjustment (models with a divorced choke)**

    Adjust the choke coil rod.

- **The choke valve and/or the linkage is sticking or binding**

    1 Clean and align the choke valve and linkage. Replace it if necessary.

    2 If you have to replace the linkage, be sure to readjust it.

- **The idle speed setting is incorrect**

    Adjust the idle speed to specifications (refer to the VECI decal in the engine compartment).

- **There's not enough fuel in the carburetor**

    1 Check the fuel pump pressure and volume.

    2 Check for a partially plugged fuel inlet filter. Replace the filter if it's dirty.

    3 Check the float level and adjust, if necessary.

-     **The engine is flooded**

    Refer to the procedure under *Engine cranks but won't start.*

The engine idles roughly and stalls

- **The idle mixture adjustment is incorrect**

    Adjust the idle mixture screws as described in the proper overhaul chapter.

- **The idle speed setting is incorrect**

    Reset the idle speed in accordance with the procedure outlined on the VECI decal in the engine compartment. Check the operation of the solenoid, if equipped.

- **The manifold vacuum lines are disconnected or improperly installed**

    Check all vacuum hoses leading to the manifold or carburetor base - make sure they're not leaking, disconnected or connected improperly. Install or replace vacuum lines as necessary.

- **The carburetor is loose on the intake manifold**

    Tighten the carburetor-to-manifold bolts to 100 in-lbs.

- **The intake manifold is loose or the gaskets are defective**

    Using a pressure oil can, spray light oil or kerosene around the manifold and the carburetor base. If the engine RPM changes, tighten the bolts or replace the manifold gaskets or carburetor base gaskets as necessary.

- **The hot-idle compensator (if equipped) is not operating**
  Normally, the hot idle compensator should be closed when the engine is running, but still cold, and open when the engine is hot (about 140-degrees F). Replace it if it's defective.

- **The carburetor is flooding**
  Refer to the procedure under *Engine cranks but won't start*.
  1 Remove the air horn and check the float adjustment.
  2 Check the float needle and seat for a good seal. If the needle is defective, replace it with a Holley matched set.
  3 Check the float for fuel contamination, a bent float hanger or binding of the float arm. Adjust to specifications.
  4 If excessive dirt is found in the carburetor, clean the fuel system and the carburetor. Replace the fuel filter(s) as necessary.

The engine runs unevenly or surges

- **There's a fuel restriction**
  Check all hoses and fuel lines for bends, kinks or leaks. Straighten and secure them if necessary. Check all fuel filters. If a filter is plugged or dirty, replace it.

- **There's dirt or water in the fuel system**
  Clean the fuel tank and lines. Remove and clean the carburetor.

- **The fuel level is too high or too low**
  Adjust the float. Verify that the float and float needle valve operate freely.

- **The main metering jet is loose or is the wrong size**
  Tighten or replace as necessary.

- **The power system in the carburetor is not functioning properly**
  1 A power valve or piston is sticking in the down position. Free it up or replace it as necessary.
  2 The power valve is loose, has the wrong gasket or is leaking around the threads. Tighten or replace as necessary.
  3 The diaphragm is leaking. Test it with a hand-operated vacuum pump and replace as necessary.

- **There's a vacuum leak somewhere**
  It is absolutely necessary that all vacuum hoses and gaskets are properly installed, with no air leaks. The carburetor and manifold should be evenly tightened to the specified torque values.

The engine hesitates on acceleration

- **The accelerator pump system is defective**
  A quick check of the pump system can be made as follows: With the engine off, remove the air cleaner, look into the carburetor bores and watch the pump stream while briskly opening the throttle valve. The pump jet should emit a full stream of fuel which strikes near the center of the venturi area.

  Piston type:
  Remove the air horn and check the pump cup. If it's cracked, scored or distorted, replace the pump plunger.

Piston and diaphragm types:
Check the pump discharge ball for proper seating and location. The pump discharge ball is located in a cavity next to the pump well. To check for proper seating, remove the air horn and gasket and fill the cavity with fuel. No "leak down" should occur. Restake and replace the check ball if it's leaking. Make sure the discharge ball, spring and retainer are properly installed.

Diaphragm type:
Check pump discharge as described above. Inspect the diaphragm and replace it if it's defective. Check the pump inlet ball valve clearance. Adjust the pump operating lever clearance.

- **There's dirt in the pump passages or in the pump jet**
  Clean and blow out the passages and jet with compressed air.

- **The fuel level is too high or too low**
  Check for a sticking float needle or binding float. Free up or replace parts as necessary. Check and reset the float level to specification.

- **The air horn-to-float bowl gasket is leaking**
  Tighten the air horn-to-float bowl fasteners.

- **The carburetor is loose on the manifold**
  Tighten the carburetor mounting bolts/nuts to 100 in-lbs.

There's no power during heavy acceleration or at high speed

- **The carburetor throttle valves aren't opening all the way**
  Push the accelerator pedal to the floor. Adjust the throttle linkage to obtain wide-open throttle.

- **The fuel filters are dirty or plugged**
  Inspect the filter(s) and replace as necessary.

- **The power enrichment system is not operating**
  Piston type:
  Check the power piston for free up-an-down movement. If the piston is sticking, check the piston and cavity for dirt or scoring. Check the power piston spring for distortion. Clean or replace as necessary.

  Piston and diaphragm types:
  Check the power-valve channel restrictions. Clean them if necessary.

- **The float level is too low**
  Check and reset the float level to specification.

- **The float is not dropping far enough into the float bowl**
  Check for a binding float hanger and for proper float alignment in the float bowl.

- **The main metering jets are dirty or plugged, or are the wrong size**
  1 If the main metering jets are plugged or dirty, or if there's a lot of dirt in the fuel bowl, the carburetor should be completely disassembled and cleaned.
  2 Check the jet sizes (see Chapter 9).

The engine is getting poor fuel economy

- **The engine needs a complete tune-up**
  Check engine compression. Examine the spark plugs. If they're dirty or improperly gapped, clean and regap or re-

place them. Check the condition of the ignition points (if equipped) and the dwell setting. Readjust the points if necessary and check and reset the ignition timing. Clean or replace the air cleaner element if it's dirty. Check for restrictions in the exhaust system and inspect the intake manifold for leaks. Make sure all vacuum hoses are properly connected. Make sure all emission systems are operating properly.

- **The choke valve is not opening all the way**
  1 Clean the choke and free up the linkage.
  2 Check the choke thermostatic bimetal coil for proper adjustment. Adjust the choke if necessary.

- **Fuel is leaking somewhere**
  Check the fuel tank, the fuel lines and the fuel pump for any fuel leakage.

- **The main metering jet is defective or loose or is the wrong size**
  Replace as necessary.

- **The power system in the carburetor is not functioning properly. The power valve or piston is sticking in the up position**
  Free up or replace as necessary.

- **There's a high fuel level in the carburetor or the carburetor is flooding.**
  1 Inspect the needle and seat for dirt. If either is damaged, replace the needle and seat assembly with a Holley matched set.
  2 Check the float for fuel contamination.
  3 Reset the float to specifications.
  4 If there's a lot of dirt in the float bowl, clean the carburetor.

- **Fuel is being pulled from the accelerator pump system into the venturi through the pump jet**
  Run the engine at a speed sufficient to squirt fuel from the nozzle and watch the pump jet. **Warning:** *Use a mirror to do this. DO NOT look directly into the carburetor while the engine is running.* If fuel is feeding from the jet, check the pump discharge ball for proper seating by filling the cavity above the ball with fuel to the level of the casting. No "leak-down" should occur with the discharge ball in place. Restake/replace the leaking check ball, defective spring or retainer as necessary.

- **The air bleeds or fuel passages in the carburetor are dirty or plugged**
  1 Clean and, if necessary, overhaul the carburetor.
  2 If gum or varnish is present in the idle or high speed air bleeds, clean them with carburetor cleaner spray.

The engine backfires

- **The choke valve is fully or partially open, or is binding or sticking**
  Free up the choke valve with carburetor cleaner spray. Realign or replace it if bent.

- **The accelerator pump is not operating properly**
  1 With the engine off, remove the air cleaner, operate the throttle and watch the pump discharge. Replace the pump cup or diaphragm as necessary.
  2 Adjust the pump stroke.
  3 Restake or replace the pump intake or discharge valve.

- **The spark plugs are old or dirty (fouled)**
  Clean or replace the spark plugs.

- **The spark plug wires are old or cracked**
  Test with a scope if possible, or watch the wires on a dark night with the engine running. Replace the wires if necessary.

- **The fuel filter is partially clogged.**
  Replace the filter.

- **The air pump diverter valve is defective (If the engine backfires on deceleration)**
  Check the hoses and fittings for tightness and leakage. Disconnect the diverter valve signal line. With the engine running, you should feel a vacuum. With the engine idling, hold your hand at the exhaust port. You shouldn't feel any air. If the valve or the hoses are defective, replace them.

The secondary throttle valves don't open

- **The throttle valves are sticking**
  1 Readjust the secondary throttle valve stop screw.
  2 The throttle valves are nicked or the throttle shaft is binding.
  3 Repair or replace the throttle valve.
  4 Check the throttle body for warpage.
  5 Tighten the throttle body screws evenly.

- **The secondary diaphragm is ruptured or leaking**
  Inspect the diaphragm. If it's damaged or defective, replace it.

- **The venturi vacuum ports are plugged**
  Try cleaning the ports with carburetor cleaner spray. It may be necessary to remove the diaphragm assembly and blow into the venturi with compressed air.

Basic vacuum troubleshooting

What is vacuum?

First, let's look at what vacuum is. In science, the term "vacuum" refers to a total absence of air; in automotive mechanics, vacuum refers a pressure level that's lower than the earth's atmospheric pressure at any given altitude. The higher the altitude, the lower the atmospheric pressure.

You can measure vacuum pressure in relation to atmospheric pressure. Atmospheric pressure is the pressure exerted on every object on earth and is caused by the weight of the surrounding air. At sea level, the pressure exerted by the atmosphere is 14.7 "pounds per square inch" (psi). We call this measurement system "pounds per square inch absolute" (psia).

But vacuum gauges don't measure vacuum in psia; instead, they measure it in "inches of Mercury" (in-Hg). Once in a while, you'll see another unit of measurement on some gauges; it's expressed in "kilopascals" (kPa). Another unit of measurement, used on manometers, is expressed in "inches of water" (in-H_2O). In some Japanese factory manuals, vacuum is referred to as "negative pressure." Don't let this term mislead you; the manufacturer is simply referring to in-Hg.

How is vacuum created in an internal combustion engine?

Positive pressure always flows to an area with a "less positive," or lower, pressure. This is a basic law of physics. Viewed from this perspective, an engine is really nothing more than an air pump. As the crankshaft rotates through two full revolutions, the engine cycles through its intake, compression, power and exhaust strokes. The first and last of these strokes - the intake and exhaust strokes - are identical to the action of the intake and exhaust strokes of any air pump: The intake "pulls" in air; the exhaust expels it.

During the intake stroke, the piston moves downward from its top dead center position. At the same time the exhaust valve closes and the intake valve opens. This downward movement of the piston in the cylinder creates a relative vacuum, drawing the air/fuel mixture into the cylinder through the open intake valve.

After the engine compression and power strokes are completed, the intake valve is still closed but the exhaust valve opens as the piston begins moving upward on its exhaust stroke. The rising piston forces the spent exhaust gases out through the open port.

The partial vacuum created by the engine's intake stroke is relatively continuous, because one cylinder is always at some stage of its intake stroke in a four, six or eight-cylinder engine. On carbureted engines, this intake vacuum is regulated, to some extent, by the position of the choke plate and the throttle valve. When the choke plate or throttle valve is in its closed position, airflow is reduced and intake vacuum is higher; as the plate or valve opens, airflow increases and vacuum decreases. The accompanying cutaways show how vacuum levels change during various engine loads **(see illustrations).**

Finding vacuum leaks

Vacuum system problems can produce, or contribute to, numerous driveability problems, including:

Deceleration backfiring
Detonation
Hard start condition
Knocking or pinging
Overheating
Poor acceleration
Poor fuel economy
Rich or lean stumbling
Rough idling
Stalling
Won't start when cold

The major cause of vacuum-related problems is damaged or disconnected vacuum hoses, lines or tubing. Vacuum leaks can cause problems such as erratic running and rough idling.

For instance, a rough idle often indicates a leaking vacuum hose. A broken vacuum line allows a vacuum leak, which allows more air into the intake manifold than the engine is calibrated for. Then the engine runs roughly due to the leaner air/fuel mixture.

Another example: Spark knock or pinging sometimes indicates a kinked vacuum hose to the EGR valve. If this hose is kinked, the EGR valve won't open when it should. The engine, which requires a certain amount of exhaust gas in the combustion chamber to cool it down, pings or knocks.

Here's another: A misfire at idle may indicate a torn or ruptured diaphragm in some vacuum-activated unit (a dashpot or EGR valve, for instance). The torn diaphragm permits air movement into the intake manifold below the carburetor. This air thins out the already lean air/fuel mixture at idle and causes a misfire. A misfire may also indicate a leaking intake manifold gasket or a leaking carburetor or throttle body base gasket. If a leak develops between the mating surfaces of the intake manifold and the cylinder head, or between the carburetor base

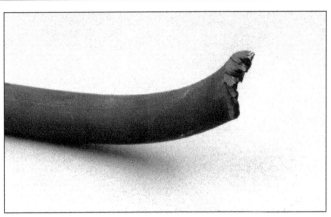

5.3 This vacuum hose was routed too close to an exhaust manifold - after being overheated repeatedly, it finally cracked and broke

gasket and the intake manifold, the extra air getting into the engine below the gasket causes a misfire.

If you suspect a vacuum problem because one or more of the above symptoms occurs, the following visual inspection may get you to the source of the problem with no further testing.

- Make sure everything is routed correctly - kinked lines block vacuum flow at first, then cause a vacuum leak when they crack and break.
- Make sure all connections are tight. Look for loose connections and disconnected lines. Vacuum hoses and lines are sometimes accidentally knocked loose by an errant elbow during an oil change or some other maintenance procedure.
- Inspect the entire length of every hose, line and tube for breaks, cracks, cuts, hardening, kinks and tears **(see illustration).** Replace all damaged lines and hoses.
- When subjected to the high underhood temperatures of a running engine, hoses become brittle (hardened). Once they're brittle, they crack more easily when subjected to engine vibrations. When you inspect the vacuum hoses and lines, pay particularly close attention to those that are routed near hot areas such as exhaust manifolds, EGR systems, reduction catalysts (often right below the exhaust manifold on modern FWD vehicles with transverse engines), etc.
- Inspect all vacuum devices for visible damage (dents, broken pipes or ports, broken tees in vacuum lines, etc.)
- Make sure none of the lines are coated with coolant, fuel, oil or transmission fluid. Many vacuum devices will malfunction if any of these fluids get inside them.

What if none of the above steps eliminates the leak? Grab your vacuum pump and apply vacuum to each suspect area, then watch the gauge for any loss of vacuum.

And if you still can't find the leak? Be sure to check the intake manifold or the base gasket between the carburetor and the manifold. To test for leaks in this area, squirt carburetor cleaner spray or WD-40 along the gasket joints with the engine running at idle. If the idle speed smoothes out momentarily, you've located your leak. Tighten the intake manifold or the carburetor fasteners to the specified torque and recheck. If the leak persists, you may have to replace the gasket. An alternative to spraying solvent is to use a short length of vacuum hose as a sort of "stethoscope," listening for the high-pitched hissing noise that characterizes vacuum leaks. Hold one end of the hose to your ear and probe close to possible sources of vacuum leakage with the other end. **Warning:** *Stay clear of rotating engine components when probing with the hose.*

TABLE 1 — VACUUM LEVELS DURING VARIOUS ENGINE LOADS

COLD START-UP, OPERATION AT FAST IDLE

The throttle plate opening uncovers the "S" then the "E" and "P" ports. Vacuum pressure at these ports and the manifold port is equal. The choke is full on.

Port	Vacuum Level
• "E" port	STRONG
• "P" port	(MAXIMUM)
• "S" port	
• Manifold	

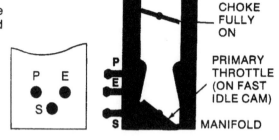

CHOKE FULLY ON

PRIMARY THROTTLE (ON FAST IDLE CAM)

MANIFOLD

COLD DRIVEAWAY, LIGHT THROTTLE

The throttle plate is farther open and vacuum decreases slightly. Vacuum will be strong with the choke plate closed and moderate when the choke starts to open.

Port	Vacuum Level
• "E" port	
• "P" port	STRONG to MODERATE
• "S" port	
• Manifold	

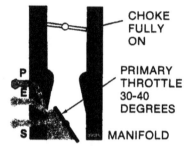

CHOKE FULLY ON

PRIMARY THROTTLE 30-40 DEGREES

MANIFOLD

WARMUP DRIVEAWAY, OR CRUISE, PART THROTTLE

The choke is partly off and the throttle plate has opened to a point where vacuum signals are equal and fairly strong.

Port	Vacuum Level
• "E" port	
• "P" port	STRONG to MODERATE
• "S" port	
• Manifold	

CHOKE PARTLY OFF

PRIMARY THROTTLE 30 DEGREES

MANIFOLD

HOT CRUISE, PART THROTTLE

Vacuum at the manifold, "P" and "S" ports is equal and moderately strong. Even though the "E" and "P" ports are closely positioned, "E" port vacuum is weakened because it "bleeds" off in the EGR valve integral transducer control.

Port	Vacuum Level
• "E" port	MODERATE to WEAK
• "P" port	
• "S" port	STRONG to MODERATE
• Manifold	

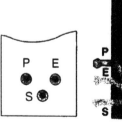

CHOKE OFF

PRIMARY THROTTLE

MANIFOLD

5.1 These cutaways (this and facing page) of a typical carburetor show how vacuum levels change during various engine loads

At this relatively "heavy" throttle positioning with the choke open, the vacuum level at all ports is weak.

"E" port ⎫
"P" port ⎬   WEAK
"S" port ⎮
Manifold ⎭

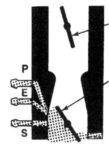

CHOKE OFF

PRIMARY THROTTLE AT 55 DEGREES

MANIFOLD

With the choke off and the throttle wide open, vacuum signals are very weak to none.

"E" port ⎫
"P" port ⎬   WEAK to ZERO
"S" port ⎮
Manifold ⎭

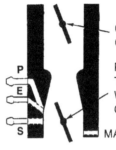

CHOKE OFF

PRIMARY THROTTLE WIDE OPEN

MANIFOLD

With the throttle closed and choke off, the "E," "P" and "S" ports are cut off from vacuum signals (below the throttle plate). Manifold vacuum is very strong.

"E" port ⎫
"P" port ⎬   ZERO
"S" port ⎭
Manifold ....MAXIMUM

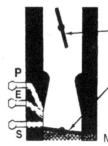

CHOKE OFF

PRIMARY THROTTLE CLOSED

MANIFOLD

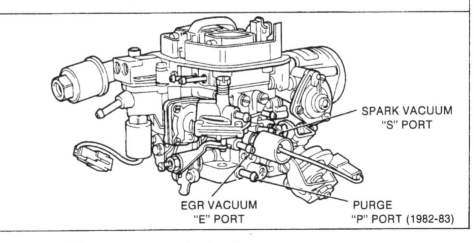

SPARK VACUUM "S" PORT

EGR VACUUM "E" PORT

PURGE "P" PORT (1982-83)

5.2 Carburetor cutaways (continued)

6 Carburetor removal and installation

This chapter takes you through removal and installation procedures step-by-step. Not all of the steps will apply to every vehicle and the carburetor you are working on may be equipped with some devices not shown here. This is a general removal and installation procedure which will otherwise work for all carburetors. The accompanying photographs illustrate the procedure on an original equipment two-barrel (2280) and also an aftermarket high-performance four-barrel (4150).

Removal

Warning: *Gasoline is extremely flammable, so take extra precautions when you work on any part of the fuel system. Don't smoke or allow open flames or bare light bulbs near the work area, and don't work in a garage where a natural gas-type appliance (such as a water heater or clothes dryer) with a pilot light is present. If you spill any fuel on your skin, rinse it off immediately with soap and water. When you perform any kind of work on the fuel system, wear safety glasses and have a Class B type fire extinguisher on hand.*

1   Remove the air cleaner **(see illustrations)**. Carefully detach all the vacuum lines going to the air cleaner, marking them with tape for easy identification. Remove the carburetor by following these steps:

2   Carefully disconnect the fuel line from the carburetor inlet

6.1 Remove the air cleaner housing cover.

6.2 Remove, and in this case throw away the air filter element. An air filter element like this can be as harmful to performance and economy as a plugged fuel filter. Replace it!

6.3a Remove the air filter housing. Shown here is a Model 4150 carburetor with mechanical choke and mechanical secondaries mounted on a Chevrolet 350 engine. Very straight forward on this model, with only one vacuum line and no electrical connections, etc. . . .

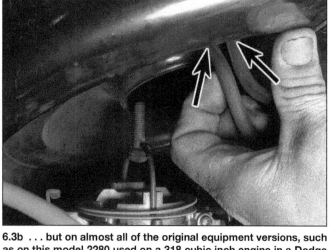

6.3b . . . but on almost all of the original equipment versions, such as on this model 2280 used on a 318 cubic inch engine in a Dodge van, you will find one or more connections under the lower air filter housing (arrows). Be sure to identify which hose goes to which connection by marking the lines or drawing a diagram. If reconnected backwards the control they operate might not work properly.

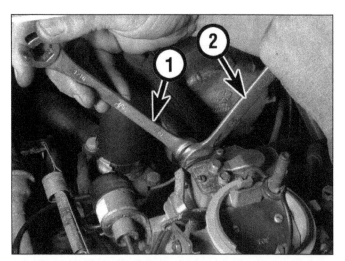

6.4 When removing the fuel line(s) always use the correct size wrench on the fuel line, preferably a "flare-nut" wrench (1) to avoid stripping the tube nut, and a "backup" wrench (2) to keep the carburetor inlet nut from turning

6.5a On models with a "divorced" style choke, which mounts on the manifold and is attached to the carburetor by a linkage rod, remove the retaining clip (arrow) from the choke rod and choke linkage

6.5b If the carburetor you're working on has a manual choke, loosen the set screw holding the cable to the linkage (right arrow) and also the clamp screw for the cable housing (left arrow)

assembly (**see illustration**). Be sure to catch any excess fuel with a shop rag.

3   Identify all the vacuum lines and electrical connectors going to the carburetor and carefully mark each one with tape or paint. Disconnect the lines and connectors.

4   Disconnect the PCV valve and hoses from the carburetor and valve cover or intake manifold.

5   Disconnect the choke rod or heat tube from the carburetor (**see illustrations**).

6   Disconnect the accelerator linkage from the carburetor (**see illustrations**). If equipped, also detach the cruise control and automatic transmission kickdown linkage.

7   Remove the carburetor nuts or bolts from the carburetor base (**see illustrations**).

8   Stuff clean shop rags or towels into the open intake manifold hole to prevent any dirt, tools, nuts or bolts from falling inside.

6.6a On carburetors with this type of linkage connections, first remove the springs and clips (arrows) from the throttle lever . . .

6.6b . . . then remove the cruise control linkage, the accelerator linkage and the transmission kickdown linkage from the throttle lever. Note: *Be sure to remove and mark the location of any spacers and/or washers used to align the linkage for smooth operation*

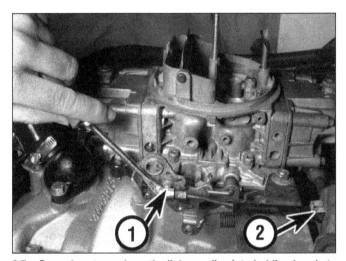

6.7a On carburetors where the linkage clips into holding brackets and onto the throttle shaft ballstud, first remove the cable from the ballstud by lightly prying from behind as shown (1). Then, using pliers, squeeze the plastic connector at the cable bracket (2) and slip the cable out, pushing it towards the firewall

6.7b Next unhook the return spring from the throttle linkage (arrow)

6.8 Remove the carburetor-to-manifold nuts or bolts

6.9 If access is difficult, use a box-end wrench (arrow) to remove the carburetor nuts

9   Remove the base plate gasket and the sealing gaskets from the intake manifold surface (**see illustration**) and from the bottom of the carburetor. It will probably be necessary to use a

6.10 Remove the base gasket from the manifold. Use a gasket scraper if necessary, but be careful not to gouge the manifold, which could cause a vacuum leak! Note: *Save as much of the gasket as possible in order to match it against one of the new ones provided in the overhaul kit*

6.11 Always use a new carburetor base gasket. Old ones become crushed and hardened due to age and heat and can't provide the kind of seal needed to prevent vacuum leaks (a very common cause of drivability problems)

good, stiff scraper to remove the old gasket material. Be careful not to gouge the surface of the intake manifold or carburetor, since this will create vacuum leaks.

10   There are always a few things to watch out for after the carburetor has been removed. If any of the carburetor studs were removed from the intake manifold along with the nut(s), it is a good idea to replace them with new ones. If none are available, try to clean-up the threads with a die or with a special thread file. Use only two nuts (back-up) to lock the stud in place while removing the seized nut from the other end. Using pliers will only damage the threads, making the original stud useless.

11   It is a good idea to locate the new base plate gaskets from your overhaul kit and compare them to the originals before scraping or replacing the base plate (if necessary). Many aftermarket kits will include several gaskets to cover different manifold designs for the same carburetor. Once the original gasket has been scraped and thrown out, it will be tough to decide which gasket is the correct one. Take a little time to get familiar with the parts from your overhaul kit and set the correct gaskets off to the side for use during installation.

Installation

12   Make sure the carburetor mounting studs are correctly installed in the intake manifold. If new studs were installed, make sure they are the correct length. If an adapter plate or spacer will be installed, make sure the studs are long enough to work properly. Many of these parts are available through Holley or automotive parts suppliers.

13   Remove the shop towels from the intake manifold and check for any objects that might have fallen into the intake manifold while the carburetor was off. Use a flashlight to look around before final approval.

14   Install the base gasket onto the intake manifold **(see illustration)**. Some models use a bakelite, phenolic or pressed paper base plate, sandwiched between two thin gaskets.

15   Install the carburetor and mounting nuts or bolts. Tighten the nuts or bolts to 100 in-lbs. Operate the throttle lever through its full range to make sure it works smoothly. If there is any binding, find out why and correct the cause before proceeding.

16   Connect the fuel line(s), tightening the tube nut(s) securely. Don't forget to hold the inlet fitting nut with a wrench to prevent it from turning.

6.12 When reconnecting the hoses, always check for hoses that are cracked or hardened. Don't waste your time doing a detailed overhaul and then putting worn-out hoses back on the connections. Make sure the hose is fairly new and soft enough to seal properly

17   Connect the vacuum hoses and electrical connectors. If any of the hoses are cracked or hardened, replace all of them at this time, since they are probably all very close to the same condition and age **(see illustration)**.

18   Finally, install a new air filter element. Following all the overhaul procedures won't do you any good if the engine can't breath.

Carburetor adjustments

Because of the many different carburetor models and types of choke mechanisms, accelerator pumps, bowl vents etc., there are quite a few of adjustment procedures, many of which are specific to an individual group of carburetors.

There are ten different overhaul Chapters in this manual. At the end of each overhaul illustration sequence there are initial assembly (or "dry") adjustments to be made as the carburetor is reassembled on the bench. Following these will be "on-vehicle" adjustments, which are made with the engine running and at operating temperature. Each adjustment will be identified by a heading preceding the procedure. **Note:** *There are many variations to each control system throughout the manual and occasionally a specific adjustment for a limited feature will not be covered. When this happens, which will be rarely, please refer to the instructions that come with each overhaul kit.*

7 Part A
Overhaul and adjustments

Overhaul tools and preparation

Regardless of which of the many one, two or four barrel carburetors you are going to overhaul, the initial preparation is the same for all of the following carburetor overhaul Chapters.

Take your time; think about what's involved. Nobody wants to or should have to do any more than is necessary. Are you going to be simply replacing a base gasket to correct a vacuum leak or is the carburetor in desperate need of a complete overhaul? Short cuts or quick fixes are nice, but only when the problem is corrected. It's always amazing how there's enough time to fix-it-right the *second* time around. Organization, cleanliness, good lighting and some advanced planning are all very important parts of getting the overhaul done quickly and correctly the first time.

Many of the parts being dealt with during an overhaul are extremely small and can be easily lost or misplaced, especially if you're using compressed air to clean out passages or dry off parts, so extra care must be taken to keep track of all of the many individual parts you will have separated during your disassembly for a carburetor overhaul.

Now that it's clear just what needs to be done it's time to put together needed parts, tools and supplies.

Note: *The numbering system for the illustrations in the overhaul procedures in this Chapter are slightly different than in the other Chapters in this book. The illustrations for each overhaul group will have a letter from A through J. The illustrations used in each Part will have their own series of illustration numbers. For example, 7C.34 will illustrate the 34th photo for the 4010/4011 model, which is Part C.*

7A.1 After you've read Chapter 4 and have identified the carburetor that you have on your vehicle, write down all the numbers that the parts department person will need to locate your parts. The decision you have to make is just how far you're going to take the overhaul. There are basic carburetor overhaul kits available from your local parts store and "Renew Kits", as Holley labels them, which contain all the necessary parts and instructions to rebuild the carburetor to original specifications. There is also a line of overhaul kits from Holley, called "Trick Kits", which contain several of each of the parts that can be changed for higher performance applications, such as power valves, accelerator pump actuating cams, etc. The choice is yours: original stock replacement parts all the way up to a competitive level of high performance.

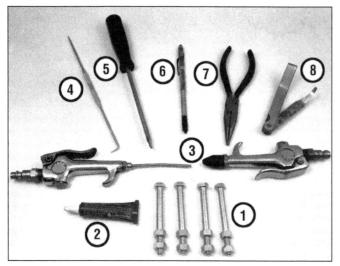

7A.2 Besides the general tools discussed in Chapter 2 some special tools are helpful to make your carburetor work easier

(1) Four 5/16 inch bolts and nuts make a very inexpensive carb stand
(2) A thread locking compound should be used on throttle body screws
(3) If using compressed air, either of these types of blow guns works well
(4) A small pick is useful for removing gaskets or small clips
(5) Clutch-head screws are used on the metering plate of some of the four-barrel models. There are clutch-head tools available, but a broken screwdriver can be ground down to the correct size
(6 & 7) A screw starter for regular and phillips screws and a needle nose pliers will come in handy in tight places
(8) A feeler gauge set is essential for clearance checks

7A.3 Using common nuts and bolts, you can make a very stable work stand for all the carburetors, with the exception of the one barrel models

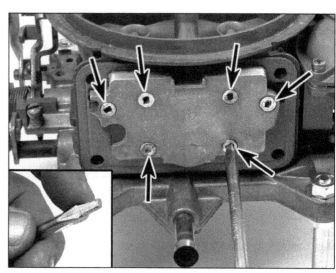

7A.4 With the use of a bench grinder, you can make your own clutch-head driver

7A.5 Before removing the carburetor from the intake manifold it will make things much easier if you loosen some of the parts while the carburetor is securely mounted to the manifold. Using a wrench on the fuel line and a back-up wrench on the carburetor nut, loosen items such as fuel line fittings . . .

7A.6 . . . fuel bowl sight plugs (if your model has them) . . .

7A.7 . . . and the lock screw for the needle-and-seat adjustment

7A.8 When removing a component where the gasket is holding the parts together, prying may become necessary. Always look for a casting protrusion or other external prying point rather than wedging a screwdriver between two parts. Besides damaging a gasket surface, which could cause fuel or vacuum leaks, there is always a possibility that hidden alignment dowels, tabs or vent tubes can be damaged or sheared off.

7A.9 When disassembling the carburetor to place it in cleaning solution or a dip tank, it is important to remember to remove all rubber and paper gasket material, electrical parts and components as well as all float assemblies from the carburetor. The cleaning solution or dip can damage non-metal parts and will swell some materials and cause them to interfere with fuel or air passages or make assembly difficult. Caution: *Remember: only metal parts can be placed into the dip tank.*

7A.10 Several different manufacturers make cleaners and dips for cleaning parts. They come in individual spray cans all the way up to bulk cans with a basket for lowering parts into the can. For thorough cleaning of metal parts, we recommend overnight soaking in a dip tank, such as the one at left. Warning: *Some cleaners and dips can be very strong caustic material that can irritate the skin. Wash your hands immediately after use of any cleaner. Read all manufacturer's label information and do not allow the dip or cleaner to contact your skin; gloves may be advisable. Also, wear safety glasses.*

7A.12 Since carburetors are complex devices, it will be difficult to remember how all the linkage on the carburetor is assembled. It's a good idea to take notes on how everything looks before disassembly, and even take instant photographs from various angles. Accelerator pump arms often have three holes (arrow) so they can be assembled one of three ways, allowing adjustment. Normally, you'll want to reassemble these parts the same way they were before disassembly.

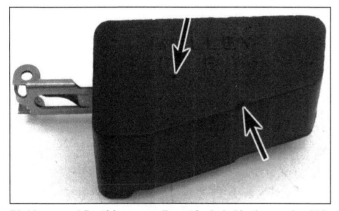

7A.14 . . . and float(s) are usually not included in the overhaul kit. This Nitrophyl float is nicked in the areas shown by the arrows and should be replaced, since damage to the surface coating will eventually result in a soggy, sinking float. Hollow plastic and metal floats can also leak, with the same results. If your float is plastic or Nitrophyl, it's a good idea to replace it routinely at overhaul time. Carefully check metal floats to be sure there's no fuel inside them and they're not dented or otherwise damaged.

7A.11 If dips, which are water soluble, are used, it is necessary to use some type of compressed air to clean and dry off the parts after they have been rinsed off with water. You must be sure, regardless of how the parts are cleaned, that all air and fuel passages are clean and not plugged before reassembly. Compressed air is the most effective way to clear passages. If you don't have access to an air compressor, you can buy an inexpensive portable air tank, which can be filled at any service station, or buy a can of compressed gas designed for cleaning electronic parts. Warning: *Always wear safety glasses whenever using compressed air. The solvents, cleaners and bits of debris can harm your eyes.*

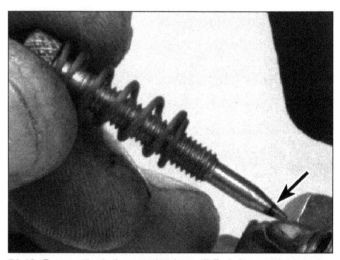

7A.13 Frequently during overhaul, you'll find damaged parts that aren't included in the overhaul kit, so be prepared to wait for special orders (If your local auto parts store can't get a part, Holley often can). It's very common to find idle mixture screws that have had their needle end damaged like this (arrow) through over-tightening. The idle mixture screw(s) . . .

7A.15 Before disassembly, check for excessive throttle shaft wear by trying to move the shaft up and down (arrows). Any movement should be almost imperceptible. If you can see and feel movement, there's enough wear to cause a vacuum leak; you'll have to find a replacement carburetor or have a carburetor overhaul shop install throttle-shaft bushings. We don't recommend removing the throttle shaft yourself, since the throttle plate screws are difficult to stake correctly on reassembly, and an improperly staked screw could vibrate loose and fall into the engine!

7 Part B
Overhaul and adjustments

Four-barrel models 4150/4160/4180/4190; 4165/4175; 4500
Three-barrel models 3150/3160
Two-barrel models 2300/2305/2380

The carburetors covered in this Part of Chapter 7 all share a similar design. The 4150, 4160, 4180, 4190 and 4500 are square-bore four-barrel carburetors (the primary and secondary throttle plates are all the same size). The 4165 and 4175 models are spread-bore four-barrel carburetors (the primary throttle plates are smaller than the secondary throttle plates); they are designed as replacements for the Rochester Quadrajet and Carter Thermoquad. The 3150 and 3160 are three-barrel models; they are designed very similarly to four-barrel models, except the area that would normally be occupied by the two secondary throttle plates has one large, oval-shaped throttle plate. The 2300, 2380 and 2305 are two-barrel carburetors whose design is basically the front half of a 4150 or 4160 four-barrel model.

The overhaul procedure is covered through a sequence of photographs laid out in order from disassembly to reassembly and finally installation and adjustment on the vehicle. The captions presented with the photographs will walk you through the entire procedure, one component at a time.

The following illustration sequence will use models 4150 and 4160 which, between the two, have enough variations that overhaul of these models will cover all the systems of the other models in this group. Specific details will vary with carburetor part numbers, because each carburetor is made to fit a specific application. Refer to the instruction sheet which comes with each overhaul kit.

Note: *Because of the large number of different models and variations of each model, it is not always possible to cover all the differences in components during the overhaul procedure. Overhaul sections only deal with disassembly, reassembly and basic adjustments of a typical carburetor in that model group. If you need further information about a specific variation, refer to the exploded views at the beginning of this Chapter or go to Chapter 3.*

Warnings:
Gasoline
Gasoline is extremely flammable, so take extra precautions when you work on any part of the fuel system. Don't smoke or allow open flames or bare light bulbs near the work area, and don't work in a garage where a natural gas-type appliance (such as a water heater or clothes dryer) with a pilot light is present. A spark caused by an electrical short circuit, by two metal surfaces striking each other, or even by static electricity built up in your body, under certain conditions, can ignite gasoline vapors. Also, never risk spilling fuel on a hot engine or exhaust component. If you spill any fuel on your skin, rinse it off immediately with soap and water.

Battery
Always disconnect the battery ground (-) cable at the battery before working on any part of the fuel or electrical system.

Fire extinguisher
We strongly recommend that a fire extinguisher suitable for use on fuel and electrical fires be kept handy in the garage or workshop at all times. Never try to extinguish a fuel or electrical fire with water. Post the phone number for the nearest fire department in a conspicuous location near the phone.

Compressed air
When cleaning carburetor parts, especially when using compressed air, be very careful to spray away from yourself. Eye protection should be worn to avoid the possibility of getting any chemicals or debris into your eyes.

Lead poisoning
Avoid the possibility of lead poisoning. Never use your mouth to blow directly into any carburetor component. Small amounts of Tetraethyl lead (a lead compound) become deposited on the carburetor over a period of time and could lead to serious lead poisoning. Check passages with compressed air and a fine-tipped blow gun or place a small-diameter tube to the component and blow through the tube to be sure all necessary passages are open.

7B.1 After removing the carburetor, mount it on a stand - four long 5/16-inch bolts, nuts and washers make an inexpensive stand

7B.2 Remove the float bowl assembly by removing the four bolts as shown here by the arrows. The bolts can have either a slotted head for a screwdriver (as shown here) or a hex head for a socket. As shown by the pointer, a gasket has remained stuck to the bowl, but they may also be stuck on the head of the bolt. Be sure the gaskets are removed before cleaning the parts.

7B.4 Removing a float bowl - the gaskets can dry out and stick and make the bowl hard to remove, so it may require a small rap with either a plastic hammer or the handle of a screwdriver to break the gasket's hold. Be careful not to wedge something between the pieces and damage the gasket surface by gouging the soft metal.

Disassembly
Float Bowls

Before removing the carburetor, be sure to loosen the items mentioned in Chapter 7A to make disassembly easier once the carburetor is removed from the vehicle.

Make notes or diagrams of levers or linkage locations, slots or holes that the linkage connects with and anything you may feel could be difficult to remember later on during re-assembly. **Note:** *Remember that on extremely dirty carburetors, overnight cleaning may be required and that reassembly may not be possible until the next day.*

7B.3 This is an example of a typical four-barrel, but with a side-pivot float design; the previous illustration was a center-pivot style. Also notice the different fuel inlet locations and types. This illustration is of the "single feed" type fuel inlet system, with a single hose fitting (arrow) for the fuel inlet; other types of float bowls have a separate fuel inlet for each bowl. Regardless of the type of fuel inlet system used on your carburetor, the inlet nut or tube, whichever is used on your model, must be removed to take out the filter and remove the gasket(s).

7B.5 On the "single feed" type fuel inlet system the float bowl for the secondaries is filled by the use of a feed tube (arrow) which connects both bowls. The tube is sealed on each end by the use of an O-ring. Be sure the old O-ring isn't left in the float bowl during cleaning. New O-rings are provided in the overhaul kit.

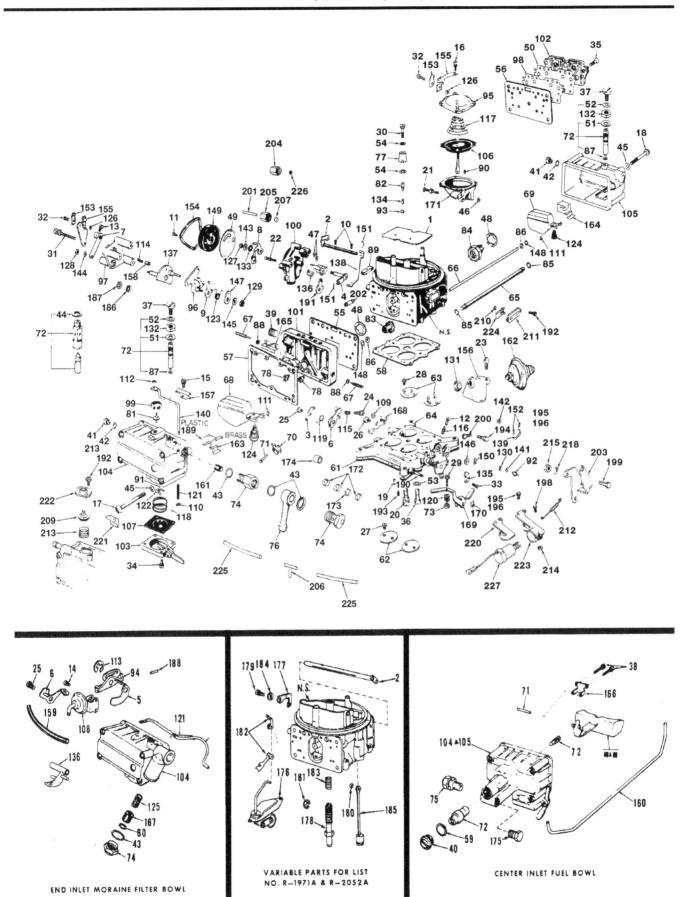

END INLET MORAINE FILTER BOWL

VARIABLE PARTS FOR LIST
NO. R-1971A & R-2052A

CENTER INLET FUEL BOWL

Model 4160 and 3160 exploded view (typical) - 4165 and 4180 similar

Index Number	Part Name	Index Number	Part Name
1	Choke Plate	71	Float Lever Shaft
2	Choke Shaft Assembly	72	Fuel Inlet Valve & Seat Assy.
3	Fast Idle Pick-up Lever	73	Pump Lever Adjusting Screw Fitting
4	Choke Housing Shaft & Lever Assy.	74	Fuel Inlet Fitting
5	Choke Control Lever	75	Fuel Transfer Tube Fitting Assy.
6	Fast Idle Cam Lever	76	Fuel Inlet Tube & Fitting Assy.
7	Choke Lever & Swivel Assy.	77	Pump Discharge Nozzle
8	Choke Therm. Lev., Link & Piston Assembly	78	Main Jet - Primary
9	Choke Rod Lev. & Bush. Assy.	81	Air Vent Valve
10	Choke Plate Screw	82	Pump Discharge Needle Valve
11	Therm. Housing Clamp Screw	83	Power Valve Assy. - Primary
12	Throttle Stop Screw	85	Fuel Line Tube "O" Ring Seal
13	Choke Lever Assembly Swivel Screw	86	Balance Tube "O" Ring Seal
14	Choke Diaph. Assy., Brkt. Scr. & Lock Washer	87	Fuel Valve Seat "O" Ring Seal
		88	Idle Needle Seal
15	Air Vent Clamp Screw & L.W.	89	Choke Rod Seal
16	Sec. Diaph. Assy. Cov. Scr. & L.W.	90	Diaphragm Housing Check Ball-Sec.
17	Fuel Bowl to Main Body Screw- Primary	91	Pump Inlet Check Ball or Value
		92	Throttle Lever Ball
18	Fuel Bowl to Main Body Screw- Secondary	93	Pump Discharge Check Ball
19	Diaph. Lever Adjusting Screw	94	Choke Diaphragm Assembly Link
20	Throt. Body Screw & Lock Washer	95	Sec. Diaphragm Housing Cover
21	Diaph. Hsg. Assy. Scr. & L.W.	96	Back-up Plate & Stud Assembly
22	Choke Housing Screw & L.W.	97	Fast Idle Cam Plate
23	Dashpot Brkt. Screw & L.W.	98	Secondary Metering Body Plate
24	Fast Idle Cam Lever Adj. Screw	99	Air Vent Cap
25	Fast Idle Cam Lev. Scr. & L.W.	100	Choke Hsg. & Plugs Assembly
26	Diaph. Lev. Assy. Scr. & L.W.	101	Main Metering Body & Plugs Assy.- Primary
27	Throt. Plate Screw - Primary	102	Metering Plate Secondary
28	Throt. Plate Screw - Secondary	103	Pump Cover Assembly
29	Pump Lever Adjusting Screw	104	Fuel Bowl & Plugs Assy.-Primary
30	Pump Discharge Nozzle Screw	105	Fuel Bowl & Plugs Assy.-Secondary
31	Fast Idle Cam Plate Scr. & L.W.	106	Secondary Diaph & Rod Assy.
32	Choke Cont. Wire Brkt. Clamp Scr.	107	Pump Diaphragm Assembly
33	Pump Cam Lock Screw	108	Choke Diaphragm Assembly - Complete
34	Fuel Pump Cov. Assy. Scr. & L.W.	109	Secondary Diaph. Link Retainer
35	Secondary Metering Body Screw	110	Air Vent Rod Spring Retainer
36	Throt. Body Screw - Special	111	Float Retainer
37	Fuel Valve Seat Lock Screw	112	Air Vent Valve Retainer
38	Float Shaft Brkt. Scr. & L.W.	113	Choke Control Lever Retainer
39	Spark Hole Plug	114	Fast Idle Cam Plunger Spring
40	Fuel Bowl Plug	115	Fast Idle Cam Lever Screw Spring
41	Fuel Level Check Plug	116	Throttle Stop Screw Spring
42	Fuel Level Check Plug Gasket	117	Secondary Diaphragm Spring
43	Fuel Inlet Fitting Gasket	118	Pump Diaphragm Return Spring
44	Fuel Valve Seat Gasket	119	Fast Idle Cam Lever Spring
45	Fuel Bowl Screw Gasket	120	Pump Lev. Adj. Screw Spring
46	Sec. Diaphragm Housing Gasket	121	Air Vent Rod Spring
47	Choke Housing Gasket	122	Pump Inlet Check Ball Ret. Spring
48	Power Valve Body Gasket	123	Choke Spring
49	Choke Thermostat Housing Gasket	124	Float Spring - Pri. & Sec.
50	Sec. Metering Body Plate Gasket	125	Fuel Inlet Filter Spring
51	Fuel Valve Seat Adj. Nut Gasket	126	Choke Cont. Wire Brkt. Clamp Scr. Nut
52	Fuel Valve Seat Lock Screw Gasket	127	Choke Housing Shaft Nut
53	Throt. Body Screw Gasket	128	Choke Control Lever Nut
54	Pump Discharge Nozzle Gasket	129	Back-up Plate Stud Nut
55	Metering Body Gasket - Primary	130	Throttle Lever Ball Nut
56	Metering Body Gasket - Secondary	131	Dashpot Locknut
57	Fuel Bowl Gasket	132	Fuel Valve Seat Adj. Nut
58	Throttle Body Gasket	133	Choke Housing Shaft Spacer
59	Fuel Bowl Plug Gasket	134	Pump Discharge Check Ball Weight
60	Fuel Inlet Filter Gasket	135	Pump Cam
61	Flange Gasket	136	Fast Idle Cam Assembly
62	Throttle Plate - Primary	137	Fast Idle Cam & Shaft Assembly
63	Throttle Plate - Secondary	138	Choke Rod
64	Throt. Body & Shaft Assembly	139	Throttle Connecting Rod
65	Fuel Line Tube	140	Air Vent Push Rod
66	Balance Tube	141	Throttle Lev. Ball Nut Washer
67	Idle Adjusting Needle	143	Choke Shaft Nut Lock Washer
68	Float & Hinge Assy. - Primary		
69	Float & Hinge Assy. - Secondary		

Parts list for models 4160 and 3160 exploded view

Index Number	Part Name	Index Number	Part Name
144	Choke Control Lev. Nut Lock Washer	186	Choke Oper. Lev. Spring Washer
145	Back-up Plate Stud Nut Lock Washer	187	Choke Oper. Lever Washer
146	Sec. Connecting Pin Washer	188	Choke Clevis Pin
147	Choke Spring Washer	189	Baffle Plate - Primary (Plastic)
148	Balance Tube Washer	***	Fuel Valve Clip
149	Choke Thermostat Assy.	190	Pump Oper. Lever Stud
150	Pin Retainer	191	Choke Thermostat Lever
151	Choke Rod Retainer	192	Vent Valve & H.I.C. Cover Screw
152	Throt. Connecting Rod Cotter Pin	193	Secondary Throttle Adjusting Screw
153	Choke Cont. Wire Brkt. Clamp	194	Solenoid Stop Screw
154	Thermostat Housing Clamp	195	Trans Spring Perch Bracket Screw & L.W.
155	Choke Control Wire Bracket	196	Solenoid Bracket Screw Lock Washer
156	Dashpot Bracket		
157	Air Vent Rod Clamp	198	Solenoid Bracket Screw
158	Fast Idle Cam Plunger	199	Throttle Lever Extension Screw
159	Choke Vacuum Tube	200	Transmission Pick-up Lever Adj. Screw
160	Fuel Transfer Tube		
161	Filter Screen	201	Tube
162	Dashpot Assembly	202	Tube & ''O'' Ring Assy.
163	Baffle Plate - Primary (Brass)	203	Throttle Lever Extension
164	Baffle Plate - Secondary	204	Choke housing Nut
165	Metering Body Vent Baffle	205	Comp Nut
166	Float Shaft Retainer Bracket	206	Hose Connector
167	Fuel Inlet Filter	207	Tube Ferrule
168	Diaphragm Lever Assembly	208	Connector
169	Pump Operating Lever	209	Vent Valve Body Assy.
170	Pump Operating Lever Retainer	210	H.I.C. Seal
171	Secondary Diaphragm Housing	211	H.I.C. Cover
172	Throt. Shaft Brg. Pri. & Sec. (Ribbon)	212	Trans kickdown Control Lever Spring
173	Throt. Shaft Brg. Pri. & Sec. (Ribbon)	213	Vent Valve Spring
174	Throt. Shaft Bearing - Pri. (Solid)	214	Solenoid Bracket Nut
		215	Extension Nut
175	Fuel Bowl Drain Plug	216	Stop Screw Lock Nut (Not Shown)
176	Choke Assy. - Complete (Divorced)	217	Pump Operating Lever Screw Spring (Not Shown)
177	Choke Shaft Lever		
178	Idle By-pass Adj. Screw	218	Extension Screw Lock Washer
179	Choke Lever Screw & L.W.	220	Trans. Spring Perch Bracket
180	Choke Piston Link Retainer	221	Retard Wire Bracket
181	Fast Idle Cam Retainer	222	Vent Valve Clamp Assy.
182	Choke Rod Clevis Clip	223	Solenoid Bracket
183	Idle By-pass Adj. Screw Spring	224	Hot Idle Compensator Assy.
184	Choke Piston Lever Spacer	225	Hose
185	Choke Piston & Link Assembly	226	Choke housing Screen
		227	Solenoid Assy.

★★★ = Not shown in illustration

Parts list for models 4160 and 3160 exploded view (continued)

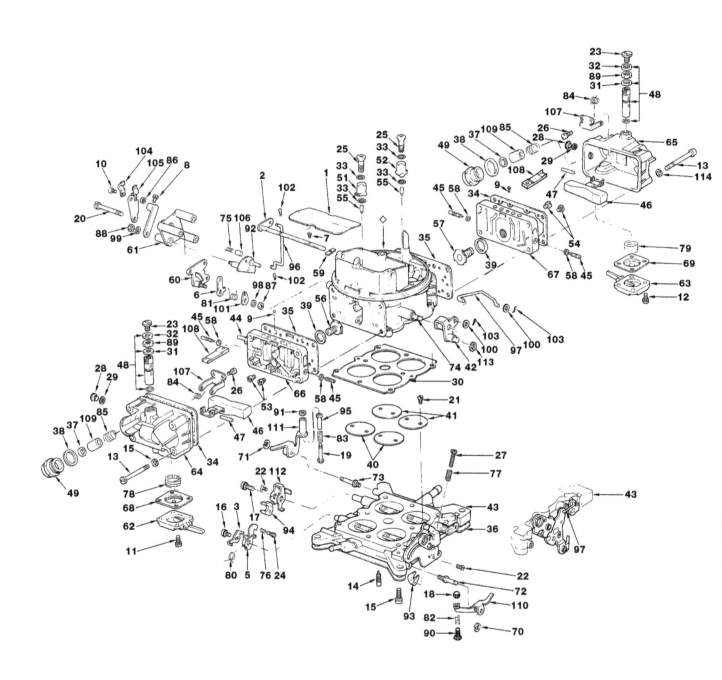

Model 4150 exploded view (typical) - 4175 and 4190 similar

Index Number	Part Name	Index Number	Part Name
		56	Power Valve Assy. Pri.
1	Choke Plate	57	Power Valve Assy. or Plug Sec.
2	Choke Shaft & Lever Assy.	58	Idle Needle Seal Pri. & Sec.
3	Fast Idle Pick-up Lever	59	Choke Rod Seal
4	Choke Lever & Swivel Assy.	60	Back-up Plate & Stud Assy.
5	Fast Idle Cam Lever	61	Fast Idle Cam Plate
6	Choke Rod Lever & Bushing Assy.	62	Pump Cover Assy. Pri.
7	Choke Plate Screw	63	Pump Cover Assy. Sec.
8	Choke Swivel Screw	64	Fuel Bowl & Plugs Assy. Pri.
9	Drive Screw	65	Fuel Bowl & Plugs Assy. Sec.
10	Clamp Screw	66	Metering Body & Plugs Assy. Pri.
11	Fuel Pump Cover Screw Pri.	67	Metering Body & Plugs Assy. Sec.
12	Fuel Pump Cover Screw Sec.	68	Pump Diaphragm Assy. Pri.
13	Fuel Bowl Screw	69	Pump Diaphragm Assy. Sec.
14	Secondary Idle Adjusting Screw	70	Pump Operating Lever Retainer Pri.
15	Throttle Body Screw & L.W.	71	Pump Operating Lever Retainer Sec.
16	Fast Idle Cam Lever Screw & L.W.	72	Pump Lever Stud Pri.
17	Secondary Pump Cam Lever Screw & L.W.	73	Pump Lever Stud Sec.
		74	Cam Follower Lever Stud
18	Pump Lever Adjusting Screw Pri.	75	Plunger Spring
19	Pump Lever Adjusting Screw Sec.	76	Fast Idle Cam Lever Screw Spring
20	Fast Idle Cam Plate Screw & L.W.	77	Throttle Stop Screw Spring
21	Throttle Plate Screw Pri. & Sec.	78	Diaphragm Return Spring Pri.
22	Pump Cam Lock Screw Pri. & Sec.	79	Diaphragm Return Spring Sec.
23	Fuel Valve Seat Lock Screw	80	Fast Idle Cam Lever Spring
24	Fast Idle Cam Lever Adj. Screw	81	Choke Spring
25	Pump Discharge Nozzle Screw Pri. & Sec.	82	Pump Lever Adj. Screw Spring Pri.
		83	Pump Lever Adj. Screw Spring Sec.
26	Float Shaft Bracket Screw & L.W. Pri. & Sec.	84	Float Spring Pri. & Sec.
		85	Fuel Inlet Filter Spring
27	Throttle Stop Screw	86	Bracket Clamp Screw Nut
28	Fuel Level Check Plug Pri. & Sec.	87	Choke Spring Nut
29	Fuel Level Check Plug Gasket Pri. & Sec.	88	Choke Lever Nut
		89	Fuel Valve Seat Adjusting Nut Pri. & Sec.
30	Throttle Body Gasket	90	Pump Lever Adjusting Screw Nut Pri.
31	Fuel Valve Seat Adjusting Nut Gasket		
32	Fuel Valve Seat Lock Screw Gasket	91	Pump Lever Adjusting Screw Nut Sec.
33	Pump Discharge Nozzle Gasket	92	Fast Idle Cam & Shaft Assy.
34	Fuel Bowl Gasket Pri. & Sec.	93	Pump Cam Pri.
35	Metering Body Gasket Pri. & Sec.	94	Pump Cam Sec.
36	Flange Gasket	95	Pump Operating Lever Screw Sleeve
37	Fuel Inlet Filter Gasket	96	Choke Rod
38	Fuel Inlet Fitting Gasket	97	Secondary Connecting Rod
39	Power Valve Gasket	98	Choke Spring Nut L.W.
40	Throttle Plate Primary	99	Choke Control Lever L.W.
41	Throttle Plate Secondary	100	Connecting Rod Washer
42	Cam Follower Lever Assy.	101	Choke Spring Washer
43	Throttle Body & Shaft Assy.	102	Choke Rod Retainer
44	Spark Tube	103	Cotter Pin
45	Idle Adjusting Needle Pri. & Sec.	104	Choke Control Wire Bracket Clamp
46	Float & Hinge Assy. Pri. & Sec.	105	Choke Control Wire Bracket
47	Float Shaft Pri. & Sec.	106	Fast Idle Cam Plunger
48	Fuel Inlet Needle & Seat Assy. Primary & Sec.	107	Float Shaft Retaining Bracket Pri. & Sec.
49	Fuel Inlet Fitting Primary	108	Fuel Bowl Vent Baffle Pri. & Sec.
50	Fuel Inlet Fitting Secondary	109	Fuel Inlet Filter Pri. & Sec.
51	Pump Discharge Nozzle Pri.	110	Pump Operating Lever Pri.
52	Pump Discharge Nozzle Sec.	111	Pump Operating Lever Sec.
53	Main Jet Pri.	112	Pump Cam Lever Sec.
54	Main Jet Sec.	113	E. Ring Retainer
55	Pump Discharge Needle Valve Pri. & Sec.	114	Fuel Bowl Screw Gasket

Parts list for the model 4150 exploded view

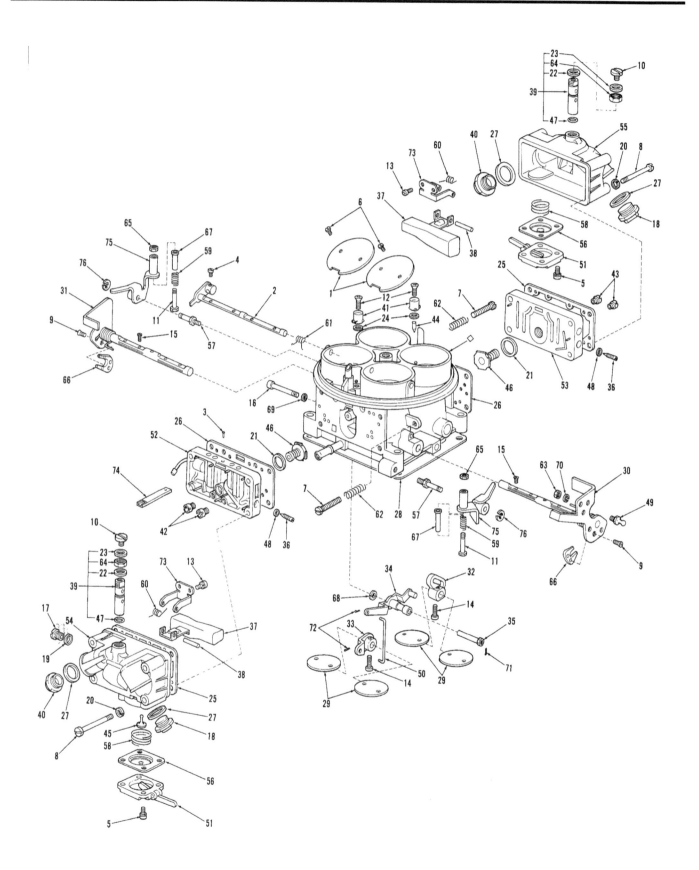

Model 4500 exploded view (typical)

Index Number	Part Name	Index Number	Part Name
1	Choke Plate	36	Idle Adjusting Needle
2	Choke Shaft Assy.	37	Float & Hinge Assy.
3	Fuel Bowl Vent Baffle Drive Screw	38	Float Shaft
		39	Fuel Valve & Seat Assy.
4	Choke Shaft Swivel Screw	40	Fuel Inlet Fitting
5	Fuel Pump Cover Screw & L.W.	41	Pump Discharge Nozzle
		42	Main Jet-Pri
6	Choke Plate Screw & L.W.	43	Main Jet - Sec.
7	Throttle Stop Screw	44	Pump Discharge Needle Valve
8	Fuel Bowl Screw	45	Pump Check Valve
9	Pump Cam Lock Screw	46	Power Valve Assy.
10	Fuel Valve Seat Lock Screw	47	Fuel Valve Seat "O" Ring Seal
11	Pump Operating Adj. Screw	48	Idle Adjusting Needle Seal
12	Pump Discharge Nozzle Screw	49	Throttle Lever Ball
		50	Connecting Link
13	Float Shaft Bracket Screw & L.W.	51	Fuel Pump Cover Assy.
		52	Metering Body & Plugs Assy. - Pri.
14	Throttle Shaft Screw	53	Metering Body & Plugs Assy. - Sec.
15	Throttle Plate Screw		
16	Pivot Screw	54	Fuel Bowl - Pri.
17	Fuel Level Check Plug	55	Fuel Bowl - Sec.
18	Fuel Inlet Plug	56	Pump Diaphragm Assy.
19	Fuel Level Check Plug Gasket	57	Pump Lever Stud
		58	Diaphragm Return Spring
20	Fuel Bowl Screw Gasket	59	Pump Operating Adj. Screw Spring
21	Power Valve Gasket		
22	Fuel Valve Seat Adj. Nut Gasket	60	Float Spring
		61	Choke Spring
23	Fuel Valve Seat Lock Screw Gasket	62	Throttle Stop Screw Spring
		63	Throttle Lever Ball Nut
24	Pump Discharge Nozzle Gasket	64	Fuel Valve Seat Adj. Nut
		65	Pump Operating Adj. Nut
25	Fuel Bowl Gasket	66	Pump Cam
26	Metering Body Gasket - Pri & Sec.	67	Pump Operating Lever Screw Sleeve
27	Fuel Inlet Fitting & Plug Gasket	68	Pivot Screw L.W.
		69	Pivot Screw Washer
28	Flange Gasket	70	Throttle Lever Ball L.W.
29	Throttle Plate	71	Pivot Screw Cotter Pin
30	Throttle Shaft Assy. - Pri.	72	Connecting Link Cotter Pin
31	Throttle Shaft Assy. - Sec.	73	Float Shaft Retaining Bracket
32	Primary Throttle Lever (Internal)	74	Fuel Bowl Vent Baffle
33	Secondary Throttle Lever & Bushing Assy.	75	Pump Operating Lever & Guide Assy.
34	Intermediate Throttle Lever Assy. (Comp.)	76	Pump Operating Lever Retainer
35	Threaded Guide Bushing		

Parts list for model 4500 exploded view

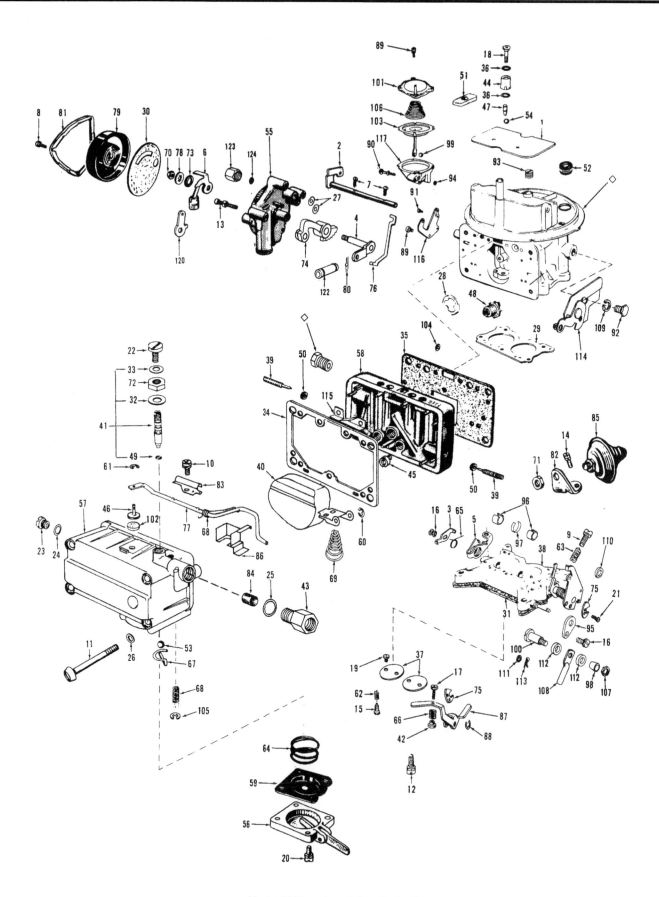

Model 2300 exploded view (typical)

Index Number	Part Name	Index Number	Part Name
1	Choke Plate	71	Dashpot Screw Nut
2	Choke Shaft Assembly	72	Fuel Valve Seat Adj. Nut
3	Fast Idle Pick-up Lever	73	Choke Therm. Lever Spacer
4	Choke Hsg. Shaft & Lev. Assy.	74	Fast Idle Cam Assembly
5	Fast Idle Cam Lever	75	Pump Cam
6	Choke Therm. Lev., Link & Piston Assembly	76	Choke Rod
		77	Air Vent Rod
7	Choke Plate Screw	78	Choke Therm. Shaft Nut L.W.
8	Therm. Hsg. Clamp Screw	79	Thermostat Housing Assembly
9	Throttle Stop Screw	80	Choke Rod Retainer
10	Air Vent Rod Clamp Scr. & L.W.	81	Thermostat Housing Clamp
11	Fuel Bowl to Main Body Screw	82	Dashpot Bracket
12	Throt. Body Scr. & L.W.	83	Air Vent Rod Clamp
13	Choke Hsg. Scr. & L.W.	84	Filter Screen Assembly
14	Dashpot Brkt. Scr. & L.W.	85	Dashpot Assembly
15	Fast Idle Cam Lever Screw	86	Baffle Plate
16	Fast Idle Cam Lev. & Throt. Lev. Screw & L.W.	87	Pump Operating Lever
		88	Pump Operating Lev. Retainer
17	Pump Oper. Lev. Adj. Screw	89	Adapter Mounting & Diaphragm Cover Assy. Screw
18	Pump Discharge Nozzle Screw		
19	Throttle Plate Screw	90	Throt. Diaphragm Hsg. Scr.
20	Fuel Pump Cov. Assy. Scr. & L.W.	91	Adapter Passage Screw
21	Pump Cam Lock Scr. & L.W.	92	Choke Bracket Screw
22	Fuel Valve Seat Lock Screw	93	Air Adapter Hole Plug
23	Fuel Level Check Plug	94	Throt. Diaphragm Hsg. Gskt.
24	Fuel Level Check Plug Gasket	95	Throttle Lever
25	Fuel Inlet Fitting Gasket	96	Throttle Shaft Bearing
26	Fuel Bowl Screw Gasket	97	Throttle Shaft Brg. (center)
27	Choke Housing Gasket	98	Throttle Connector Pin Bushing
28	Power Valve Body Gasket	99	Diaphragm Check Ball
29	Throttle Body Gasket	100	Throttle Connector Pin
30	Choke Therm. Housing Gasket	101	Diaphragm Housing Cover
31	Flange Gasket	102	Air Vent Cap
32	Fuel Valve Seat Adj. Nut Gskt.	103	Diaphragm Assembly
33	Fuel Valve Seat Lock Scr. Gskt.	104	Diaphragm Link Retainer
34	Fuel Bowl Gasket	105	Air Vent Rod Spg. Retainer
35	Metering Body Gasket	106	Diaphragm Spring
36	Pump Discharge Nozzle Gasket	107	Throttle Link Connector Pin Nut
37	Throttle Plate	108	Throttle Connector Bar
38	Throt. Body & Shaft Assy.	109	Choke Brkt. Scr. Lock Washer
39	Idle Adjusting Needle	110	Throt. Link Connector Pin Washer
40	Float & Hinge Assy.	111	Throt. Connector Pin Washer
41	Fuel Inlet Valve & Seat Assy.	112	Throttle Connector Pin Spacer
42	Pump Oper. Lev. Adj. Scr. Fitting	113	Throt. Connector Pin Retainer
43	Fuel Inlet Fitting	114	Choke Control Lever Bracket
44	Pump Discharge Nozzle	115	Metering Body Vent Baffle
45	Main Jet	116	Throt. Diaphragm Adapter
46	Air Vent Valve	117	Diaphragm Housing
47	Pump Discharge Needle Valve or Check Ball Weight	119	Pump Oper. Lever Stud
		*	Vent Tube
48	Power Valve Assembly	*	Heat Tube Nut
49	Fuel Valve Seat "O" Ring Seal or Gasket	*	Heat Tube Ferrule
		*	Fuel Tube Hose Clamp
50	Idle Needle Seal	*	Retainer
51	Choke Rod Seal	*	Air Cleaner Stud (Long)
52	Choke Code Air Tube Grommet	*	Air Cleaner Stud (Short)
53	Pump Inlet Check Ball or Valve	*	Choke Heat Tube
54	Pump Discharge Check Ball	*	Fuel Line Hose
55	Choke Hsg. & Plugs Assy.		
56	Fuel Pump Cover Assy.	*	Fresh Air Hose
57	Fuel Bowl & Plugs Assy.	*	Spring
58	Main Metering Body & Plugs Assy.	120	Choke Thermostat Lever
59	Pump Diaphragm Assembly	121	Vacuum Tube Plug
60	Float Spring Retainer	122	Tube & "O" Ring Assy.
61	Air Vent Retainer	123	Nut
62	Fast Idle Cam Lev. Scr. Spring	124	Screen
63	Throttle Stop Screw Spring		Parts not shown on T.V. III. 3-3
64	Pump Diaphragm Return Spring		Solenoid Bracket Screw & L.W.
65	Fast Idle Cam Lev. Spring		Bracket Screw
66	Pump Oper. Lev. Adj. Spring		Bracket Nut
67	Pump Inlet Check Ball Retainer		Solenoid Bracket Assy.
68	Air Vent Rod Spring		Solenoid Assy.
69	Float Spring		Solenoid Part apply to list R-7333A
70	Choke Thermostat Shaft Nut		

* = Not shown in illustration

Parts list for model 2300 exploded view

CENTER TRIPLE INST. UNIT

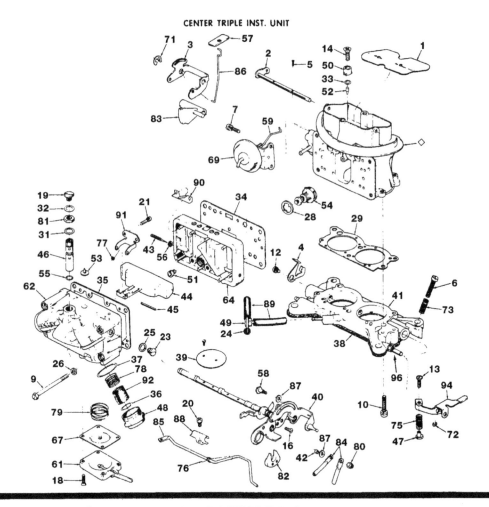

FRONT AND REAR TRIPLE INST. UNIT

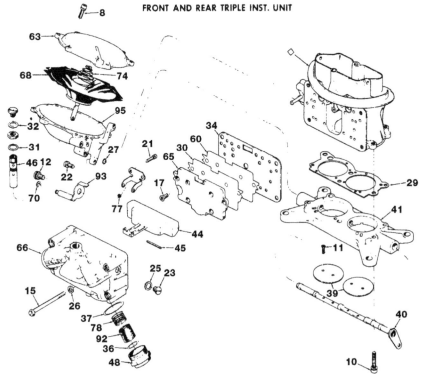

Model 2300-C exploded view (typical)

Index Number	Part Name	Index Number	Part Name
1	Choke Plate	46	Fuel Inlet Valve & Seat Assy.
2	Choke Shaft & Lever Assembly	47	Pump Operating Lever Adj. Screw Fitting
3	Choke Control Lever	48	Fuel Inlet Fitting
4	Fast Idle Cam Lever	49	Tee Connector
5	Choke Plate Screw	50	Pump Discharge Nozzle
6	Throttle Stop Screw	51	Main Metering Jet
7	Choke Diaphragm Assy. Bracket Screw & Lock Washer	52	Pump Discharge Needle Valve
		53	Air Vent Valve
8	Diaph. Cover Assy. Scr. & L.W.	54	Power Valve Assy.
		55	Fuel Valve Seat "O" Ring
9	Fuel Bowl to Main Body Screw	56	Idle Adjusting Needle Seal
10	Throttle Body Screw & L.W.	57	Choke Rod Seal
11	Throttle Plate Screw	58	Throttle Connector Pin
12	Fast Idle Cam Lever Screw & L.W.	59	Choke Diaphragm Assy. Link
		60	Metering Body Plate
13	Pump Operating Lever Adj. Screw	61	Fuel Pump Diaphragm Cover Assy.
14	Pump Discharge Nozzle Screw	62	Fuel Bowl & Plugs Assy.
15	Fuel Bowl To Main Body Scr. Sec.	63	Diaphragm Housing Cover Sec.
		64	Metering Body & Plugs Assy.
16	Pump Cam Lock Screw	65	Sec. Metering Body
17	Metering Body Screw Sec.	66	Sec. Fuel Bowl
18	Fuel Pump Cover Screw	67	Pump Diaphragm Assy.
19	Fuel Valve Seat Lock Screw	68	Sec. Diaphragm Assy.
20	Air Vent Rod Clamp Screw	69	Choke Diaphragm Assy. Complete
21	Float Shaft Bracket Scr. & L.W.	70	Sec. Diaphragm Link Retainer
		71	Choke Control Lever Retainer
22	Diaphragm Mounting Scr. & L.W.	72	Pump Operating Lever Retainer
23	Fuel Level Check Plug	73	Throttle Stop Screw Spring
24	Tee Connector Plug	74	Diaphragm Spring Sec.
25	Fuel Level Check Plug Gasket	75	Pump Operating Lever Adj. Spring
26	Fuel Bowl Screw Gasket	76	Air Vent Rod Spring
27	Diaphragm Housing Gasket Sec.	77	Float Spring
28	Power Valve Body Gasket	78	Fuel Inlet Filter Spring
29	Throttle Body Gasket	79	Diaphragm Return Spring
30	Metering Body Plate Gasket Sec.	80	Throttle Connector Pin Nut
31	Fuel Inlet Adjusting Nut Gasket	81	Fuel Valve Seat Adj. Lock Nut
32	Fuel Valve Seat Screw Gasket	82	Pump Cam
33	Pump Discharge Nozzle Gasket	83	Fast Idle Cam
		84	Throttle Connector Bar
34	Metering Body Gasket	85	Air Vent Rod
35	Fuel Bowl Gasket	86	Choke Rod
36	Fuel Inlet Filter Gasket	87	Throttle Connector Pin Spacer
37	Fuel Inlet Fitting Gasket		
38	Flange Gasket	88	Air Vent Rod Clamp
39	Throttle Plate	89	Choke Vacuum Hose
40	Throttle Lever & Shaft Assy.	90	Metering Body Vent Baffle
41	Throttle Body & Shaft Assy.	91	Float Shaft Retaining Bracket
42	Throttle Connector Pin Bushing	92	Fuel Inlet Filter
		93	Diaphragm Lever & Pin Assy.
43	Idle Adjusting Needle	94	Pump Operating Lever
44	Float & Hinge Assy.	95	Sec. Diaphragm Housing
45	Float Lever Shaft	96	Pump Oper. Lever Stud

Parts list for model 2300-C exploded view

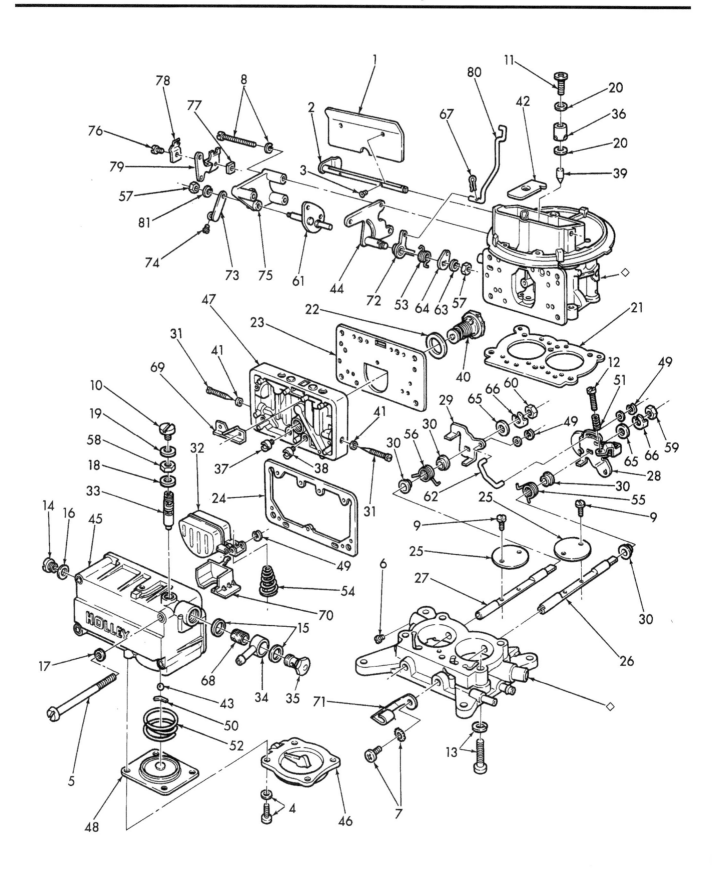

Model 2305 exploded view (typical)

Index Number	Part Name	Index Number	Part Name
1	Choke Plate	65	Washer
2	Choke Shaft & Lever Assembly	66	Throttle Shaft Nut Lockwasher
3	Choke Plate Screw	67	Choke Rod Retainer
4	Pump Dia. Cover Screw & L.W.	68	Fuel Inlet Filter Screen
5	Fuel Bowl Screw	69	Metering Body Vent Baffle
6	Secondary Idle Adj. Screw	70	Fuel Inlet Baffle
7	Pump Operating Lever Screw & L.W.	71	Pump Operating Lever
		72	Choke Rod Lever & Bushing Assy.
8	Fast Idle Cam Plate Screw & L.W.	73	Choke Lever & Swivel Assy.
9	Throttle Plate Screw	74	Swivel Screw
10	Needle & Seat Lock Screw	75	Fast Idle Cam Plate
11	Pump Discharge Nozzle Screw	76	Bracket Clamp Screw
12	Throttle Stop Screw	77	Clamp Screw Nut
13	Throttle Body Screw & L.W.	78	Bracket Clamp
14	Fuel Level Check Plug	79	Choke Wire Bracket
15	Fuel Inlet Fitting Gasket	80	Choke Rod
16	Fuel Level Plug Gasket	81	Choke Control Lever Lockwasher
17	Fuel Bowl Screw Gasket	*	Float Hinge Pin
18	Needle & Seat Adj. Nut Gasket	*	Flange Gasket
19	Seat Lock Screw Gasket	*	Air Cleaner Gasket
20	Pump Discharge Nozzle Gasket		
21	Throttle Body Gasket		
22	Power Valve Gasket		
23	Metering Body Gasket		
24	Fuel Bowl Gasket		
25	Throttle Plate		
26	Primary Throttle Shaft		
27	Secondary Throttle Shaft		
28	Primary Throttle Lever		
29	Secondary Throttle Lever		
30	Return Spring Bushing		
31	Idle Adjusting Needle		
32	Float & Hinge Assembly		
33	Fuel Inlet Needle & Seat Assy.		
34	Fuel Inlet Tube & Fitting Assy.		
35	Fuel Inlet Fitting		
36	Pump Discharge Nozzle		
37	Primary Main Jet		
38	Secondary Main Jet		
39	Pump Discharge Needle Valve		
40	Power Valve Assembly		
41	Idle Adj. Needle Seal		
42	Choke Rod Seal		
43	Check Ball		
44	Back-Up Plate & Stud Assembly		
45	Primary Fuel Bowl & Plugs Assembly		
46	Pump Diaphragm Cover Assy.		
47	Metering Body & Plugs Assy.		
48	Pump Diaphragm Assy.		
49	Retainer		
50	Pump Check Ball Retainer		
51	Idle Speed Screw Spring		
52	Pump Diaphragm Return Spring		
53	Choke Lever Spring		
54	Float Spring		
55	Primary Throttle Return Spring		
56	Secondary Throttle Return Spring		
57	Hex Nut		
58	Needle & Seat Adjusting Nut		
59	Primary Throttle Lever Nut		
60	Secondary Throttle Lever Nut		
61	Fast Idle Cam & Shaft Assy.		
62	Secondary Connecting Rod		
63	Choke Rod Lever Nut Lockwasher		
64	Choke Spring Washer		

✱ = Not shown in illustration

Parts list for model 2305 exploded view

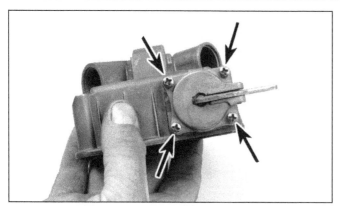

7B.6 The accelerator pump is located on the bottom of the float bowl, attached by four phillips head screws (arrows). Some models have two (dual) accelerator pumps, one in each float bowl. Both have the same components and are disassembled the same way. But be careful for capacity differences. There may be some instances where a larger 50cc pump is used in one bowl, but a standard pump is used in the other bowl. If in doubt, mark the pump so it can be reinstalled in the original location. Remove the screws and the pump cover.

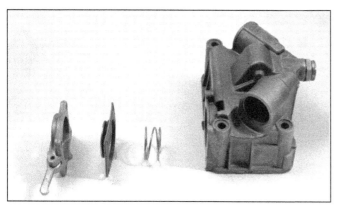

7B.8 As the pieces are removed note the location of the spring and how it sits in the diaphragm. Even though the diaphragm is being replaced during the overhaul with the new part from the kit, always examine the pieces to help pinpoint carburetor problems.

7B.10 The accelerator pump is operated by a lever pressing against the pump arm to cause the fuel shot. The lever is controlled by a plastic cam installed on the throttle shaft (1) and fastened by a screw in the throttle linkage (2). This cam can be changed to others that operate the accelerator pump(s) at different rates.

7B.7 Remove the pump diaphragm and the spring from the float bowl

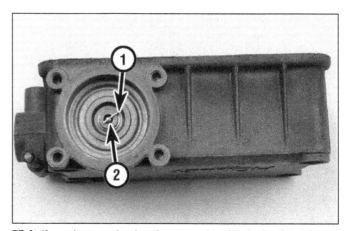

7B.9 If you have a check-valve assembly that looks like this, remove the clip (1) and accelerator pump ball (2). If you have an assembly with a bar-type retainer (see illustration 7B.52), DO NOT disassemble the check-valve - the bar-type retainer is permanently staked in place. Some models may have a plastic umbrella-type check valve, which you can simply pull out of the float bowl.

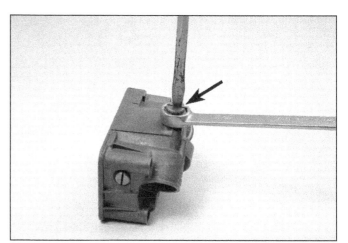

7B.11 There are a variety of needle-and-seat assemblies used for different float bowls. On the externally adjustable type there is a center-pivot style, as shown here, and a side-pivot style. Regardless of center or side pivot float styles the removal and installation is the same for all externally adjustable floats: first loosen the lock screw (arrow) on top of the adjustment nut. It isn't necessary to remove the screw. Loosen it enough to be able to turn the adjustment nut . . .

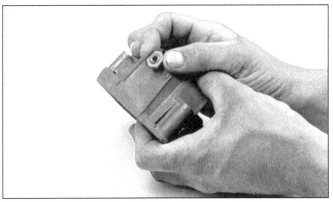

7B.12 . . . then use the adjustment nut to unscrew the needle and seat from the float bowl. This eliminates the need to place a wrench or pliers on the assembly, and possibly damaging threads, when removing it.

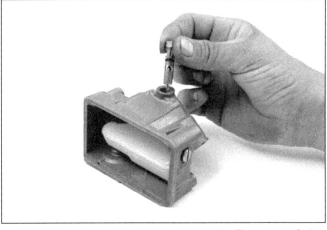

7B.13 Remove the needle-and-seat assembly. The rubber O-ring may cause it to stick slightly while pulling it straight up.

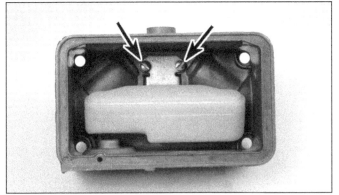

7B.14 The float assembly will mount differently, depending on whether it is a center- or side-pivot float style. To remove the center-pivot float shown here, simply remove the two screws shown (arrows) and remove the float assembly. **Note:** *Floats often become saturated with fuel and become much heavier than normal. This will cause the float to sink and raise the fuel level in the bowl, causing an overly rich mixture or fuel leaks. Not all floats become saturated, but pinholes or breaks in the solder can develop over time, even on metal or plastic floats. If in doubt about the condition of your float(s) it's recommended that the float be replaced at the time of the overhaul.*

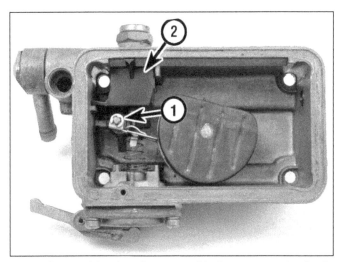

7B.15 On the side-pivot float assembly first remove the C-clip from the float hinge pin (1). The float and spring can be removed from the bowl and set aside. It's not necessary or advised to place the float assembly into a cleaning solution. Next, using a needle-nose pliers, pull the needle and seat baffle (2) straight out from the bowl.

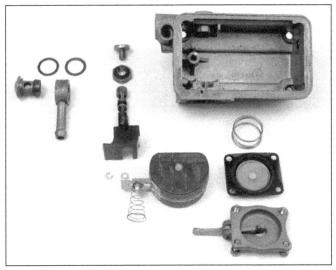

7B.16 An exploded view of the parts in a side-pivot, externally adjustable float bowl assembly

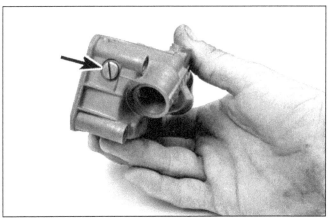

7B.17 Regardless of whether the floats are center- or a side-pivot style, as long as they are externally adjustable, there will be a sight plug/screw and gasket (arrow). Remove the screw so the gasket can be removed before cleaning.

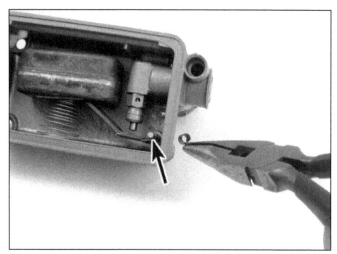

7B.18 On float bowls with no external adjustments, the float must be removed before the needle and seat. Using a needle-nose pliers, remove the small clip that holds the float in the bowl by sliding it straight off the hinge pin (arrow). Then slide the float with the spring, which is attached, off the pin and out of the bowl.

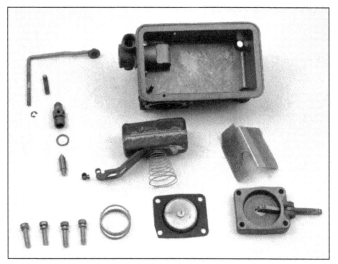

7B.20 Here's an exploded view of the parts needed to make up one complete side-pivot float bowl assembly, with accelerator pump and no external adjustment.

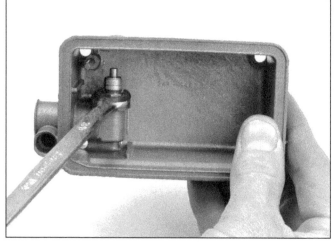

7B.19 Next, using a 3/8-inch wrench, unscrew the needle and seat and remove the assembly from inside the float bowl. The needle on this style isn't captured in the seat housing, so be careful you don't loose the needle when removing it.

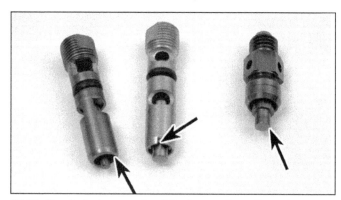

7B.21 Here's just a sample of the types of needle-and-seat assemblies you may find during your overhaul. On the left and in the center are similar types for externally adjustable float levels. The only difference is how the needle is retained in the seat housing (left and center arrows). The assembly on the right is used in float bowls with side-pivot floats with no external adjustments; the needle in this type is not retained in the seat.

7B.22 On models with removable needles, check the needle tip (arrow) for evidence of sticking or leakage, which would cause serious driveability problems.

Metering bodies and plates

7B.23 The gaskets can have a very strong hold on the parts they connect, especially after they have been in a carburetor for years. To remove the body, use a screwdriver and lightly pry, using the capped hole on the top of the metering body for the tip of the screwdriver to sit in and the ring around the air horn to pry against. Doing this, you can avoid trying to wedge something between the two pieces and possibly damaging the gasket surfaces.

7B.24 After the metering body has been removed, part of the reason for the overhaul will become clear. As gaskets dry up over time, they shrink. This can cause everything from leaking vacuum or fuel to actually blocking passages between the metering body and the main carburetor body. In this illustration, you can see how the gasket has blocked several passages. Be sure to remove all the old gasket material and blow out all the passages in the metering body.

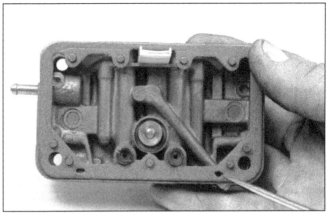

7B.25 The gasket on the float-bowl side of the metering body, although just around the outer edge, can cause some problems also. The hole shown by the pointer in this illustration shows the accelerator pump fuel passage, where it comes from the float bowl. Incorrect gasket selection or placement, as well as an overtightened gasket, can cause this passage to become either partially or completely blocked. So check this area out carefully.

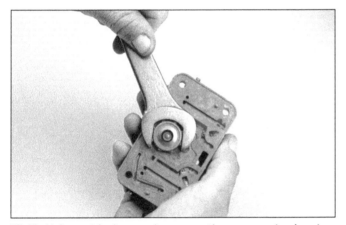

7B.26 Using a 1-inch wrench, remove the power valve (or plug, which is used on some four-barrel models in the secondary metering body); the valve and plug look very much alike, so be sure you note which one you have in the secondary body for correct re-assembly

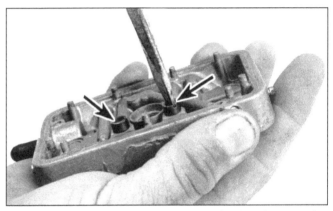

7B.27 Since there are at least two and possibly four fuel metering jets (arrows) to be removed during an overhaul, it is important to write down information on parts as they are removed so there is no confusion later during reassembly. The metering jets have their numbering system stamped on the side of the jet. Be sure to write down which jet came from which location BEFORE removal. To remove the main metering jets use a flat-blade screwdriver large enough to cover the entire slot in the jet. Unscrew the jet from the metering body and set it aside to be cleaned with the other parts. The jets don't use a gasket; they sit directly against the metering body.

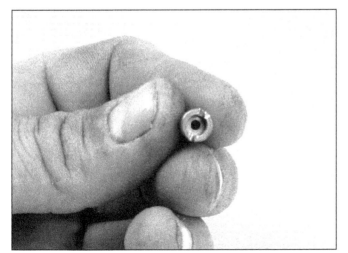

7B.28 The wrong screwdriver was used to remove this jet. Since the fuel passage is very small, even minor damage or small bits of debris can interfere with fuel flow.

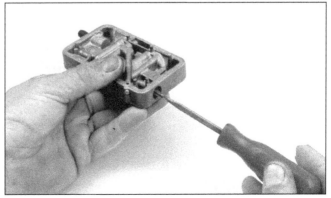

7B.29 Remove the two idle mixture screws, one on either side of the metering body, and the small cork gaskets behind the screws

7B.30 If equipped with a secondary metering plate, unscrew the clutch-head screws and remove the plate from the main body. The gasket may have this plate stuck very securely, so it may be necessary to lightly pry it loose. **Note:** *If it necessary to use a tool of any kind to pry the metering plate loose, be careful not to damage the gasket surfaces of the metering plate and main body.*

7B.31 The secondary metering plate assembly has a metal plate between it and the main carburetor body, with gaskets between each. Be certain the plate is separated from the body and the gasket removed before cleaning the parts. This exploded view of the secondary metering body assembly shows more clearly the order in which the parts are assembled.

Choke assemblies

Manual

Step 1: Referring to illustration 7B.32, loosen the set screw where the choke cable attaches to the choke lever swivel assembly (1) and the clamp screw (2) which holds the cable housing in place. Pull the cable out of the choke linkage. **Note:** *Since there are no gaskets or rubber parts in the manual linkage assembly it's not necessary go further and remove this component for cleaning. This removal is being described for anyone who wants to change their choke system over to an automatic choke, which is available as a kit from Holley.*

Step 2: Using a needle-nose pliers or a tool with a fine point pull the c-clip or cotter pin-type clip from the choke plate linkage arm. This is located behind the fast idle cam plate (3).

Step 3: Remove the three screws (4) and remove the linkage from the main body.

7B.32 The manual choke linkage is simple to remove - refer to the text at left for the procedure

Automatic

To remove the automatic choke, refer to illustrations 7B.34 through 7B.36.

7B.33 Using needle-nose pliers or a pick-type tool, pull the clip from the linkage

7B.34 Mark the relationship of the plastic cover to the thermostat housing, then remove the three cover clamp screws (1) and retaining ring (2). Remove the choke thermostat cover and gasket. Once the thermostat cover and gasket have been removed locate the three housing mounting screws (3) and remove the screws and housing.

7B.35 Both of these choke covers use a bi-metal spring in the cover. The one on the left uses engine heat to control the movement of the bi-metal spring, while the one on the right uses an electric heater element behind the spring to operate it. Be sure to check the vacuum piston (arrows) for free movement.

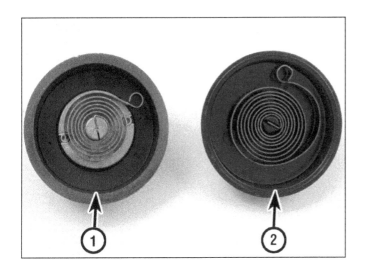

7B.36 Whether your carburetor has an electrically heated spring element (1) or a manifold heat-stove-operated spring (2) the movement of the choke plate still relies on a bi-metal spring. If the spring is broken or corroded or looks deformed, like the one on the right, obtain a replacement.

Secondary Linkage - mechanical and vacuum

The secondary throttle plates are operated one of two ways: either mechanically, by a rod that connects the primary and secondary throttle shafts or by a vacuum diaphragm that opens the secondary throttles in response to high-airflow conditions.

Mechanical Secondaries

7B.37 The mechanical linkage is nothing more than a linkage rod (1) connecting the throttle mechanism to the secondary throttle shaft (2). Pull off the clip and remove the rod.

Warning: *When assembling carburetor components, it is important to ensure that you are able to get full movement of each part to ensure maximum performance. Linkage rods can be bent slightly to adjust overall length to be sure components have their full range of movement. Be careful that an adjustment doesn't cause any binding between throttle shafts or the throttles could stick open, possibly resulting in a traffic collision!*

Vacuum secondaries

Carburetors with vacuum secondaries have a housing with a rubber diaphragm attached to the secondary throttle shaft. Vacuum is applied to the assembly through a passage from the main carburetor body. This opens and closes the secondary throttle plates in response to engine rpm (which determines vacuum above the throttle plates in the main body).

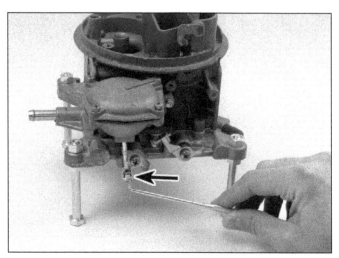

7B.38 The secondary vacuum diaphragm must be disconnected from the secondary throttle shaft by removing the C-clip as shown (arrow) and sliding the vacuum diaphragm rod off the stud

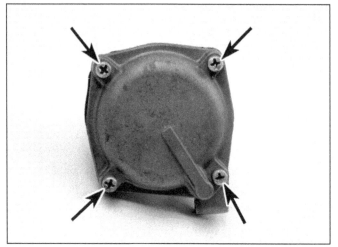

7B.40 To disassemble the diaphragm housing, remove the four screws that hold the upper and lower halves together (arrows)

7B.42 Here's the secondary vacuum diaphragm housing disassembled - note the locations of the three mounting screws (arrows). The spring, which controls secondary opening rate, can be changed for a lighter or heavier one to improve throttle response (see Chapter 8).

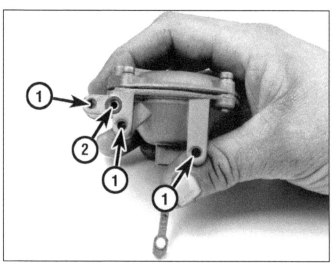

7B.39 There are three (four on large-diaphragm 2300-C models) screws that mount the diaphragm assembly to the carburetor main body (1). **Note:** *The view shown here is from the side which sits against the main body. This view is used to show the vacuum passage and gasket (2) where vacuum is applied to the diaphragm.*

7B.41 In the bottom half of the assembly housing some models have a check ball in the vacuum passage. Be sure to remove this check ball before tipping the lower housing half over or placing it into the cleaning solution

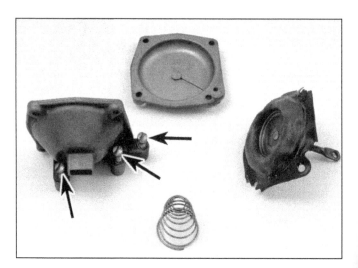

Throttle body

The throttle bodies of this group of carburetors are all based upon similar designs. Some four-barrel models have four identically sized throttle plates referred to as "square bore". Others are the "spread bore" type, with two smaller primaries and two larger secondaries

Throttle bodies are fastened to the main body by five screws on two-barrel models, or by eight screws on three- and four-barrel models.

There will be differences in the size, shape and number of passages cast into the throttle body. These are determined by the intended application of the carburetor and will not affect the disassembly procedure.

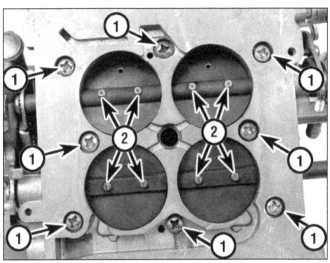

7B.43 Using a large Phillips screwdriver, remove the screws (1) from the throttle body.

Caution: *When removing any screws or bolts, always keep track of any differences such as length and fine or coarse thread. Reinstalling screws or bolts in the wrong location can damage the carburetor or, at the very least, make reassembly and adjustment difficult or impossible.*

The throttle plate screws(2) are "staked" during assembly. The purpose is to be sure they don't vibrate loose and fall into the carburetor and ultimately the engine.

Warning: *Although the throttle plates are removable by grinding off the ends of the screws and unscrewing them, it's not recommended they be removed at any time. It's not necessary for cleaning the parts. Reinstallation of the screws may not secure them properly or the shaft may be bent while trying to stake the screws again, causing a possible binding of the throttle plates in a wide-open-throttle position.*

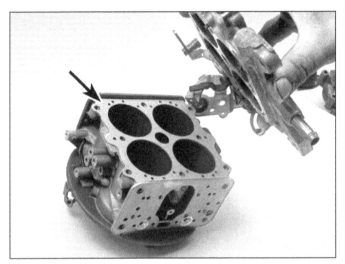

7B.44 When removing the throttle body from the main body, be sure to carefully remove the gasket (arrow) so you can to match it with the new one to make sure all necessary passages are matched between components.

Main body

The main body is essentially the same for all the models in this group. Some differences in venturi size, discharge nozzle configuration, the number of accelerator pump "shooters", their sizes and shape will be apparent. These differences will not affect the overhaul procedure.

7B.45 The main body shown for this example is from a model 4150 carburetor with a "double pumper" set-up, and so two accelerator pump discharge nozzle locations ("shooters") are necessary (arrows); other models in this group have one or two shooters, depending upon application. Using a large Phillips screwdriver, remove the screw holding the shooter in place in the main body. Then, with small needle-nose pliers, lift the assembly out of the carburetor. There are gaskets between all the parts of this assembly, so be sure they get removed before cleaning.

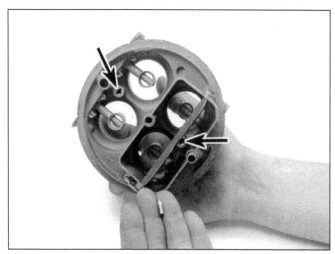

7B.46 After the "shooter(s)" have been removed, be sure to re-move the accelerator pump discharge valve, sometimes referred to as a "pill" or "anti-siphon valve", from the fuel passage

7B.47 A disassembled view of the accelerator pump discharge nozzle, screw, discharge valve and gaskets. Note: *The arrow indi-cates a gasket stuck inside the shooter. Always check for gaskets stuck to components before cleaning.*

Cleaning

7B.48 Assembly of the choke plate is done with "staked" screws, as were the throttle plates. The purpose is to be sure they don't vi-brate loose and fall into the carburetor and ultimately the engine.

Although the choke plate is removable by grinding off the ends of the screws and unscrewing them, it's not recommended. It's not necessary for cleaning and reinstallation of the screws may not properly secure the choke plate or the shaft may be bent while try-ing to stake the screws again, causing a possible binding condi-tion.

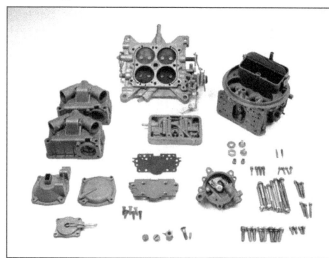

7B.49 Here's a disassembled 4160, ready to go into the dip tank for an overnight soak - note that plastic and rubber parts are NOT shown here, since they can be damaged by harsh solvents - use soap and water to clean plastic and rubber parts that will be re-used.

Reassembly

Reassembly is basically the reverse of disassembly. Sev-eral assembly checks, tips and adjustments are shown in the following illustrations to help you with areas of special concern. Read the special instructions in the carburetor kit and follow the illustrations here when making adjustments.

> **Tip:**
> When reassembling the carburetor, lightly spray gas-kets with silicone spray lubricant. This will help elimi-nate the kind of sticking you saw as you disassem-bled your carburetor. There are also special rubber gaskets available through Holley for those who expect frequent disassembly of their carburetor, especially for performance modifications and adjustments.

7B.50 Quality of parts is as important as the quality of reassembly, so use a quality overhaul kit.

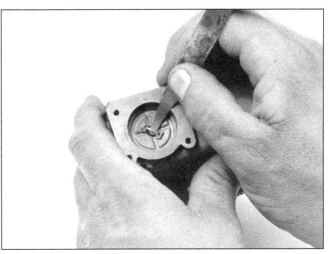

7B.51 On the type of accelerator pump shown here, the clearance between the check ball and the retainer bar must be checked. Turn the fuel bowl over and measure the clearance with a narrow feeler gauge (here a wide gauge has been trimmed to the correct size). It must be 0.011 to 0.015 inch. Bend the retainer bar slightly to get the correct clearance.

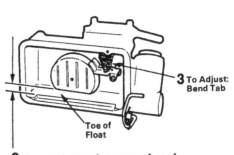

1 With fuel bowl inverted

3 To Adjust: Bend Tab

Toe of Float

2 Measure distance between surface of fuel bowl and float, at toe of the float

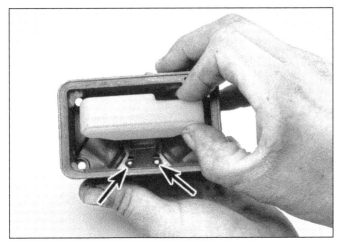

7B.52 On center-pivot floats, put the float assembly into the float bowl as shown and install the two screws (arrows) that hold the float assembly in place

7B.53 A few carburetors do not have externally adjustable floats, so you'll have to set the float level during assembly, as shown. The distance between the arrows should be as listed in the overhaul kit instructions.

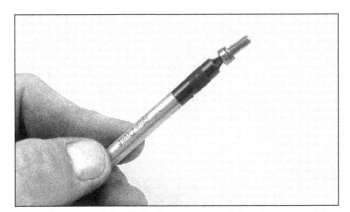

7B.54 Because of the limited space to work in when assembling parts, if can be very difficult to get smaller screws into a recessed area without dropping them. The thin locking screwdriver shown here is a great help in both removing and installing these parts. It's inexpensive and most tool supply outlets carry a version of this type of tool.

7B.55 Before inserting the needle-and-seat assembly into the float bowl, always apply lithium-base grease to the o-ring to make installation easier and provide a better seal.

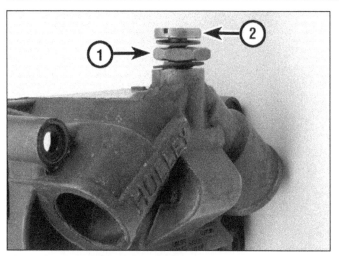

7B.56 On externally adjustable floats, once the needle and seat are installed, place a gasket on first, install the adjusting nut (1) install another gasket, then screw in the lock screw (2).

7B.57 On externally adjustable floats, with the float, needle and seat installed, turn the float bowl upside down and adjust the float level by turning the adjusting nut so it moves the float up or down. The initial setting for the float should set the top of the float parallel with the float bowl. Using a common pencil as a gauge (arrow), will get you very close to the actual running float height, which you'll set after the carburetor is installed (center-hung float shown, side-hung uses the same technique).

Metering bodies and plates

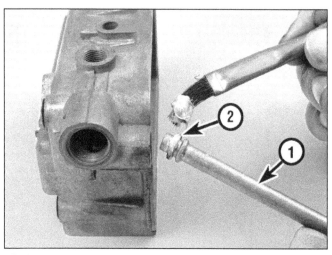

7B.58 If your carburetor has this type of fuel inlet system, before placing the fuel balance tube (1) and O-ring (2) into the float bowl for assembly, always place some grease on the O-ring to ease the assembly of the parts. Very little is needed, so don't overdo it.

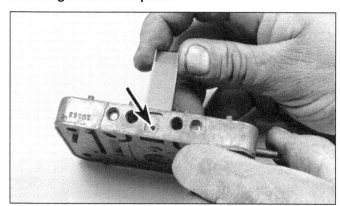

7B.59 When installing the pin to hold the new vent tube in place (arrow), look at the vent tube itself first. A small hole may need to be drilled in the plastic so that the pin can be pushed through to hold the tube in place.

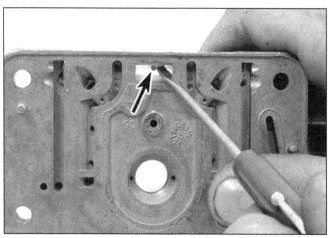

7B.60 Take a look at the installed vent and retaining pin (arrow). If installation of the pin has pushed the vent closed, take a small screwdriver and open the vent tube back up to fit the opening in the metering body.

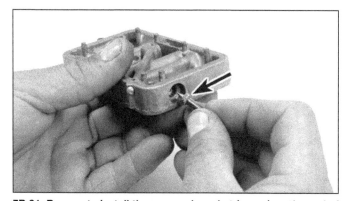

7B.61 Be sure to install the new cork gasket (arrow) on the end of each idle mixture screw before reinstalling it into the metering body. The idle mixture screws need to be installed with an initial setting for the engine to idle after starting. Turn the idle mixture screws all the way into the seat. Caution: *Don't force the screws into the seat; just turn the screw in until it lightly touches it.* **Then back each screw out one-and-one half to two turns**

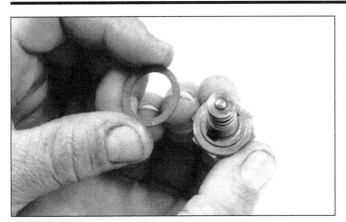

7B.62 Install the gasket on the power valve, then . . .

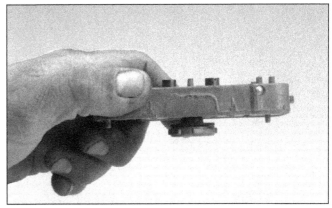

7B.63 . . . hold the metering body horizontal and screw in the valve until it is tight. This method will make sure the gasket stays centered during assembly.

7B.64 Here's an example of a power valve that came off of a metering body when the body was held vertically during assembly. The gasket dropped before the power valve could be tightened all the way. This caused a fuel leak through the valve, making the power valve less effective.

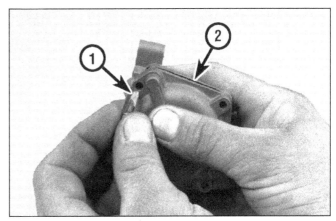

7B.65 When reassembling the vacuum-secondary diaphragm, be sure to use a small amount of grease on the screws (1) so that the screw doesn't grab the rubber diaphragm and damage it as the screws are being installed. Visually look for the diaphragm around the edges of the assembly (2). This lets you know the diaphragm is sitting between the two halves properly and a proper vacuum seal will be made.

7B.66 If you have an automatic choke, your fast-idle cam and choke rod arm should look like this before you install the choke housing on the carburetor - the arrow shows the first step of the fast-idle cam, which is where you'll be placing the fast-idle screw when you're setting the fast-idle speed later in this procedure

7B.67 When reinstalling the choke housing, make sure the choke rod (arrow) is positioned UNDER the fast-idle cam arm - also, don't forget to reinstall the clip that retains the choke rod to the choke arm

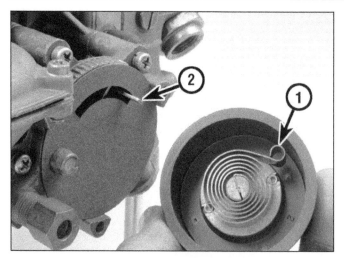

7B.68 When reinstalling the automatic choke, be sure the coiled end (1) hooks over the protruding tang (2) that extends out from the choke housing

7B.69 Reinstall the retainer ring (arrow) so that, when the screws are reinstalled, the retainer applies the most tension that it can. If the retainer ring is turned over on installation, you'll find that there isn't enough tension to secure the cover and the choke setting can easily change.

7B.70 Look very closely at all gaskets used during reassembly. Notice the differences in the number and placement of holes between the upper two gaskets. This is an example of why the old parts and gaskets should be kept until the overhaul is complete.

Caution: *Carburetor kits generally come with gaskets and some parts for more than one application, so extreme care must be taken when matching up replacement pieces. Use of the wrong replacement part could lead to poor performance or even an inoperative condition.*

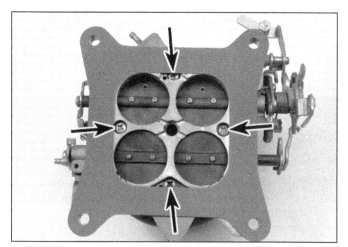

7B.71 By placing the carburetor base gasket in place, you can clearly see that not all the screws are covered by the gasket. There is a possibility the screws could come loose and fall into the engine.

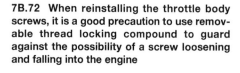

7B.72 When reinstalling the throttle body screws, it is a good precaution to use removable thread locking compound to guard against the possibility of a screw loosening and falling into the engine

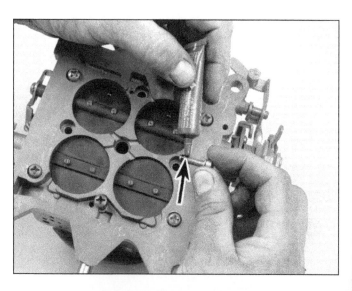

Fast-idle speed setting

7B.73 Before installing the carburetor on the vehicle, open the throttle and set the fast-idle screw (2) on the first step of the fast-idle cam (see illustration 7B.67). Then, using drill bits as measuring tools (1), check that the primary throttle plates are open approximately 0.035-inch (0.035-inch is between 1/32-inch and 3/64-inch, but closer to 1/32-inch). If the clearance is not correct, adjust it with the fast-idle screw. After the carburetor is reinstalled, all on-vehicle adjustments have been made and the engine is at normal operating temperature, place the fast-idle screw back on the first step of the fast-idle cam and check that the idle speed is approximately 1700 rpm. If the speed is not correct, adjust it with the screw. If you notice the engine idling slowly and "chugging" when the engine is cold and the choke is fully applied, try turning the screw clockwise a bit to increase the fast-idle speed. If the idle speed seems too fast with the choke applied, turn the screw counterclockwise slightly.

Secondary idle speed setting

7B.76 Secondary idle speed on three- and four-barrel models must be set before carburetor installation. Tighten the screw (arrow) until it just touches the secondary throttle lever (you'll be able to see the throttle plates open slightly), then tighten the screw approximately one more turn

Tip:
If you are having a hard time getting idle speed and mixture adjustment on the primaries, open the secondary idle speed screw just slightly more. This will allow you to close the primaries enough to cover the transfer slots and give a better signal to the idle circuit for more responsive adjustments.

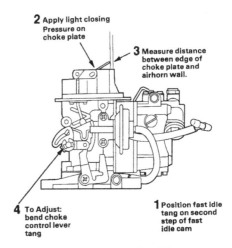

7B.74 A few carburetors do not have fast-idle screws, and you adjust the fast idle by bending the choke control lever tang until the choke plate-to-airhorn wall clearance is as listed in the overhaul kit specifications

Accelerator pump setting

7B.75 To make the accelerator pump adjustments, with the throttle held in the wide open position, you should be able to move the accelerator pump arm approximately 0.010 inch from the operating lever (arrow) before it stops moving. This will prevent any binding problems at the full stroke of the pump. On the other end of the stroke, at idle, be sure you also have about 0.010 inch clearance. Holley recommends that the operating lever contact the pump arm (0 clearance), but operating with no clearance can sometimes cause an unnecessary "drip" at the discharge nozzle, because every movement or bump will cause very slight movements of the accelerator pump arm. Not enough for a shot, but more than enough to trickle small amounts of fuel into the venturi.

Automatic choke setting

Most models are equipped with a choke thermostat coil and housing mounted on the carburetor. To set the choke, realign the marks you made on disassembly of the choke cover from the housing. You may not need as much choke as before now that the carburetor is working properly, so readjustment may be necessary.

To readjust, rotate the thermostat cover counter-clockwise until the choke plate is closed, then rotate it clockwise until the plate just starts to move away from the closed position. Tighten the three retaining screws to hold the cover in place.
Note: *The location of the mark on the cover may not line up with any marks on the housing, but it is still set correctly for your vehicle. Spring tension can deteriorate with age and use,*

so the location of the correct setting can change.

On some models, the choke is operated by a thermostat coil mounted in the intake manifold, which actuates the choke via a rod. With the carburetor installed and the engine cold, the coil should hold the choke almost (but not quite) completely closed. If the choke is closed tightly or open too far, remove the coil from the intake manifold; the center mount for the thermostat coil should be slotted so you can adjust the coil's tension with a screwdriver. If there's no adjustment at the choke coil's center mount, bend the rod to adjust the choke. If the coil is broken, corroded or deformed, replace it.

Once the reassembly is complete, reinstall the carburetor on the vehicle. Refer to Chapter 6.

On-vehicle adjustments

After the carburetor is reinstalled on the vehicle (see Chapter 6), start the engine and warm it to normal operating temperature. Make sure it is on a level surface, then proceed with the adjustments described in illustrations 7B.77 through 7B.80. **Warning:** *Be sure the vehicle is parked on a level surface and the parking brake is set, the wheels are blocked and the transmission is in Neutral (manual) or Park (automatic).*

Float level setting

7B.77 The same procedure is used for setting the float level in both the front and rear float bowls; first, remove the sight plug . . .

Idle mixture adjustment

7B.79 With the float level correct, it's time to make the final adjustments on the idle mixture screws. Hook up a tachometer in accordance with the manufacturer's instructions. On a standard idle system, with the engine idling at the proper speed, working with one idle screw at a time, turn the screw clockwise until the idle speed drops a noticeable amount (on models with reverse-idle systems, turn the screw counterclockwise). Now slowly turn the idle screw counterclockwise (clockwise on reverse-idle systems) until the maximum rpm is reached, but no further or the mixture will be overly rich. Repeat the procedure on the other idle mixture screw. Keep track of the number of turns each screw is off the seat to be sure they are each out approximately the same number of turns. This will verify that the idle circuit is operating properly and a balance is maintained.

7B.78 . . . then loosen the lock screw slightly to allow the adjustment nut to turn so you can raise or lower the float. By setting the dry float level as described previously, you should be very close to the correct setting. Now turn the adjustment nut so the fuel level just comes up to and possibly "trickles" over the threads of the sight hole. Once the adjustment is reached, and while holding the adjustment nut in place, tighten the lock screw down to maintain the float level. Now repeat this procedure for the rear float bowl (three- and four-barrel models).

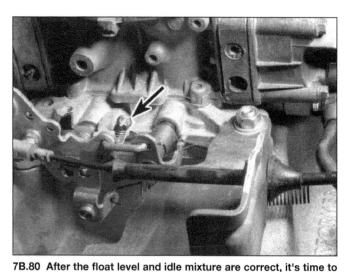

7B.80 After the float level and idle mixture are correct, it's time to set the curb idle speed. Check the Vehicle Emissions Control Information (VECI) label under the hood for the proper specification. Hook up a tachometer in accordance with the manufacturer's instructions and turn the idle speed screw (arrow) until the idle rpm is as specified (turning the screw clockwise increases the idle speed, while turning it counterclockwise decreases idle speed). Note that some models (primarily the 4180) have idle speed control solenoids. If your carburetor is so equipped, adjust the idle speed by turning the screw that contacts the solenoid plunger.

7 Part C
Overhaul and adjustments
Four-barrel models 4010/4011

This group of carburetors includes a variety of models all based on a similar design. The models included in this group are the four-barrel square-bore model 4010 and four-barrel spreadbore model 4011.

The overhaul procedure is covered through a sequence of photographs laid out in order from disassembly to reassembly and finally installation and adjustment on the vehicle. The captions presented with the photographs will walk you through the entire procedure one component at a time.

The following illustration sequence will use a model 4010. Specific details will vary with carburetor part numbers, because each carburetor is made to fit a specific application. Refer to the instruction sheet which comes with each overhaul kit. **Note:** *Because of the large number of different models and variations of each model, it is not always possible to cover all the differences in components during the overhaul procedure. Overhaul sections only deal with disassembly, reassembly and basic adjustments of a typical carburetor in that model group. If you need further information about a specific variation, refer to the exploded views at the beginning of this Chapter or go to Chapter 3.*

Warnings:
Gasoline

Gasoline is extremely flammable, so take extra precautions when you work on any part of the fuel system. Don't smoke or allow open flames or bare light bulbs near the work area, and don't work in a garage where a natural gas-type appliance (such as a water heater or clothes dryer) with a pilot light is present. A spark caused by an electrical short circuit, by two metal surfaces striking each other, or even by static electricity built up in your body, under certain conditions, can ignite gasoline vapors. Also, never risk spilling fuel on a hot engine or exhaust component. If you spill any fuel on your skin, rinse it off immediately with soap and water.

Battery

Always disconnect the battery ground (-) cable at the battery before working on any part of the fuel or electrical system.

Fire extinguisher

We strongly recommend that a fire extinguisher suitable for use on fuel and electrical fires be kept handy in the garage or workshop at all times. Never try to extinguish a fuel or electrical fire with water. Post the phone number for the nearest fire department in a conspicuous location near the phone.

Compressed air

When cleaning carburetor parts, especially when using compressed air, be very careful to spray away from yourself. Eye protection should be worn to avoid the possibility of getting any chemicals or debris into your eyes.

Lead poisoning

Avoid the possibility of lead poisoning. Never use your mouth to blow directly into any carburetor component. Small amounts of Tetraethyl lead (a lead compound) become deposited on the carburetor over a period of time and could lead to serious lead poisoning. Check passages with compressed air and a fine-tipped blow gun or place a small-diameter tube to the component and blow through the tube to be sure all necessary passages are open.

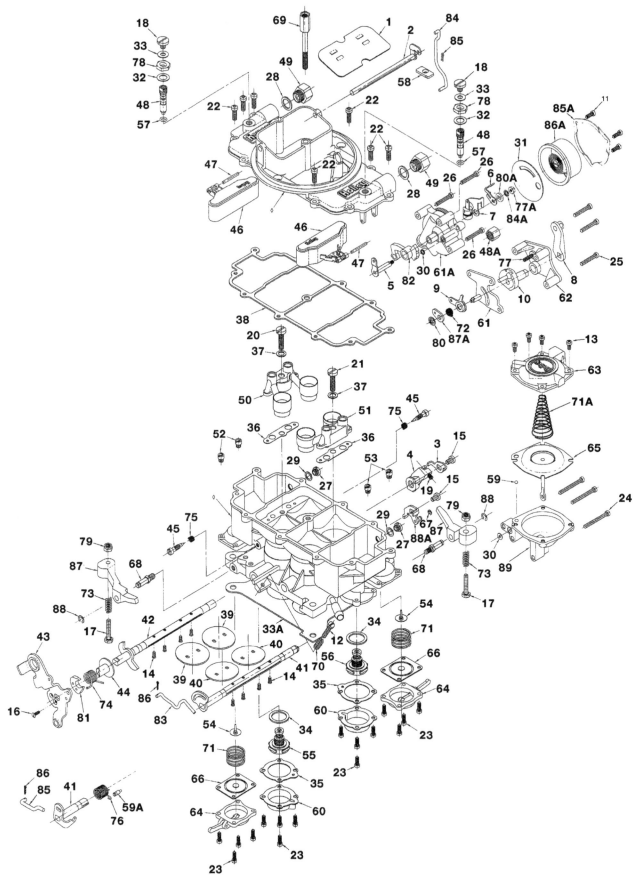

Model 4010 exploded view with a typical Holley square-bore throttle-body flange

Index Number	Part Name	Index Number	Part Name
◇	Airhorn Assembly	51	Secondary Nozzle Bar Assy.
1	Choke Plate	52	Primary Main Jet
2	Choke Shaft and Lever Assembly	53	Secondary Main Jet
3	Cam Pick-Up Lever	54	Pump Check Valve
4	Fast Idle Cam Lever†	55	Power Valve - Primary
5	Choke Hsg. Shaft & Lever Assy. ††	56	Power Valve - Secondary
6	Choke Thermostat Lever ††	57	O-Ring Seal
7	Thermostat Lever and Piston Assy.††	58	Choke Rod Seal
		59	Check Ball
8	Choke Lever and Swivel Assy. †	59A	Throttle Return Spring Pin
9	Choke Rod Lever Molded †	60	Power Valve Cover
10	Fast Idle Cam †	61	Choke Lever Back-up Plate †
11	Thermostat Cap Clamp Screw ††	61A	Choke Housing & Plugs Assy. ††
12	Throttle Stop Screw	62	Fast Idle Cam Plate †
13	Secondary Diaphragm Cover Screw	63	Secondary Diaphragm Cover
		64	Acc. Pump Diaphragm Cover Assy.
14	Throttle Plate Screw	65	Secondary Diaphragm & Link Assy.
15	Fast Idle Cam Lever & Sec. Diaphragm Lever Screw & L.W.	66	Acc. Pump Diaphragm Assy.
		67	E-Ring Retainer
16	Pump Cam Screw	68	Pump Lever Stud
17	Pump Lever Adj. Screw	69	Air Cleaner Stud
18	Fuel Valve Seat Screw	70	Throttle Stop Adjusting Screw Spring
19	Fast Idle Adj. Screw	71	Pump Diaphragm Return Spring
*	Secondary Idle Adj. Screw		
20	Nozzle Bar Screw - Primary	71A	Secondary Diaphragm Spring
21	Nozzle Bar Screw - Secondary	72	Choke Lever Spring
22	Airhorn Screw	73	Pump Lever Adj. Screw Spring
23	Pump Diaph. Cover & Power Valve Cover Screw	74	Primary Throttle Return Spring
		75	Idle Adj. Needle Spring
24	Secondary Diaphragm Housing Screw	76	Secondary Throttle Return Spring
25	Fast Idle Cam Plate Screw †	77	Fast Idle Cam Spring †
26	Choke Housing Screw ††	77A	Hex Nut ††
27	Fuel Level Check Plug	78	Fuel Valve Seat Lock Nut
28	Fuel Inlet Fitting Gasket	79	Fibre Locking Nut
29	Fuel Level Plug Gasket	80	Push Nut Fastener †
30	Sec. Diaphragm Housing & Choke Housing Gasket	80A	Choke Thermostat Lever Spacer ††
31	Thermostat Cap Gasket ††	81	Pump Cam
32	Needle and Seat Adj. Nut Gasket	82	Fast Idle Cam Assy. ††
33	Seat Lock Screw Gasket	83	Secondary Connecting Rod
33A	Flange Gasket	84	Choke Rod
34	Power Valve Gasket	84A	Lockwasher ††
35	Power Valve Cover Gasket	85	Cotter Pin -Choke Rod
36	Nozzle Bar Gasket	85A	Thermostat Cap Clamp ††
37	Nozzle Bar Screw Gasket	86	Cotter Pin - Throttle Connecting Rod
38	Airhorn Gasket	86A	Thermostat & Cap Assy. Indexed ††
39	Primary Throttle Plate		
40	Secondary Throttle Plate	87	Pump Operating Lever Assy.
41	Secondary Throttle Shaft & Lever Assy.	87A	Choke Spring Ret. Washer †
		88	E-Ring Retainer
42	Throttle Shaft and Stop Lever Assy.	88A	Sec. Diaphragm Lever Assy.
		89	Secondary Diaphragm Housing
43	Primary Throttle Lever	*	J - Tube Bowl Vent
◇	Carburetor Body	*	Fuel Inlet Filter Screen
*	Secondary Throttle Shaft Bushing		
44	Throttle Lever Bushing		
45	Idle Adj. Needle		
46	Float and Tab Assy.		
47	Float Shaft		
48	Fuel Inlet Needle and Seat Assy.		
48A	Compression Nut ††		
49	Fuel Inlet Fitting		
50	Primary Nozzle Bar Assy.		

Parts list for model 4010 exploded view

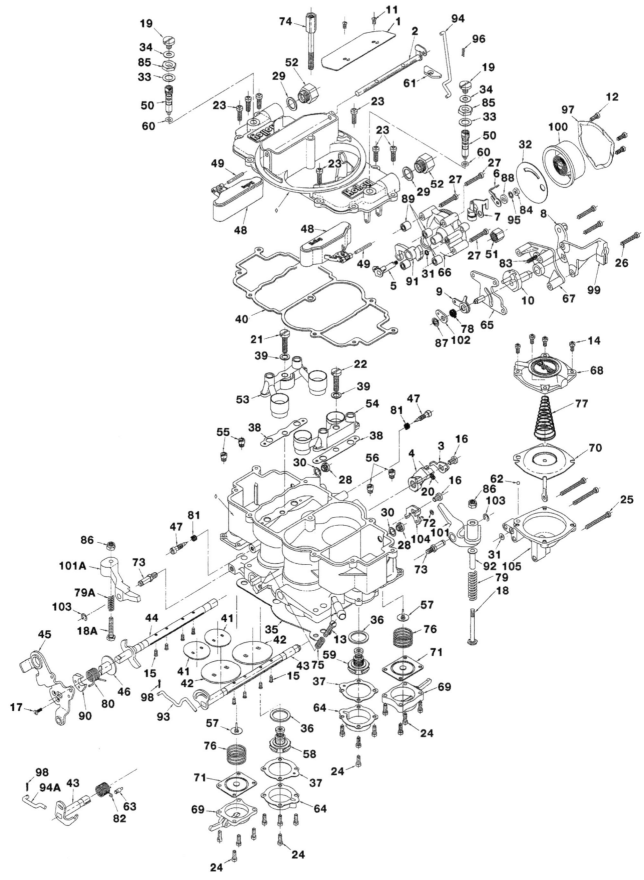

Model 4011 exploded view with the Rochester Quadrajet spread-bore flange

Index Number	Part Name	Index Number	Part Name
◇	Airhorn Assembly	54	Secondary Nozzle Bar Assy.
1	Choke Plate	55	Primary Main Jet
2	Choke Shaft and Lever Assembly	56	Secondary Main Jet
3	Cam Pick-Up Lever	57	Pump Check Valve
4	Fast Idle Cam Lever†	58	Power Valve - Primary
5	Choke Hsg. Shaft & Lever Assy. ††	59	Power Valve - Secondary
6	Choke Thermostat Lever ††	60	O-Ring Seal
7	Thermostat Lever and Piston Assy.††	61	Choke Rod Seal
8	Choke Lever and Swivel Assy. †	62	Check Ball
9	Choke Rod Lever Molded †	63A	Throttle Return Spring Pin
10	Fast Idle Cam †	64	Power Valve Cover
11	Choke Plate Screw	65	Choke Lever Back-up Plate †
12	Thermostat Cap Clamp Screw ††	66A	Choke Housing & Plugs Assy. ††
13	Throttle Stop Screw	67	Fast Idle Cam Plate. †
14	Secondary Diaphragm Cover Screw	68	Secondary Diaphragm Cover
15	Throttle Plate Screw	69	Acc. Pump Diaphragm Cover Assy.
16	Fast Idle Cam Lever & Sec. Diaphragm Lever Screw & L.W.	70	Secondary Diaphragm & Link Assy.
17	Pump Cam Screw	71	Acc. Pump Diaphragm Assy.
18	Pump Lever Adj. Screw	72	E-Ring Retainer
18A	Pump Lever Adj. Screw	73	Pump Lever Stud
19	Fuel Valve Seat Screw	74	Air Cleaner Stud
20	Fast Idle Adj. Screw Secondary Idle Adj. Screw	75	Throttle Stop Adjusting Screw Spring
21	Nozzle Bar Screw - Primary	76	Pump Diaphragm Return Spring
22	Nozzle Bar Screw - Secondary	77	Secondary Diaphragm Spring
23	Airhorn Screw	78	Choke Lever Spring
24	Pump Diaph. Cover & Power Valve Cover Screw	79	Pump Lever Adj. Screw Spring
25	Secondary Diaphragm Housing Screw	79A	Pump Lever Adj. Screw Spring
26	Fast Idle Cam Plate Screw †	80	Primary Throttle Return Spring
27	Choke Housing Screw ††	81	Idle Adj. Needle Spring
28	Fuel Level Check Plug	82	Secondary Throttle Return Spring
29	Fuel Inlet Fitting Gasket	83	Fast Idle Cam Spring †
30	Fuel Level Plug Gasket	84	Hex Nut ††
31	Sec. Diaphragm Housing & Choke Housing Gasket	85	Fuel Valve Seat Lock Nut
32	Thermostat Cap Gasket ††	86	Fibre Locking Nut
33	Needle and Seat Adj. Nut Gasket	87	Push Nut Fastener †
34	Seat Lock Screw Gasket	88A	Choke Thermostat Lever Spacer ††
35	Flange Gasket	89	Choke Housing Spacer
36	Power Valve Gasket	90	Pump Cam
37	Power Valve Cover Gasket	91	Fast Idle Cam Assy. ††
38	Nozzle Bar Gasket	92	Pump Lever Adj. Screw Sleeve
39	Nozzle Bar Screw Gasket	93	Secondary Connecting Rod
40	Airhorn Gasket	94	Choke Rod
41	Primary Throttle Plate	94A	Secondary Connecting Rod
42	Secondary Throttle Plate	95	Lockwasher ††
43	Secondary Throttle Shaft & Lever Assy.	96	Retainer-Choke Rod
44	Throttle Shaft and Stop Lever Assy.	97	Thermostat Cap Clamp ††
45	Primary Throttle Lever	98	Cotter Pin - Throttle Connecting Rod
◇	Carburetor Body	99	Choke Control Bracket
*	Secondary Throttle Shaft Bushing	100	Thermostat & Cap Assy. Indexed ††
46	Throttle Lever Bushing	101	Pump Operating Lever Assy.
47	Idle Adj. Needle	101A	Pump Operating Lever Assy.
48	Float and Tab Assy.	102	Choke Spring Ret. Washer †
49	Float Shaft	103	E-Ring Retainer
50	Fuel Inlet Needle and Seat Assy.	104	Sec. Diaphragm Lever Assy.
51A	Compression Nut ††	105	Secondary Diaphragm Housing
52	Fuel Inlet Fitting	*	J - Tube Bowl Vent
53	Primary Nozzle Bar Assy.	*	Fuel Inlet Filter Screen

Parts list for model 4011 exploded view

Disassembly

7C.1 Using a needle-nose pliers, or something similar, remove the clip from the linkage that holds the lower end of the choke rod (1) to either the manual or automatic choke assembly, whichever is used on your model. Next, remove the three Torx screws (2) that hold the assembly to the main body of the carburetor and remove the assembly and set it aside. This assembly is normally cleaned by hand, with spray cleaner, but placing this type of hard plastic into a cleaning solution will not harm it.

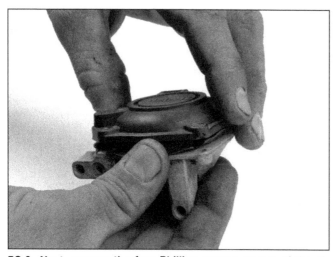

7C.3 Next, remove the four Phillips screws on top of the diaphragm assembly and remove the top so you can remove the rubber diaphragm and spring inside; these parts can't go into any strong cleaning solutions, but the housing body and cover should be cleaned thoroughly.

7C.2 Remove the C-clip (1) that attaches the vacuum secondary assembly to the secondary throttle shaft linkage. Remove the three Torx screws (2) attaching the assembly to the main body and remove the secondary diaphragm from the main body of the carburetor.

7C.4 Remove the brass fuel inlet nuts at both float bowls, then take the filter screens out of the cover and throw them away; new ones are normally provided in the overhaul kit.

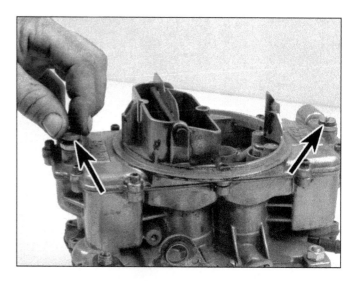

7C.5 Remove the locking screws from each needle and seat, . . .

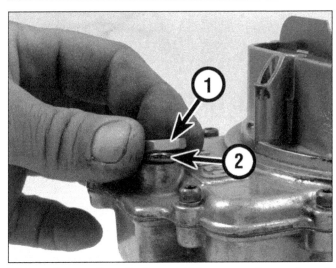

7C.6 ... then use the adjusting nut (1) to unscrew the needle-and-seat assembly (2) from the cover. Lift the needle and seat assembly from the cover. New needle-and-seat assemblies will be provided in the overhaul kit.

7C.7 There are two idle mixture screws, one on either side of the carburetor. Remove the screws and set them aside to be cleaned. Check the tips to be sure there are no grooves or pit marks. Any damage or wear of this type would cause poor idle mixture control. Replace them, if necessary. Also, remove the fuel-level sight plugs (arrow)

7C.8 Remove the ten Torx screws (arrows) from the cover of the carburetor. . .

7C.9 . . .and remove the cover, and attached floats, from the carburetor main body

7C.10 Remove the four Torx screws (1) from the accelerator pump cover. If your model has dual accelerator pumps, one located on each of the float bowls, be sure to remove both assemblies.

Remove the C-clip (2) from the shaft and remove the accelerator pump operating lever. Here again, there will be an operating lever for each accelerator pump. The shafts screw into the main body, but, for cleaning, they don't need to be removed.

7C.11 Remove the accelerator pump diaphragm and spring. The diaphragm will be replaced by a new one provided in the overhaul kit. **Note:** *Even when new parts are provided in the overhaul kit, it's always a good idea to wait until the overhaul is complete to throw away any parts or gaskets. Examination of the old parts is a good way to help diagnose problems.*

7C.12 The inlet side of the accelerator pump circuit uses a rubber check valve to open and close the fuel inlet passage from the fuel bowl. Pull the valve from the pump casting. The valve will be replaced after cleaning with a new one provided in the overhaul kit.

7C.13 Remove the four Torx screws holding each power valve cover onto the main body. Each float bowl will have a separate power valve, . . .

7C.14 . . . so be sure to remove both covers

7C.15 Unscrew and remove the power valves from each float bowl - note that the carburetor is being held horizontal while the power valve is being unscrewed. When installing a power valve, it is essential that the carburetor be held horizontal so the gasket will not slip out of place, causing the power valve to malfunction.

7C.16 Remove the screw that attaches each booster venturi cluster assembly to the main body and . . .

7C.17 . . . lift out the two cluster assemblies, noting which one is installed on which side - although they look similar, they are frequently not the same and installing them in the wrong locations can result in driveability problems

7C.18 Remove the pump check ball weight . . .

7C.19 . . . and check ball - if you don't have a magnet that will reach the ball, turn the carburetor over, but be careful not to lose the ball. **Note:** *Be sure to get these parts from both discharge holes if your carburetor has dual accelerator pumps.*

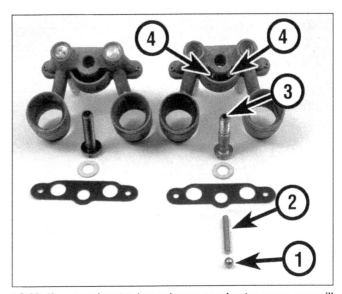

7C.20 If your carburetor has only one accelerator pump, you will find these boosters in your model. The one on the right is the one used on the side of the carburetor with the accelerator pump. You can see the part of the casting near the screw hole is made to flow fuel through it, while the assembly on the left does not have this provision. If your model has dual accelerator pumps, you will have two assemblies similar to the one on the right with:

1　Accelerator pump check ball
2　Check ball weight
3　Drilled anchor screw
4　Accelerator pump discharge nozzles

7C.21 Using the proper-size screwdriver, remove the four metering jets, two from each fuel bowl. **Note:** *There is a possibility that the jets can be different sizes, especially if the carburetor has been overhauled before, so be sure to write down the jet sizes (stamped on the sides of the jets) and record which positions they came from to avoid confusion on reassembly.*

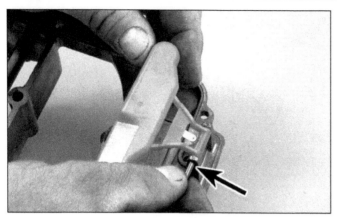

7C.22 Take the carburetor cover, removed earlier, and turn it over so the floats are up. Slide each pivot pin (arrow) out and remove the floats. Note: *Floats often become saturated with fuel and become much heavier than normal. This will cause the float to sink and raise the fuel level in the bowl, causing an overly rich mixture or fuel leaks. Not all floats become saturated, but pinholes or breaks in the solder can develop over time, even on metal or plastic floats. If in doubt about the condition of your float(s) it's recommended that the float be replaced at the time of the overhaul.*

7C.24 Turn the adjustment nut until each upper float surface is approximately parallel with the cover, as shown here.

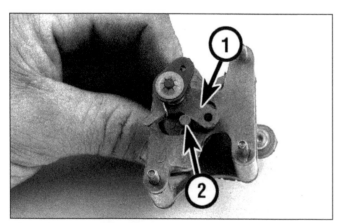

7C.26 Before you reinstall the choke linkage, be sure it looks like this, with the choke lever (1) and rod (2) in this relationship.

7C.27 When reassembling the vacuum secondary diaphragm, the spring tension will make it tricky to keep the rubber diaphragm between the upper and lower halves. To make it easier, keep light pressure on the connecting rod (arrow) to offset the spring pressure. Note: *Be sure to use some white grease on the threads of the screws to keep the rubber diaphragm from being grabbed and torn as the screws are tightened.*

Reassembly

Several assembly checks, tips and adjustments are shown in the following illustration sequence, but, if nothing is mentioned here, reassembly is in the reverse order of disassembly. Read the special instructions in the carburetor kit and follow the illustration sequence in making adjustments.

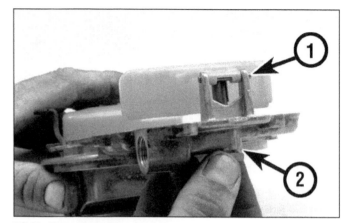

7C.23 First reinstall the both floats and pivot pins (1). Then, holding the cover as shown, thread the needle-and-seat assembly (2) into the cover. Note: *Use of a small amount of grease on the O-ring on the needle-and-seat assembly to make installation easier and provide a better seal.*

7C.25 Reinstall the accelerator pump check valve into the hole in the pump cavity by very lightly pulling the valve through the hole until the thick portion of the stem comes into the fuel bowl. You can cut off the excess stem, but this isn't necessary.

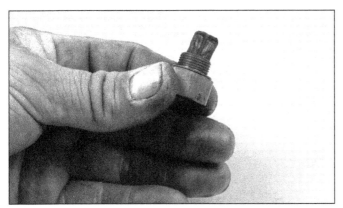

7C.28 It is easier to install the filter basket into the inlet nut first, then screw it into the carburetor cover. If you place the filter into the carburetor first, the inlet nut can crush the filter if it doesn't center itself in the hole.

7C.30 On the other end of the stroke, at idle, be sure you have about 0.010 inch clearance. Holley reccomends that the operating lever contact the pump arm (0 clearance), but operating with no clearance can sometimes cause an unnecessary "drip" at the discharge nozzle because every movement or bump will cause very slight movements of the accelerator pump arm. Not enough for a shot, but more than enough to trickle small amounts of fuel into the venturi.

Secondary idle speed adjustment

Accelerator pump setting

7C.29 With the throttle held in the wide-open position, you should be able to move the accelerator pump arm approximately 0.010 inch away from the operating lever. This will prevent any binding problems at the full stroke of the pump.

Idle mixture screw setting

7C.31 Make the initial idle screw settings during reassembly. The idle mixture screws, one on either side on the primary side, have an initial setting between 1 1/2 to 2 turns out from the lightly seated position. This will get the vehicle running, then the final settings need to done while the engine is running.

> **Tip:**
> If you are having a hard time getting idle speed and mixture adjustment on the primaries, open the secondaries just slightly more. This will allow you to close the primaries enough to cover the transfer slots and give a better signal to the idle circuit for more responsive adjustments.

7C.32 Secondary idle speed must be set before carburetor installation. Adjust the screw (arrow) until it just touches the secondary throttle lever (you'll be able to see the throttle plates open slightly), then tighten the screw approximately one more turn.

Fast-idle speed setting

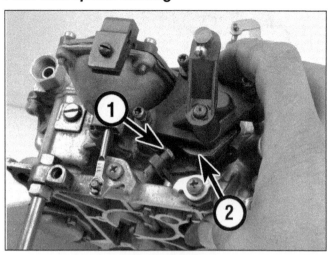

7C.33 Set the fast-idle speed using the procedure described at the end of Chapter 7B. For the adjustment, the fast-idle screw (1) should be positioned on the first step of the fast-idle cam (2).

Automatic choke setting

Most models have an integral choke thermostat housing mounted on the carburetor, although some models have a "divorced" choke coil mounted to the intake manifold. To set the choke, refer to the procedures at the end of Chapter 7B.

On-vehicle adjustments

The procedures for adjusting the float level, idle speed and idle mixture on the vehicle are the same as those described in Chapter 7B. Refer to the procedures in Chapter 7B, but note that the idle mixture screws are in different locations **(see illustrations 7C.7 and 7C.31)**.

7 Part D
Overhaul and adjustments

Four-barrel model 4360
Two-barrel models 2360/6360

This group of carburetors includes a variety of models all based on a similar design. The models included in this group are the four-barrel model 4360, two-barrel model 2360 and it's feedback version, model 6360.

The overhaul procedure is covered through a sequence of photographs laid out in order from disassembly to reassembly and finally installation and adjustment on the vehicle. The captions presented with the photographs will walk you through the entire procedure, one component at a time.

The following illustration sequence will use model 4360. Specific details will vary with carburetor part numbers because each carburetor is made to fit a specific application. Refer to the instruction sheet which comes with each overhaul kit. **Note:** *Because of the large number of different models and variations of each model, it is not always possible to cover all the differences in components during the overhaul procedure. Overhaul chapters only deal with disassembly, reassembly and basic adjustments of a typical carburetor in that model group. If you need to know about specific differences, refer to the exploded views beginning on the next page or go to Chapter 3 for more details.*

Warnings:
Gasoline

Gasoline is extremely flammable, so take extra precautions when you work on any part of the fuel system. Don't smoke or allow open flames or bare light bulbs near the work area, and don't work in a garage where a natural gas-type appliance (such as a water heater or clothes dryer) with a pilot light is present. A spark caused by an electrical short circuit, by two metal surfaces striking each other, or even by static electricity built up in your body, under certain conditions, can ignite gasoline vapors. Also, never risk spilling fuel on a hot engine or exhaust component. If you spill any fuel on your skin, rinse it off immediately with soap and water.

Battery

Always disconnect the battery ground (-) cable at the battery before working on any part of the fuel or electrical system.

Fire extinguisher

We strongly recommend that a fire extinguisher suitable for use on fuel and electrical fires be kept handy in the garage or workshop at all times. Never try to extinguish a fuel or electrical fire with water. Post the phone number for the nearest fire department in a conspicuous location near the phone.

Compressed air

When cleaning carburetor parts, especially when using compressed air, be very careful to spray away from yourself. Eye protection should be worn to avoid the possibility of getting any chemicals or debris into your eyes.

Lead poisoning

Avoid the possibility of lead poisoning. Never use your mouth to blow directly into any carburetor component. Small amounts of Tetraethyl lead (a lead compound) become deposited on the carburetor over a period of time and could lead to serious lead poisoning. Check passages with compressed air and a fine-tipped blow gun or place a small-diameter tube to the component and blow through the tube to be sure all necessary passages are open.

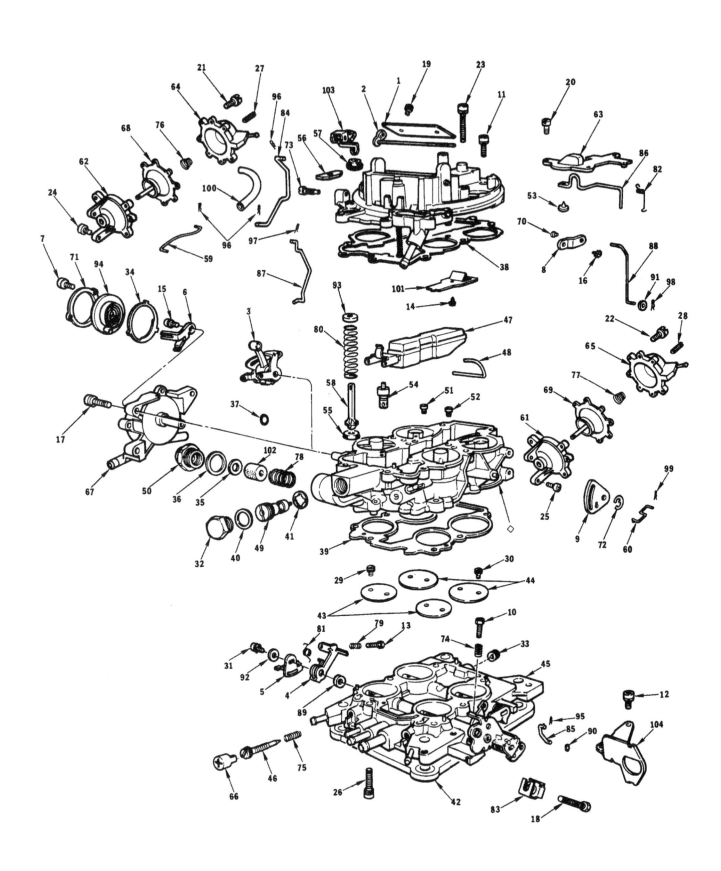

Four-barrel model 4360 exploded view (typical)

Index Number	Part Name	Index Number	Part Name
1	Choke Plate	50	Fuel Inlet Fitting
2	Choke Shaft & Lever Assembly	51	Main Jet Primary
3	Choke Control Lever, Cam & Shaft Assy.	52	Main Jet Secondary
4	Fast Idle Cam Lever	53	E.C.S. Vent Seal
5	Accelerator Pump Driver	54	Power Valve Assy.
6	Choke Thermostat Lever	55	Pump Cup
7	Clamp Screw Coke Thermostat	56	Choke Rod Seal
8	Choke Lever Quick Pull	57	Pump Steam Seal
9	Choke Quick Pull Intermediate Lever	58	Accelerating Pump Assy.
10	Kill Idle Adjusting Screw	59	Choke Diaphragm Link
11	Air Horn to Main Body Screw Short	60	Quick Pull Diaphragm Link
12	Solenoid Bracket Screw & L.W.	61	Quick Pull Diaphragm Housing
13	Fast Idle Adjusting Screw	62	Choke Diaphragm Housing
14	Fuel Bowl Baffle Screw	63	E.C.S. Vent Cover
15	Choke Control Leverl Screw & L.W.	64	Choke Diaphragm Housing Cover Tube Assy.
16	Choke Lever Screw & Lockwasher	65	Quick Pull Diaphragm Housing Cover Tube Assy.
17	Choke Thermostat Housing Mounting Screw	66	Idle Adjusting Needle Limiter
18	Adj. Screw Trans. Kickdown	67	Choke Thermostat Housing & Bushing Assy.
19	Choke Plate Screw	68	Choke Diaphragm Assy.
20	E.C.S. Vent Cover Screw	69	Quick Pull Diaphragm Assy.
21	Choke Diaphragm Cover Hold Down Screw	70	Choke Quick Pull Rod Retainer
22	Quick Pull Diaphragm Cover Hold Down Screw	71	Thermostat Housing Retainer Ring
23	Air Horn to Main Body Screw Long	72	Lever Retainer Interm.
24	Choke Diaphragm Housing Mounting Screw	73	Pump Lever Stud
25	Quick Pull Diaphragm Housing Mounting Screw	74	Kill Idle Screw Spring
26	Throttle Body to Main Body Screw & L.W.	75	Idle Needle Spring
27	Choke Diaphragm Adjusting Screw	76	Choke Diaphragm Spring
28	Quick Pull Diaphragm Adjusting Screw	77	Quick Pull Diaphragm Spring
29	Throttle Plate Screw Pri.	78	Fuel Inlet Filter Spring
30	Throttle Plate Screw Sec.	79	Fast Idle Screw Spring
31	Fast Idle Lever Screw & L.W.	80	Drive Spring
32	Fuel Inlet Plug	81	Fast Idle Cam Lever Return Spring
33	Power Brake Plug	82	E.C.S. Vent Return Spring
34	Thermostat Housing Gasket	83	Locknut Trans. Kickdown
35	Fuel Inlet Filter Gasket	84	Choke Rod
36	Fuel Inlet Fitting Gasket	85	Secondary Connecting Rod
37	Choke Housing Gasket	86	E.C.S. Vent Rod
38	Main Body Gasket	87	Accelerating Pump Rod
39	Throttle Body Gasket	88	Choke Pull Off Quick Pull
40	Fuel Inlet Plug Gasket	89	Fast Idle Cam Lever Spacer
41	Fuel Valve Gasket	90	Connecting Rod Washer
42	Flange Gasket	91	Choke Pull Off Rod Washer
43	Throttle Plate Pri.	92	Accelerating Pump Lever Retaining W.
44	Throttle Plate Sec.	93	Spring Perch Washer
45	Throttle Body & Shaft Assy.	94	Thermostat & Cover Assy.
46	Idle Adjusting Needle	95	Throttle Connecting Rod Retainer
47	Float & Hinge Assy.	96	Choke Rod Retainer
48	Float Hinge Shaft & Retainer	97	Pump Rod Retainer
49	Fuel Inlet Valve Assy.	98	Quick Pull Choke Rod Lever Rod Retainer
		99	Quick Pull Rod Retainer
		100	Choke Vacuum Hose
		101	Fuel Bowl Baffle
		102	Fuel Inlet Filter
		103	Accelerating Pump Lever
		104	Solenoid Bracket

Parts list for model 4360 exploded view

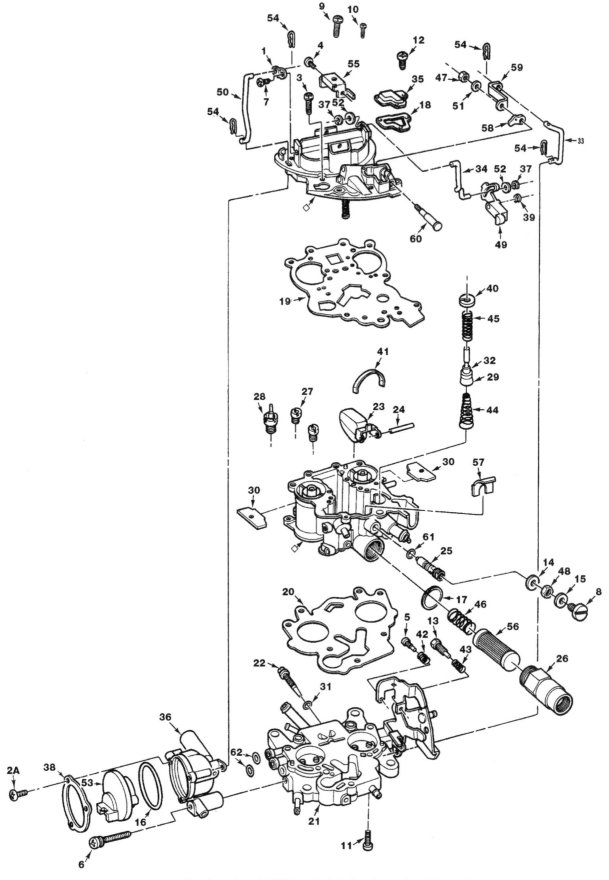

Two-barrel model 2360 exploded view (typical) - 6360 similar

Index Number	Part Name	Index Number	Part Name
*	Choke Plate	49	Fast Idle Cam Assy.
*	Choke Shaft & Lever Assy.	50	Choke Rod
1	Choke Rod Lever	*	Pump Discharge Weight
*	Choke Hsg. Lever & Shaft Assy.	51	Pump Operating Lever L.W.
*	Choke Thermostat Lever & Piston Assy.	*	Pump Link Washer
		52	Fast Idle Cam Link Washer
2A	Bracket Clamp Screw	53	Choke Thermostat Assy.
3	Airhorn Screw & L.W.	54	Retainer
4	Air Cleaner Bracket Screw	55	Air Cleaner Bracket
5	Fast Idle Adj. Screw	*	Idle Stop Solenoid Bracket
6	Choke Housing Screw & L.W.	56	Fuel Inlet Filter
7	Choke Lever Screw & L.W.	57	Fuel Inlet Baffle
8	Needle & Seat Lock Screw	58	Pump Lever
9	Fuel Bowl Screw & L.W.	59	Pump Operating Lever
10	Airhorn Screw & L.W.	60	Pump Operating Shaft
11	Throttle Body Screw & L.W.	*	Choke Wire Assy.
*	Throttle Plate Screw & L.W.	*	Idle Stop Solenoid
*	Choke Plate Screw	61	Fuel Inlet Needle & Seat O-Ring
12	Pump Lever Cover Screw		
13	Curb Idle Adj. Screw	62	Choke Housing O-Ring Seal
*	Power Brake Fitting Plug		
14	Needle & Seat Adj. Nut Gasket		
15	Needle & Seat Lock Screw Gasket		
16	Thermostat Housing Gasket		
*	Fuel Inlet Filter Gasket		
17	Fuel Inlet Fitting Gasket		
18	Pump Lever Cover Gasket		
*	Air Cleaner Gasket		
*	Flange Gasket		
19	Main Body Gasket		
20	Throttle Body Gasket		
*	Throttle Plate		
21	Throttle Body Assy.		
22	Idle Adj. Needle		
23	Float & Hinge Assy.		
24	Float Lever Shaft		
25	Fuel Inlet Needle & Seat Assy.		
26	Fuel Inlet Fitting		
27	Main Jet		
28	Power Valve Assy.		
*	Pump Discharge Nozzle O-Ring		
29	Pump Piston Cup		
30	Choke Rod Seal		
31	Idle Adj. Needle O-Ring Seal		
*	Check Ball		
*	Drive Screw		
32	Pump Piston Assembly		
33	Pump Link		
34	Fast Idle Cam Link		
35	Pump Lever Cover		
36	Choke Housing & Plugs Assy.		
*	E-Ring Retainer		
37	E-Ring Retainer		
38	Thermostat Retainer Ring		
39	Fast Idle Cam Retainer		
40	Pump Spring Retainer		
41	Float Shaft Retainer		
*	Air Cleaner Stud		
42	Fast Idle Adj. Screw Spring		
43	Curb Idle Adj. Screw Spring		
44	Pump Return Spring		
45	Pump Operating Spring		
46	Fuel Inlet Filter Spring		
47	Pump Operating Lever Nut		
48	Needle & Seat Adj. Nut		

Parts list for model 2360 exploded view

Disassembly

7D.1 After removing the carburetor and mounting it on a stand (see Chapters 6 and 7A), remove the fuel inlet nut, gasket and filter (1) and the fuel inlet plug and gasket (2)

7D.2 Remove the two screws that hold the choke diaphragm assembly to the main body

7D.3 Lift the diaphragm assembly as it is being removed so that the end of the linkage rod (arrow) can be pulled through the slot in the choke control lever

7D.4 Remove the clip from the accelerator pump rod (1) and remove the rod from the accelerator pump lever (2). **Note:** *Make a diagram or a note as to which hole the linkage goes into; there are three possible positions.*

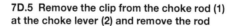

7D.5 Remove the clip from the choke rod (1) at the choke lever (2) and remove the rod

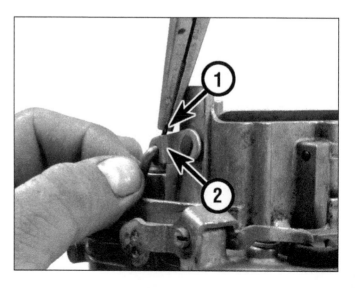

7D.6 Remove the Phillips screws (arrows) holding the cover to the carburetor main body

7D.7 Lift the air horn straight up - be sure the accelerator pump and the power valve piston clear the main body

7D.8 Once clear of the main body, turn the cover to allow the choke rod to come through the hole in the plastic link guide in the cover (arrow). Be careful not to lose the plastic insert when removing the rod. **Note:** *Observe the direction the plastic guide goes into the cover. If installed in reverse, it can interfere with the proper operation of the choke.*

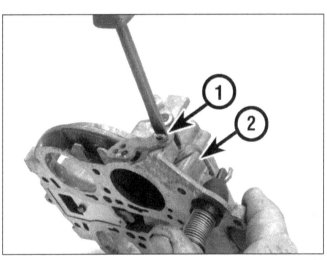

7D.9 Remove the screw (1) holding the accelerator pump lever (2) to the cover and remove the lever

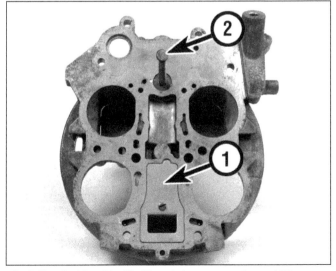

7D.10 The fuel bowl baffle (1) and the power valve rod (2) can remain in the air horn cover while cleaning

7D.11 Unscrew and remove the needle-and-seat assembly (arrow)

7D.12 Lift out the float assembly by the hinge pin (arrow). This pin, or shaft as Holley refers to it, also keeps the needle and float properly seated when the cover is installed on reassembly. **Note:** *Floats often become saturated with fuel and become much heavier than normal. This will cause the float to sink and raise the fuel level in the bowl, causing an overly rich mixture or fuel leaks. Not all floats become saturated, but pinholes or breaks in the solder can develop over time, even on metal or plastic floats. If in doubt about the condition of your float(s) it's recommended that the float be replaced at the time of the overhaul.*

7D.13 Using the proper-size screwdriver, remove the power valve assembly (arrow). **Note:** *Older models have the power valve as three separate pieces, but newer models and kits for all models have replaced these with an assembly that installs as one piece.*

7D.14 Remove the main metering jets (arrows). **Note:** *The jets normally have numbers stamped on their sides to indicate their size. Write down where each jet is located so they can be returned to their original locations on reassembly.*

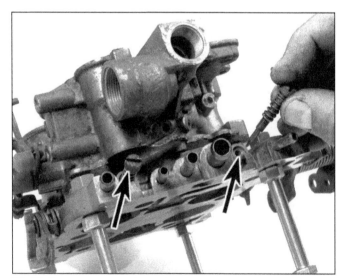

7D.15 Remove the two idle mixture screws (arrows). Check the tips of the mixture screws for signs of pitting or a grooved surface. Replace them if either of these conditions is present. When installing the screws after cleaning, turn them clockwise until they're lightly seated (**DO NOT** tighten them!), then turn them counterclockwise 2 turns - this will provide an initial setting so the engine can run for fine-tuning.

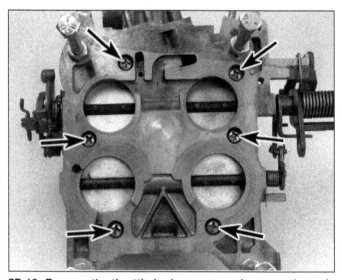

7D.16 Remove the throttle body screws and remove the main body assembly from the throttle body

Reassembly

Several assembly checks, tips and adjustments are shown in the following illustration sequence, but, unless specifically detailed here, reassembly is in the reverse order of disassembly. Follow the illustration sequence in making adjustments. Read the special instructions in the carburetor kit for specific details on your carburetor.

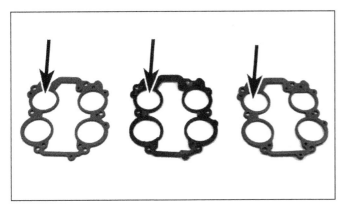

7D.17 After cleaning (see Chapter 7A), begin reassembly by selecting the correct throttle body gasket. Be sure to match the gasket that came off the carburetor with the selection that comes in the overhaul kit. The gasket in the center is the original and the ones on either side came in the kit. Notice the difference in the primary bores. The bore in the original is round, while one of the selections given in the kit has a flat front edge; the other, the correct one for us to use, has the round bore and matches the original.

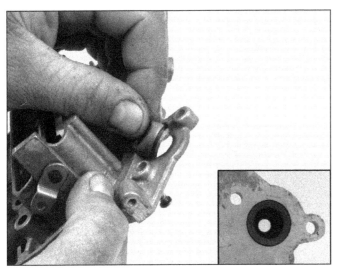

7D.19 Replace the protective boot in the cover. Be sure the boot fits properly into the cover as shown here in the inset photo.

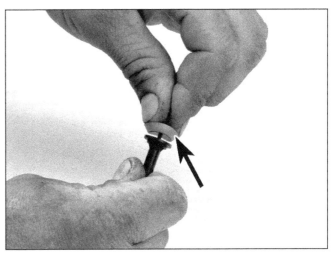

7D.18 Remove the old accelerator pump cup and replace it with the new one from the kit (arrow)

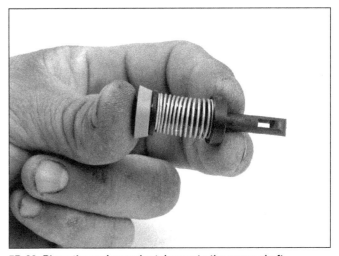

7D.20 Place the spring and retainer onto the pump shaft

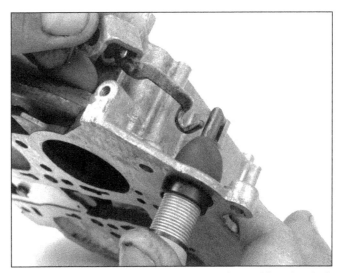

7D.21 Reassemble the accelerator pump by pushing the shaft through the boot and hooking the operating lever through the slot in the shaft. Then reattach the operating lever to the cover with the screw that was removed earlier.

7D.22 Place a new gasket on the needle-and-seat assembly (arrow)

7D.23 Reinstall the power valve (1) and the metering jets (2). Be certain the fuel passages are clean before the parts are reinstalled. Also, note the location of the idle speed adjusting screw (3).

7D.24 Reinstall the float. Caution: *The float can be installed upside down and appear to be correct. Be sure the float tang (1) makes contact with the needle (2) when the float is raised.*

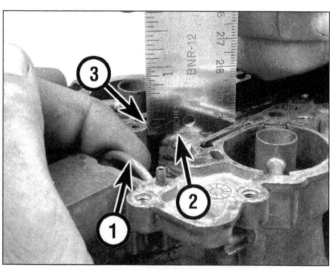

7D.25 Check the float level setting before assembly of the air horn gasket and cover. Lightly hold the float tang against the needle (1). Be careful not to let the hinge pin move up in the slot. Measure, using the specification from the overhaul kit, from the end of the float closest to the needle (2) to the gasket surface (3). Adjust by gently bending the tang until the correct dimension is reached.

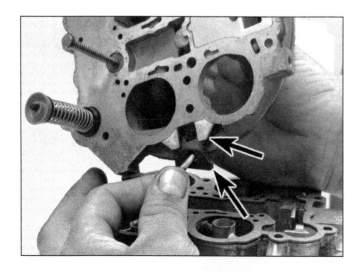

7D.26 When reassembling the air horn cover, be sure to install the plastic guide over the choke rod and into it's location in the cover. Note: *If the guide is installed backwards, it may interfere with the operation of the choke.*

Accelerator pump adjustment

7D.27 Measure the accelerator pump travel by checking the height of the pump rod (arrow) when it's all the way up . . .

7D.28 . . . then checking the height when the accelerator pump is fully applied by the linkage; DO NOT make your measurements by moving the plunger by hand, as you can move the pump plunger further than the travel allowed by the linkage.

7D.29 You should have 7/16 inch of total travel. If not, bend the linkage rod (arrow) to get the correct amount.

Automatic choke setting

Most models have an integral choke thermostat housing mounted on the carburetor, although some models have a "divorced" choke coil mounted to the intake manifold. To set the choke, refer to the procedures at the end of Chapter 7B.

On-vehicle adjustments

The procedures for adjusting the idle speed and idle mixture on the vehicle are the same as those described in Chapter 7B. Refer to the procedures in Chapter 7B, but note that the idle speed and mixture screws are in different locations (see illustrations 7D.15 and 7D.23).

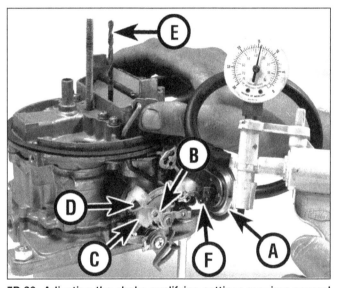

7D.30 Adjusting the choke qualifying settings requires several steps:

1 Close the choke plate
2 Apply vacuum to the choke diaphragm (A) and slight closing force to the choke control lever (B)
3 Open the throttle to allow the fast idle cam to drop (C). The fast idle screw should now be on the second step of the cam. If it is not, re-index the cam by bending the lever (D)
4 While applying slight pressure to the choke control lever (B), measure the clearance at the lower edge of the choke plate using a drill bit (E). Compare the drill bit size to the correct Specification (see Chapter 9). Adjust of necessary.
5 Adjust by bending the linkage rod (F)

7D.31 After adjusting the idle speed and mixture, adjust the fast-idle speed. With the engine running at normal operating temperature, open the primary throttle plates slightly and position the screw (arrow) on the highest step of the fast-idle cam, as shown; the idle speed should be approximately 1700 rpm. If it's not, turn the screw as necessary until it's correct (turning it clockwise increases the speed and turning it counter-clockwise decreases the speed). After adjustment, if you notice the engine runs slowly and "chugs" when the engine is cold and the choke is fully applied, increase the fast-idle speed

7 Part E
Overhaul and adjustments

Two-barrel models 5200/5210/5220 and 6500/6510/6520

This group of carburetors includes a variety of models all based on a similar design. The models included in this group are the non-feedback two-barrel models 5200, 5210, 5220 and feedback two-barrel models 6500, 6510 and 6520.

The overhaul procedure is covered through a sequence of photographs laid out in order from disassembly to reassembly and finally installation and adjustment on the vehicle. The captions presented with the photographs will walk you through the entire procedure, one component at a time.

The following illustration sequence will use model 5200. Specific details will vary with carburetor part numbers because each carburetor is made to fit a specific application. Refer to the instruction sheet which comes with each overhaul kit. **Note:** *Because of the large number of different models and variations of each model, it is not always possible to cover all the differences in components during the overhaul procedure. Overhaul Chapters only deal with disassembly, reassembly and basic adjustments of a typical carburetor in that model group. If you need further information about specific differences, refer to the exploded views beginning on the next page or refer to Chapter 3.*

Warnings:
Gasoline

Gasoline is extremely flammable, so take extra precautions when you work on any part of the fuel system. Don't smoke or allow open flames or bare light bulbs near the work area, and don't work in a garage where a natural gas-type appliance (such as a water heater or clothes dryer) with a pilot light is present. A spark caused by an electrical short circuit, by two metal surfaces striking each other, or even by static electricity built up in your body, under certain conditions, can ignite gasoline vapors. Also, never risk spilling fuel on a hot engine or exhaust component. If you spill any fuel on your skin, rinse it off immediately with soap and water.

Battery

Always disconnect the battery ground (-) cable at the battery before working on any part of the fuel or electrical system.

Fire extinguisher

We strongly recommend that a fire extinguisher suitable for use on fuel and electrical fires be kept handy in the garage or workshop at all times. Never try to extinguish a fuel or electrical fire with water. Post the phone number for the nearest fire department in a conspicuous location near the phone.

Compressed air

When cleaning carburetor parts, especially when using compressed air, be very careful to spray away from yourself. Eye protection should be worn to avoid the possibility of getting any chemicals or debris into your eyes.

Lead poisoning

Avoid the possibility of lead poisoning. Never use your mouth to blow directly into any carburetor component. Small amounts of Tetraethyl lead (a lead compound) become deposited on the carburetor over a period of time and could lead to serious lead poisoning. Check passages with compressed air and a fine-tipped blow gun or place a small-diameter tube to the component and blow through the tube to be sure all necessary passages are open.

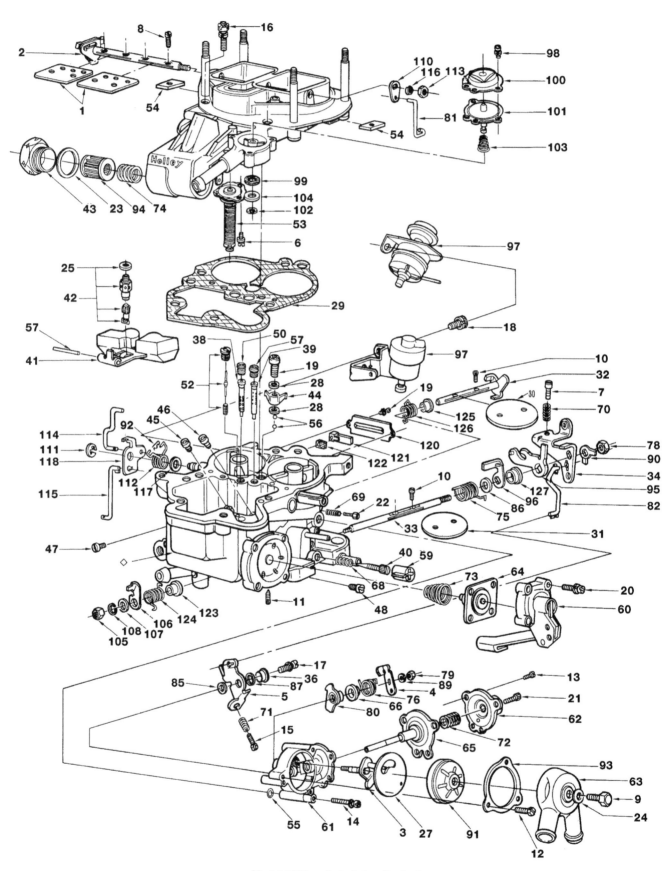

Model 5200 exploded view (typical)

Index Number	Part Name	Index Number	Part Name
1	Choke Plate	51	High Speed Bleed Restriction Primary
2	Choke Shaft & Lever Assy.	52	Power Valve Assembly
3	Choke Housing Shaft & Lever Assy.	53	Power Valve Economizer Assy.
4	Choke Lever	54	Choke Rod Seal
5	Fast Idle Adj. Lever	55	Choke Housing Seal "O" Ring
6	Economizer Screw & L.W.	56	Pump Discharge Ball
7	Solenoid Adj. Screw	57	Float Hinge Pin
8	Choke Plate Screw	58	Secondary Return Spring Pin*
9	Water Housing Retaining Screw	59	Idle Needle Limiter Cap
10	Throttle Plate Screw	60	Pump Cover Assembly
11	Secondary Idle Adj. Screw	61	Choke Housing & Plugs Assy.
12	Retaining Ring Screw	62	Choke Diaphragm Cover Assy.
13	Diaphragm Cover Screw & L.W.	63	Choke Water Housing (Mach.)
14	Choke Housing Screw & L.W.	64	Pump Diaphragm Assy. Comp.
15	Fast Idle Adj. Screw	65	Diaphragm & Shaft Assy.
16	Air Horn Screw & L.W.	66	Spring Retainer (Choke)
17	Fast Idle Lever Screw & L.W.	67	Air Cleaner Stud
18	Solenoid Bracket Screw	68	Idle Adj. Needle Spring
19	Pump Discharge Screw	69	Idle Stop Screw Spring
20	Pump Cover Screw & L.W.	70	Solenoid Adj. Screw Spring
21	Choke Diaphragm Adj. Screw	71	Fast Idle Adj. Screw Spring
22	Idle Stop Screw	72	Diaphragm Return Spring
23	Fuel Inlet Fitting Gasket	73	Pump Return Spring
24	Water Housing Retaining Screw Gasket	74	Fuel Inlet Filter Spring
25	Fuel Inlet needle Seat Gasket	75	Throttle Return Spring Pri.
26	Choke Water Housing Gasket	76	Fast Idle Cam Spring
27	Thermostat Housing Gasket	77	Sec. Operating Lever Return Spring*
28	Pump Discharge Nozzle Gasket	78	Throttle Lever Retaining Nut Primary
29	Air Horn Gasket	79	Choke Shaft Nut
30	Throttle Plate Sec.	80	Fast Idle Cam & Hub Assy.
31	Throttle Plate Pri.	81	Choke Rod
32	Throttle Shaft & Lever Assy. Secondary	82	Fast Idle Rod
33	Throttle Shaft & Cam Assy. or Throttle Shaft Machined	83	Sec. Return Spring Pin L.W.*
34	Throttle Lever Assy. Pri.	84	Solenoid Bracket Screw L.W.
36	Fast Idle Lever Bushing	85	Fast Idle Adj. Lever Washer
37	Secondary Operating Lever Bushing	86	Shaft Leveling Washer Pri.
38	Main Well Tube Sec.	87	Fast Idle Lever Spring Washer
39	Main Well Tube Pri.	89	Choke Shaft Nut L.W.
40	Idle Adjusting Needle	90	Throttle Lever Nut L.W. Pri.
41	Float Assembly	91	Thermostat Housing Assy.
42	Fuel Inlet Needle & Seat Assy.	92	Choke Rod Retainer
43	Fuel Inlet Fitting	93	Thermostat Housing Retaining Ring
44	Pump Discharge Nozzle	94	Fuel Inlet Filter Assy.
45	Main Metering Jet Sec.	95	Secondary Operating Lever Assy.
46	Main Metering Jet Pri.	96	Idle Stop Lever
47	Idle Jet Sec.	97	Solenoid, Solepot Assy.
48	Idle Jet Pri.	98	ECS Cover Screw & L.W.
49	Idle Jet Retainer Pri. & Sec.	99	ECS Seal
50	High Speed Bleed Restriction Secondary		
100	ECS Diaphragm Cover	115	Choke Pull-Off Rod
101	ECS Diaphragm & Stem Assy.	116	Choke Lever Washer
102	ECS Seal & Washer Retainer	117	Spring Retaining Washer
103	ECS Spring	118	Choke Pull-Off Intermediate Lever
104	ECS Washer	119	H.I.C. Cover Screw
105	Pump Cam Nut	120	H.I.C. Valve Cover
106	Pump Cam	121	H.I.C. Valve Assy.
107	Pump Cam Washer	122	H.I.C. Valve Gasket
108	Pump Cam Lock Washer	123	Pri. Return Spring Bushing— Pump Side
109	Pump Nozzle Check Spring	124	Pri. Throttle Return Spring— Pump Side
110	Choke Pull-Off Lever	125	Sec. Return Spring Bushing
111	"E" Clip Retainer	126	Sec. Throttle Return Spring
112	Choke Closure Spring	127	Pri. Return Spring Bushing— Throttle Lever Side
113	Choke Pull-Off Lever Nut		
114	Choke Pull-Off Rod (Solenoid)		

Parts list for model 5200 exploded view

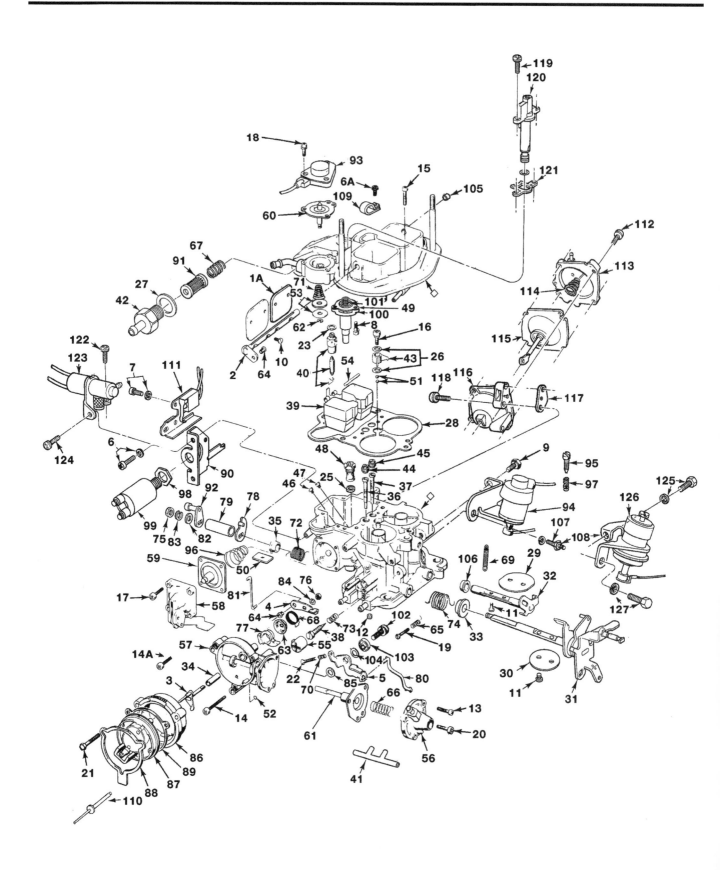

Model 6500 exploded view (typical)

Index Number	Part Name	Index Number	Part Name
1	Choke Plate-Primary	66	Choke Diaphragm Return Spring
1A	Choke Plate-Secondary	67	Fuel Filter Spring
2	Choke Shaft & Lever Assembly	68	Fast Idle Cam Spring
3	Choke Housing Shaft & Lever Assembly	69	Secondary Operating Lever Spring
4	Fast Idle Cam Pick-up Lever	70	Fast Idle Adjusting Screw Spring
5	Fast Idle Adjusting Lever	71	E.A.V. Diaphragm Spring
6	T.P.T. Bracket Screw & L.W.	72	Primary Throttle Return Spring —Pump Side
6A	Bracket Screw & L.W.		
7	A/C Switch Bracket Screw & L.W.	73	Idle Adjusting Needle Spring
8	Economizer Diaphragm Screw & L.W.	74	Primary Throttle Return Spring —Throttle Side
9	Idle Solenoid Bracket Screw & L.W.		
10	Choke Plate Screw	75	T.P.T. Lever Nut
11	Throttle Plate Screw	76	Choke Housing Shaft Nut
12	Secondary Idle Adjusting Screw	77	Fast Idle Cam
13	Choke Diaphragm Cover Screw & L.W.	78	Pump Cam
14	Choke Housing Screw & L.W.	79	Pump Cam Sleeve
14A	Choke Housing Screw & L.W.	80	Fast Idle Rod
15	Airhorn Screw & L.W.	81	Choke Rod
16	Pump Discharge Nozzle Screw	82	T.P.T. Lever Washer
17	Pump Cover Screw & L.W	83	T.P.T. Lever Lockwasher
18	E.A.V. Solenoid Cover Screw & L.W.	84	Choke Housing Shaft Lockwasher
19	Idle Stop Screw	85	Fast Idle Adj. Lever Washer*
20	Choke Diaphragm Adjusting Screw	86	Thermostat Cap & Well Cover
21	Thermostat Retaining Ring Screw	87	Thermostat & Cap Assembly
22	Fast Idle Adjusting Screw	88	Thermostat Retaining Ring
23	Fuel Inlet Needle & Seat Gasket	89	Thermostat Ground Ring
*	Choke Housing Gasket	90	T.P.T. Bracket
25	Power Valve Gasket	91	Fuel Filter Assembly
26	Pump Discharge Nozzle Gasket	92	T.P.T. Lever & Pin Assembly
27	Fuel Inlet Fitting Gasket	93	E.A.V. Solenoid Assembly
28	Air Horn Gasket	94	Idle Stop Solenoid
29	Secondary Throttle Plate	95	Solenoid Adjusting Screw
30	Primary Throttle Plate	96	Pump Return Spring
31	Primary Throttle Shaft & Lever Assembly	97	Solenoid Adjusting Screw Spring
		98	T.P.T. Adjusting Nut
32	Secondary Throttle Shaft & Lever Assembly	99	T.P.T. Assembly
		100	Power Valve Diaphragm Body
33	Primary Throttle Shaft Bushing— Throttle Side	101	Power Valve Diaphragm Spring
		102	Fast Idle Lever Screw & L.W.*
34	Choke Housing Shaft & Lever Bushing	103	Fast Idle Lever Bushing*
		104	Fast Idle Lever Washer*
35	Primary Throttle Shaft Bushing— Pump Side	105	Choke Shaft & Lever Assy. Bushing
36	Primary Main Well Tube	106	Secondary Throttle Shaft Bushing
37	Secondary Main Well Tube	107	Secondary Return Spring Pin Washer
38	Idle Adjusting Needle		
39	Float & Hinge Assembly	108	Secondary Return Spring Pin
40	Fuel Inlet Needle & Seat Assembly	109	Wire Hold-Down Clamp
41	Vacuum Hose Tee Connector	110	Thermostat Retaining Ring Rivet
42	Fuel Inlet Fitting	111	A/C Switch & Connector Assy.
43	Pump Discharge Nozzle	112	Secondary Diaphragm Cover Screw & L.W.
44	Primary High Speed Bleed		
45	Secondary High Speed Bleed	113	Secondary Diaphragm Cover & Tube Assembly
46	Primary Main Jet		
47	Secondary Main Jet	114	Diaphragm Spring
48	Power Valve Assembly	115	Secondary Diaphragm & Link Assembly
49	Economizer Diaphragm & Stem Assembly		
		116	Secondary Diaphragm Housing
50	Choke Rod Seal	117	V.O.S. Gasket
51	Pump Discharge Check Ball	118	Secondary Diaphragm Housing Screw
52	Choke Housing "O" Ring Seal		
53	E.A.V. Seal & Washer Assembly	119	Duty Cycle Solenoid Screw & L.W.
54	Float Hinge Pin	120	Duty Cycle Solenoid Assembly
55	Idle Needle Limiter Cap	121	Duty Cycle Solenoid Gasket
56	Choke Diaphragm Cover	122	Dump Valve Solenoid Screw & L.W.
57	Choke Housing & Plugs Assembly	123	Dump Valve Solenoid Assembly
58	Pump Cover Assembly	124	Dump Valve Solenoid Screw & L.W.
59	Pump Diaphragm Assembly	125	A/C Speed-up Solenoid Screw & L.W.
60	E.A.V. Diaphragm & Stem Assembly	126	A/C Speed-up Solenoid Assembly
61	Choke Diaphragm & Shaft Assembly	127	A/C Speed-up Solenoid Screw & L.W.
62	E.A.V. Seal & Washer Retainer		
63	Fast Idle Cam Spring Retainer		
64	Choke Rod Retainer		
65	Adjusting Screw Spring		

* This part is not used on units using a stud to retain the lever

Parts list for model 6500 exploded view

Disassembly

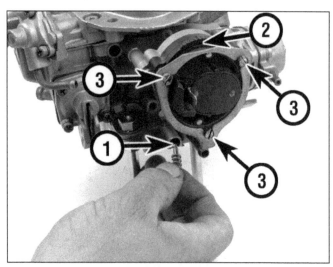

7E.1 After removing the carburetor and mounting it on a stand (see Chapters 6 and 7A), remove the idle mixture screw. The idle mixture screw is adjusted and sealed with a plug from the factory on most models. To remove the idle mixture screw, the plug must be removed. There are several ways to dig the plug out, but most people drill a small hole or cut a groove with a hacksaw from underneath and behind the plug (see Chapter 7F, illustrations 7F.17 through 7F.19 for more details). The plug isn't going to be replaced upon reassembly, so don't worry about saving the pieces removed. After you've got a hole or groove in the plug, drive the plug out from behind with a hammer and punch or chisel. Then unscrew the idle mixture screw (1) and check for a groove or any signs of pitting. Replace it if it's worn or damaged.

Make a note detailing which adjustment marks on the choke cover and housing (2) are aligned. Remove the three screws (3) that hold the cover to the housing and remove the cover.

7E.2 Remove the ground ring and choke housing and set them aside. This plastic part doesn't need to go into the parts cleaner. Just clean it off with a rag and, if necessary, a mild cleaner.

7E.3 Remove the three screws (arrows) holding the choke housing to the carburetor main body.

7E.4 Remove the linkage rod from the plastic clip that holds it in the choke lever (arrow).

7E.5 Also remove the top of the choke rod from the connector at the air horn. This will make disassembly and reassembly of the carburetor much easier.

7E.6 When removing the choke housing assembly, you will have to turn it to get the fast-idle rod (arrow) lined up with the notch in the lever for removal.

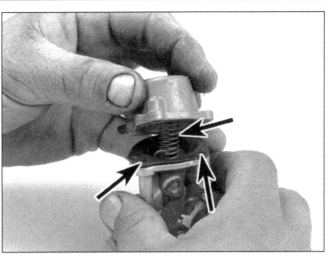

7E.7 Remove the three screws from the choke pull-off diaphragm cover and remove the spring and the diaphragm with the plunger.

7E.8 Remove the four Phillips screws for the accelerator pump cover and remove the diaphragm and the spring.

7E.9 Remove both bolts (arrows) and take off the solenoid assembly (if equipped).

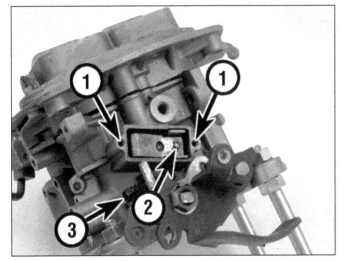

7E.10 Take out the two screws (1) that hold the hot idle compensator cover, remove it, and remove the C-clip (2) and the metal valve - note the location of the curb idle speed adjusting screw (3).

7E.11 Remove the cork gasket that is beneath the hot idle compensator valve.

7E.12 Hopefully you've loosened the fuel inlet nut while the carburetor was still attached to the manifold. Remove the nut (arrow) and gasket . . .

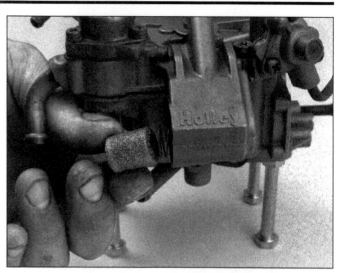

7E.13 . . . and then the filter and spring. Always replace the filter on reassembly. Many professionals prefer a filter that's mounted in the fuel line, which has a much larger filter surface area.

7E.14 Remove the five screws that hold the air horn cover to the main body and . . .

7E.15 . . . remove the cover. Be careful when you remove the cover; the power valve (arrow) and float extends down into the main body of the carburetor. Raise the cover high enough to clear all the components.

7E.16 Slide out the hinge pin (1), or float shaft as Holley refers to it, and remove the float (2) and needle (3). Check the needle tip for wear or damage. Since a new needle and seat comes in the overhaul kit, it would be foolish not to replace them when doing a rebuild. **Note:** *Floats often become saturated with fuel and become much heavier than normal. This will cause the float to sink and raise the fuel level in the bowl, causing an overly rich mixture or fuel leaks. Not all floats become saturated, but pinholes or breaks in the solder can develop over time, even on metal or plastic floats. If in doubt about the condition of your float(s) it's recommended that the float be replaced at the time of the overhaul.*

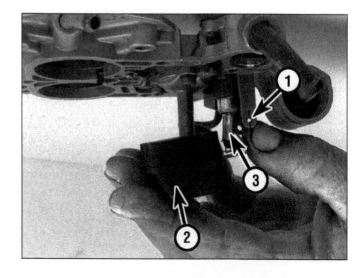

7E.17 Using a box-end wrench or a socket and ratchet, remove the seat from the carburetor cover.

7E.18 Remove the C-clip (arrow) on the E.A.V (Electronic Air Valve) shaft (if equipped).

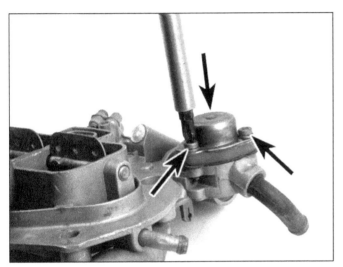

7E.19 Take out the three screws (arrows) holding the E.A.V and remove the cover, diaphragm and spring.

7E.20 Take out the three screws and remove the power valve, sometimes referred to as an economizer valve. Be careful not to lose the spring that is under the valve.

7E.21 Remove the accelerator pump discharge nozzle (arrow). Be sure to remove the gaskets both above and below the nozzle.

7E.22 There will usually be two check balls in the fuel well, which can be removed with a magnetic tool. Note: *Some very early models may have only a single check ball under the discharge nozzle.*

7E.23 Remove the power valve assembly from the bottom of the fuel bowl.

7E.24 Remove the main metering jets. The primary jet is on the right and the secondary jet is on the left; be sure to write down the size number of each jet and which side it came from so you can return them to their original locations on reassembly.

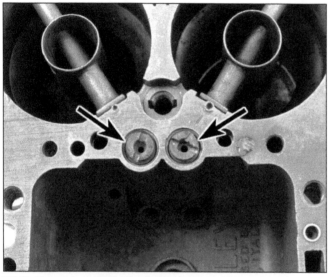

7E.25 The main well air bleeds and the main well tubes are already marked to indicate primary and secondary. Notice the mark on the primary (right) air bleed and carburetor body. These notches indicate the primary parts and which side they go into. Unscrew and remove the two air bleeds (arrows).

7E.26 If turning the carburetor over to dislodge the tubes from the carburetor doesn't work, a finely pointed pick or scribe works well to lift them from the carburetor. Try to avoid nicking or damaging the passages inside these tubes when removing them.

Reassembly

Several assembly checks, tips and adjustments are shown in the following illustration sequence, but, unless otherwise covered, reassembly is in the reverse order of disassembly. Read the special instructions in the carburetor kit and follow the illustration sequence in making adjustments.

7E.27   When reinstalling the power valve (economizer valve), be sure the correct spring is in place (arrow). The 5200 model carburetor has several diaphragm-operated components, and they all use a small spring in their operation.

7E.28 The idle feedback valve has a spring (arrow) very similar to the power valve. Compare lengths, sizes, etc. to be sure the proper spring is being used with each valve.

Tip:

If the springs for the power valve (economizer valve) and the idle feedback valve get confused, use the longer spring for the power valve.

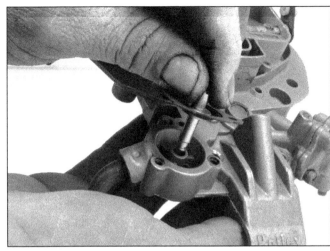

7E.29 Reinstall the E.A.V. diaphragm through the cover . . .

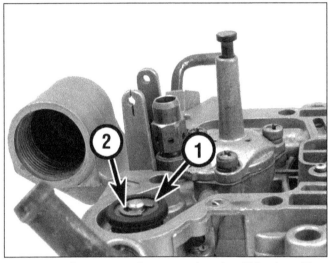

7E.31 . . . before you put the vent seal washer (1) in place. Then insert the C-clip (2).

7E.30 . . . be sure to reinstall the spring . . .

7E.32 After the new needle and seat have been installed and the float has been attached to the cover by the hinge pin, it's time to adjust the float height and drop. **Caution:** *Be sure the wire clip on the needle is attached to the float (arrow).*

Using the specifications given for your model in the overhaul kit instructions, measure the distance as shown here and adjust as needed by gently bending the tang on the float assembly at the needle. . .

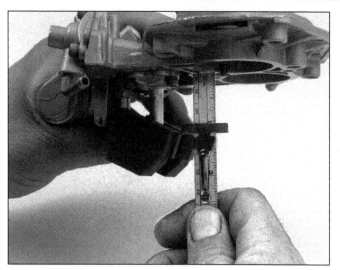

7E.33 . . . turn the cover over and measure the float drop as shown here - it should be 1" ± 1/8".

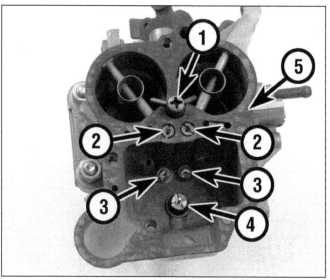

7E.34 Make sure the following are all reinstalled in the carburetor main body:

1   The accelerator discharge check ball(s) and nozzle
2   The main well tubes and air bleeds
3   The main metering jets
4   The power valve
5   The new gasket

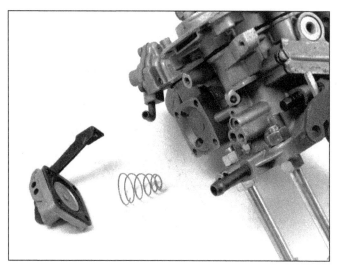

7E.35 Reassemble the accelerator pump with the spring in the correct direction.

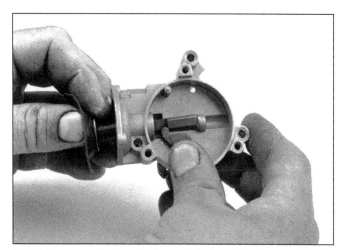

7E.36 Reassemble the choke and choke pull-off assembly. Be sure the spring is put in before the cover goes on (refer to illustration 7E.7).

7E.37 After attaching the assembly to the carburetor main body, install the choke thermostatic control. Make sure the coil or tang from the coil on the cover connects to the arm inside the housing (arrow). Then install the retaining ring and the three screws.

Accelerator pump setting

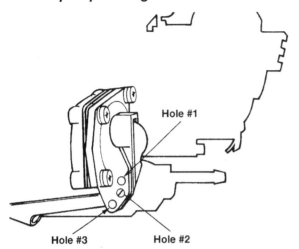

7E.38 When reassembling the accelerator pump, note the three holes in which the pump lever pin can be inserted. Most models come from the factory with the pin in Hole #2; however, if you want more of a fuel "shot" from the accelerator pump (for example, to correct a tendency to hesitate on acceleration), change the pin to Hole #3. To decrease the fuel "shot," move the pin to Hole #1.

Automatic choke setting

These models use automatic chokes with either an electrically heated or coolant-heated choke housing assembly that's mounted on the carburetor. Refer to the procedure at the end of Chapter 7B to set the choke.

Fast idle setting

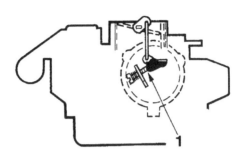

7E.39 Before installing the carburetor on the vehicle, open the throttle and position the fast idle adjustment screw on the first step of the fast idle cam (1), then close the throttle. Using a drill bit, measure the clearance between the front of the primary throttle plate and the throttle body bore, which should be approximately 1/32-inch. If the throttle is not open the correct amount, adjust it with the screw. After the carburetor has been installed, all other adjustments have been made and the engine is running at normal operating temperature, open the throttle and place the adjusting screw back on the first step of the fast-idle cam. Check the idle speed, which should be approximately 1700 rpm. If the speed is not correct, adjust it with the screw. If you notice the engine idling slowly and "chugging" when the engine is cold and the choke is fully applied, try turning the screw clockwise a bit to increase the fast-idle speed. If the idle speed seems too fast with the choke applied, turn the screw counterclockwise slightly.

4 Take slack out of linkage in open choke direction

3 Insert 1/4-inch drill or gauge between lower edge of choke plate and airhorn wall

5 Turn screw to adjust

2 Push diaphragm stem against stop

1 Place fast idle screw on high step of fast idle cam

Choke pull-down adjustment

7E.40 To adjust the choke pull-down, follow numerical sequence shown in this illustration.

On-vehicle adjustments

The procedures for adjusting the idle speed and idle mixture on the vehicle are the same as those described in Chapter 7B. Refer to the procedures in Chapter 7B, but note that the idle speed and mixture screws are in different locations (see illustrations 7E.1 and 7E.10). Also note that these models have only one idle mixture screw.

7 Part F
Overhaul and adjustments
Two-barrel models 2280 and 6280

This group of carburetors includes a variety of models all based on a similar design. The models covered in this group are the two-barrel model 2280 and feedback model 6280.

The overhaul procedure is covered through a sequence of photographs laid out in order from disassembly to reassembly and finally installation and adjustment on the vehicle. The captions presented with the photographs will walk you through the entire procedure one component at a time.

The following illustration sequence will use model 2280. Specific details will vary with carburetor part numbers because each carburetor is made to fit a specific application. Refer to the instruction sheet that comes with each overhaul kit. **Note:** *Because of the large number of different models and variations of each model, it is not always possible to cover all the differences in components during the overhaul procedure. Overhaul sections only deal with disassembly, reassembly and basic adjustments of a typical carburetor in that model group. If explanation about individual differences of parts are needed to clearly understand or choose the correct part for your application, refer to the exploded views at the beginning of this Chapter or go to Chapter 3, Carburetor Fundamentals for more details on components not discussed in the photograph sequence.*

Warnings:
Gasoline

Gasoline is extremely flammable, so take extra precautions when you work on any part of the fuel system. Don't smoke or allow open flames or bare light bulbs near the work area, and don't work in a garage where a natural gas-type appliance (such as a water heater or clothes dryer) with a pilot light is present. A spark caused by an electrical short circuit, by two metal surfaces striking each other, or even by static electricity built up in your body, under certain conditions, can ignite gasoline vapors. Also, never risk spilling fuel on a hot engine or exhaust component. If you spill any fuel on your skin, rinse it off immediately with soap and water.

Battery

Always disconnect the battery ground (-) cable at the battery before working on any part of the fuel or electrical system.

Fire extinguisher

We strongly recommend that a fire extinguisher suitable for use on fuel and electrical fires be kept handy in the garage or workshop at all times. Never try to extinguish a fuel or electrical fire with water. Post the phone number for the nearest fire department in a conspicuous location near the phone.

Compressed air

When cleaning carburetor parts, especially when using compressed air, be very careful to spray away from yourself. Eye protection should be worn to avoid the possibility of getting any chemicals or debris into your eyes.

Lead poisoning

Avoid the possibility of lead poisoning. Never use your mouth to blow directly into any carburetor component. Small amounts of Tetraethyl lead (a lead compound) become deposited on the carburetor over a period of time and could lead to serious lead poisoning. Check passages with compressed air and a fine-tipped blow gun or place a small-diameter tube to the component and blow through the tube to be sure all necessary passages are open.

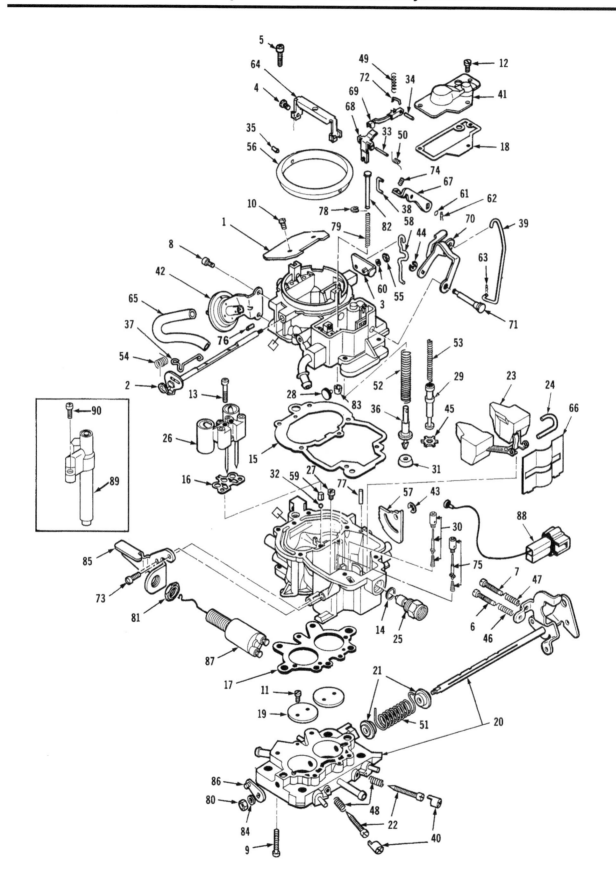

Models 2280 and 6280 exploded view (typical)

Index Number	Part Name	Index Number	Part Name
1	Choke Plate	43	Fast Idle Cam Retainer
2	Choke Shaft & Lever Assembly	44	Pump Operating Lever Retainer
3	Dechoke Lever	45	Power Valve Piston Retainer
4	Air Cleaner Bracket Screw	46	Throttle Stop Screw Spring
5	Air Horn to Main Body Screw & L.W.	47	Fast Idle Adjusting Screw Spring
6	Throttle Stop Screw	48	Idle Adjusting Needle Spring
7	Fast Idle Adjusting Screw	49	Vent Valve Operating Lever Spring
8	Bracket Retaining	50	Vent Valve Spring
9	Throttle Body to Main Body Screw & L.W.	51	Throttle Return Spring
10	Choke Plate Screw	52	Pump Drive Spring
11	Throttle Plate Screw	53	Power Valve Piston Spring
12	Cover Screw	54	Choke Shaft Spring
13	Nozzle Bar Screw	55	Dechoke Lever Nut
14	Fuel Inlet Fitting Gasket	56	Air Cleaner Ring
15	Main Body Gasket	57	Fast Idle Cam
16	Nozzle Bar Gasket	58	Fast Idle Rod
17	Throttle Body Gasket	59	Pump Discharge Weight
18	E.C.S. Cover Gasket	60	Dechoke Lever Nut L.W.
19	Throttle Plate	63	Pump Link Retainer
20	Throttle Body & Shaft Assembly	64	Air Cleaner Bracket
21	Throttle Return Spring Bushings	65	Vacuum Hose
		66	Float Baffle
22	Idle Adjusting Needle	67	Pump Lever
23	Float & Hinge Assembly	68	Vent Valve Lever
24	Float Hinge Shaft & Retainer	69	Vent Valve Operating Lever
25	Fuel Inlet Needle & Seat Assembly	70	Pump Operating Lever
		71	Pump Operating Lever Shaft
26	Nozzle Bar Assembly	72	Vent Valve Retainer
27	Main Jet	73	T.P.T. Bracket Screw
28	Vent Valve	74	Power Valve Adjusting Screw
29	Power Valve Piston	75	Power Valve Assembly (Staged)
30	Power Valve Assembly	76	Choke Diaphragm Link Pin
31	Pump Cup	77	Roll-Pin
32	Pump Discharge Ball	78	Power Valve Stem Retainer
33	Vent Valve Lever Pin	79	Power Valve Spring
34	Vent Valve Operating Lever Pin	80	T.P.T. Lever Nut
35	Air Cleaner Ring Retaining Pin	81	T.P.T. Assembly Locknut
36	Pump Steam & Head Assembly	82	Mechanical Power Valve Stem
37	Choke Diaphragm Link	83	Mechanical Power Valve Stem Cap
38	Pump Stem Link	84	T.P.T. Lever Lockwasher
39	Pump Link	85	T.P.T. Bracket
40	Idle Needle Limiter	86	T.P.T. Lever
41	Accelerator Pump Housing Cover	87	T.P.T. Assembly
42	Choke Diaphragm Assembly Complete	88	T.P.T. Wire Assembly
		89	Duty-Cycle Solenoid

Parts list for models 2280 and 6280 exploded view

Disassembly

Make notes or diagrams of levers of linkage locations, slots or holes that the linkage connects with and anything you may feel could be difficult to remember later on during re-assembly. **Note:** *Remember that, on extremely dirty carbure-tors, overnight cleaning may be required and reassembly may not be possible until the next day.*

7F.1 After removing the carburetor, mount it on a stand - long 5/16-inch bolts, nuts and washers make an inexpensive stand.

7F.2 Remove the two screws (arrows) and remove the choke pull-off diaphragm and linkage arm.

7F.3 Remove the screw (arrow) and the idle control solenoid.

7F.4 Remove the accelerator pump linkage by pulling the clip (arrow) . . .

7F.5 . . . and removing the rod from the pump lever.

7F.6 Remove the nut (arrow) and arm from the choke plate shaft.

7F.7 Remove the three screws, cover and gasket for the vent valve assembly.

7F.8 Remove the six screws holding the carburetor top cover to the main body. Remove the cover by raising it high enough to clear the main-well tubes from the body. Be careful not to set the cover down after removal and bend or otherwise damage pieces attached to the cover assembly.

7F.9 Disconnect the accelerator pump from the operating lever. **Note:** *It may be simpler to leave the accelerator pump attached to the cover and only cut off the pump cup. On reassembly a new one can be pushed onto the end of the shaft.*

7F.10 Remove the accelerator pump (arrow) from the cover.

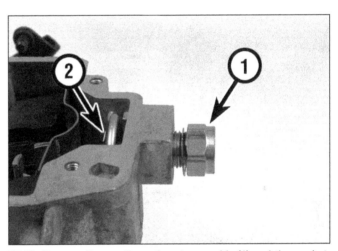

7F.11 Remove the needle-and-seat assembly (1) and the gasket. Identified also is the adjustment tang (2) on the float assembly, that makes contact with the needle.

7F.12 Take out the two screws (arrows) and remove the booster assembly. Under the booster is the discharge check ball and weight - be careful not to loose these small pieces.

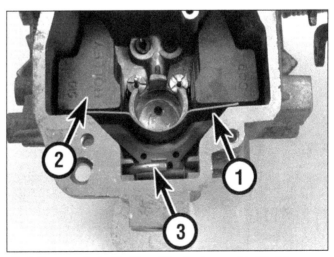

7F.13 Remove the baffle (1) and the float assembly (2) at the same time by lifting the assembly up by the float hinge pin (3). **Note:** *After removing the float, inspect it as described in Chapter 7, Part A.*

7F.14 Remove the two power valve assemblies (note the notch filed into the blade of the screwdriver to clear the stem).

7F.15 Remove the two main metering jets.

7F.16 Turn the carburetor over and remove the four screws (arrows) holding the throttle body to the main body. Set the main body aside to be cleaned.

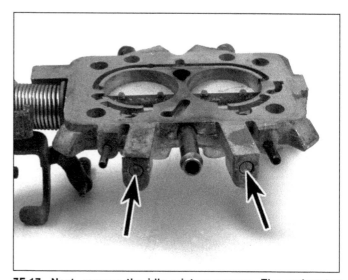

7F.17 Next, remove the idle mixture screws. The carburetor comes from the manufacturer with plugs inserted (arrows) to keep the initial settings from being tampered with. During an overhaul, you can remove the plugs and thoroughly clean this critical area.

7F.18 The easiest and cleanest way to remove the plugs is to cut off the ends of the housing where the plugs are located, but just ahead of the idle mixture screws. **Caution:** *Do not cut off any more of the casting than necessary. You should not have to go beyond the large end of the casting (arrows).* **Note:** *Instead of concealment plugs, some models have plastic limiter caps on the idle mixture screws. These can be removed with a pair of pliers or a screwdriver.*

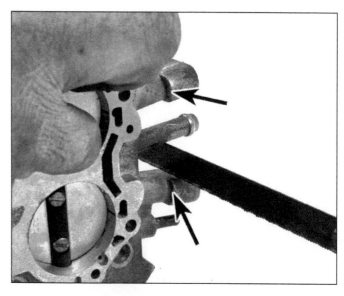

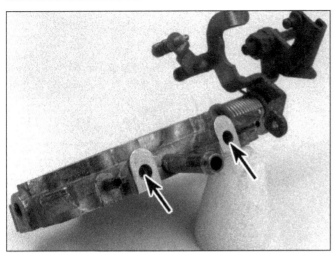

7F.19 Once cut off, you will be able to remove the two idle mixture screws (arrows) using a small Allen-head wrench to turn the screws out.

7F.20 Reinstall the two idle mixture screws. Turn the screws in until they lightly seat. Then back the screws out 1 to 1-1/2 turns as a base setting. This will get the vehicle to run, although final settings will have to be done on the vehicle.

Cleaning

Refer to Chapter 7A for information on "dips" and solvents used in cleaning. Remember not to place the float or any other plastic or rubber parts into a dip tank, since the chemicals will damage the parts. After cleaning (which usually requires an overnight soak in a dip tank), blow out all passages in the carburetor body with compressed air. **Warning:** *Wear eye protection!* Inspect the cleaned parts to be sure there's no gasket material remaining on any surfaces and the surfaces are not damaged or warped. Inspect the float, as shown in illustration 7A.14. Make sure all parts are completely dry before beginning reassembly.

Reassembly

Reassembly is basically the reverse of disassembly, but several assembly checks, tips and adjustments are shown in the following illustrations to help you with areas of special concern. Be sure to follow any special instructions that may be included in the carburetor kit and follow the illustration sequence in making adjustments.

7F.21 Install the pump discharge check ball first . . .

7F.22 . . . then install the check ball weight.

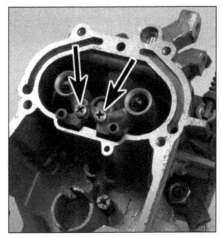

7F.23 Install the pump discharge nozzle assembly and screws. Be certain the gasket is in place beneath the discharge assembly.

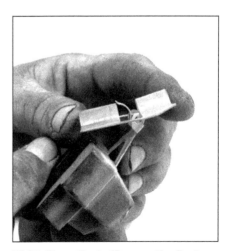

7F.24 Place the baffle over the float assembly as shown and lower both parts into the float bowl at the same time.

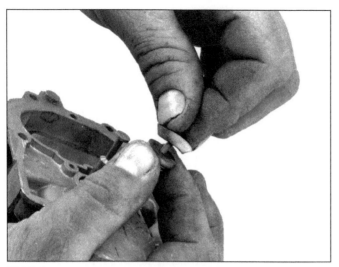

7F.25 A new accelerator pump cup must be installed on the pump shaft during reassembly. Lubricate the lips of the cup with a light film of clean engine oil.

7F.26 Reinstall a new vent valve (arrow) by pushing the stem through the hole in the cover . . .

7F.27 . . . then turn the cover over and pull the valve the rest of the way through the cover.

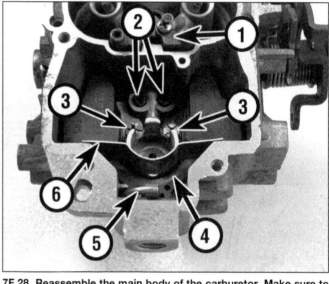

7F.28 Reassemble the main body of the carburetor. Make sure to install:

1	Discharge nozzle cluster	4	Float assembly
2	Main metering jets	5	Float hinge/retainer
3	Power valve assemblies	6	Baffle

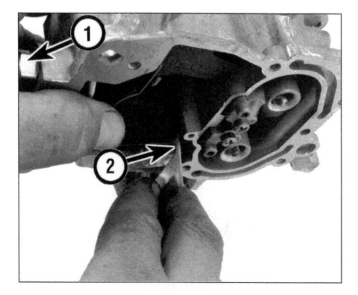

7F.29 After reinstalling the needle-and-seat assembly (1) measure the float height. Turn the carburetor over so the float hangs down, holding the float assembly in place with your thumb on the hinge pin and baffle. Measure the distance from the tip of the float to the gasket surface of the of the fuel bowl (2). Adjust the distance to the specification given in the overhaul kit instructions by gently bending the tang on the float where it contacts the fuel inlet needle (refer to illustration 7F.11). **Note:** *Measure the float height on each individual float. Bend either float arm to equalize the float positions.* Put the gasket in place and reinstall the cover assembly and all remaining components in the reverse order of disassembly.

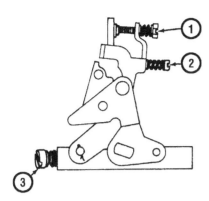

7F.30 Idle mixture and speed adjusting screw locations.

1   Curb idle adjusting screw   3   Idle mixture adjusting
2   Fast idle adjusting screw   screws

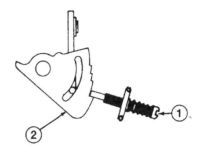

7F.31 Place the fast idle screw on the second highest step of the fast idle cam.

1   Fast idle adjusting screw   2   Fast idle cam

On-vehicle adjustments

After the carburetor is reinstalled and the fuel line and vacuum and vacuum and electrical connectors (if equipped) have been connected, there are several on-vehicle adjustments to be made. Hook up a tachometer in accordance with the manufacturer's instructions, start the engine and allow it to reach normal operating temperature. It may be necessary to turn the idle speed screw in to maintain sufficient rpm so the engine doesn't stall. Check the ignition timing, adjust if necessary, and follow any special instructions found on the Vehicle Emissions Control Information label. **Warning:** *Be sure the vehicle is parked on a level surface, the parking brake is set, the wheels are blocked and the transmission is in Neutral (manual) or Park (automatic).*

Idle mixture adjustment

After the engine has reached normal operating temperature, turn the idle speed adjusting screw out as far as possible without the engine stalling, since the throttle plates must be closed when adjusting the idle mixture.

Working with one idle mixture screw at a time, turn the screw clockwise until the idle speed drops a noticeable amount. Now slowly turn the screw out until the maximum rpm is reached. Repeat the procedure for the other idle mixture screw. Keep track of the number of turns each screw is off the seat to be sure they are each out approximately the same number of turns. This will verify that the idle circuit is operating properly and balance is maintained. **Note:** *If it was necessary to turn the idle speed screw in (to keep the engine running) before setting the idle mixture, turn the idle screw out as far as possible without the engine stalling, then perform the mixture adjustment procedure again. This will ensure that the idle mix-ture is set with the throttle plates fully closed, so the engine is drawing the fuel mixture only from the idle circuit.*

Curb idle speed adjustment

With the idle mixture properly adjusted, the curb idle speed can now be set. Turn the curb idle screw **(see illustration)** clockwise to increase the idle speed and counterclockwise to decrease it. Set the curb idle speed to the specifications found on the Vehicle Emissions Control Information label or on the sheet furnished with the overhaul kit.

Fast idle speed adjustment

After adjusting the idle mixture and curb idle speed, adjust the fast idle speed. Place the fast idle screw on the second highest step of the cam **(see illustration).** By turning the fast idle speed screw, adjust the fast idle to the specifications on the Vehicle Emissions Control Information label or on the sheet furnished with the overhaul kit.

Choke pulloff and choke unloader adjustments

Adjustments to the choke pulloff and choke unloader are made by bending the linkage rods. These adjustments usually do not change from the factory settings but should be checked and adjusted if necessary. Refer to your carburetor kit instruction sheet for the procedures and specifications.

Throttle control solenoids and transducers

Various model carburetors may be equipped with different types of throttle control solenoids or transducers. Again, refer to your carburetor kit instruction sheet for the procedures and specifications for adjusting the throttle control solenoid or transducer.

Notes

7 Part G
Overhaul and adjustments
Two-barrel models 2210/2211/2245

This group of carburetors includes a variety of models all based on a similar design. The models included in this group are the two-barrel models 2210, 2211 and 2245.

The overhaul procedure is covered through a sequence of photographs laid out in order from disassembly to reassembly and finally installation and adjustment on the vehicle. The captions presented with the photographs will walk you through the entire procedure, one component at a time.

The following illustration sequence will use model 2210. Specific details will vary with carburetor part numbers because each carburetor is made to fit a specific application. Refer to the instruction sheet which comes with each overhaul kit. **Note:** *Because of the large number of different models and variations of each model, it is not always possible to cover all the differences in components during the overhaul procedure. Overhaul Chapters only deal with disassembly, reassembly and basic adjustments of a typical carburetor in that model group. If explanation about individual differences of parts are needed to clearly understand or choose the correct part for your application,refer to the exploded views or go to Chapter 3.*

Warnings:
Gasoline

Gasoline is extremely flammable, so take extra precautions when you work on any part of the fuel system. Don't smoke or allow open flames or bare light bulbs near the work area, and don't work in a garage where a natural gas-type appliance (such as a water heater or clothes dryer) with a pilot light is present. A spark caused by an electrical short circuit, by two metal surfaces striking each other, or even by static electricity built up in your body, under certain conditions, can ignite gasoline vapors. Also, never risk spilling fuel on a hot engine or exhaust component. If you spill any fuel on your skin, rinse it off immediately with soap and water.

Battery

Always disconnect the battery ground (-) cable at the battery before working on any part of the fuel or electrical system.

Fire extinguisher

We strongly recommend that a fire extinguisher suitable for use on fuel and electrical fires be kept handy in the garage or workshop at all times. Never try to extinguish a fuel or electrical fire with water. Post the phone number for the nearest fire department in a conspicuous location near the phone.

Compressed air

When cleaning carburetor parts, especially when using compressed air, be very careful to spray away from yourself. Eye protection should be worn to avoid the possibility of getting any chemicals or debris into your eyes.

Lead poisoning

Avoid the possibility of lead poisoning. Never use your mouth to blow directly into any carburetor component. Small amounts of Tetraethyl lead (a lead compound) become deposited on the carburetor over a period of time and could lead to serious lead poisoning. Check passages with compressed air and a fine-tipped blow gun or place a small-diameter tube to the component and blow through the tube to be sure all necessary passages are open.

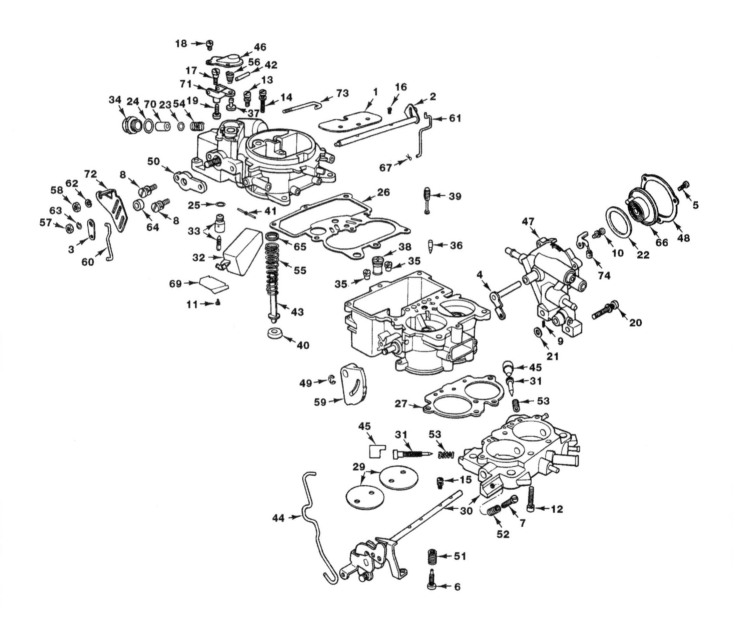

Models 2210 and 2211 exploded view (Model 2245 similar)

Index Number	Part Name	Index Number	Part Name
1	Choke Plate	39	Power Valve Piston Assembly
2	Choke Shaft & Lever Assembly	40	Pump Piston Cup
3	Fast Idle Lever	41	Float Hinge Pin
4	Choke Housing Lever & Shaft Assembly	42	E.C.S. Lever Hinge Pin
5	Thermostat Cap Clamp Screw	43	Pump Stem Assembly
6	Fast Idle Adjusting Screw	44	Pump Link
7	Curb Idle Adjusting Screw	45	Idle Needle Limiter Cap
8	Pump Shaft Lever Screw	46	E.C.S. Vent Cover
9	Choke Piston Adjusting Screw	47	Choke Housing Assembly
10	Choke Lever screw & L.W.	48	Thermostat Cap Retainer Ring
11	Fuel Bowl Baffle Screw & L.W.	49	"E" Ring Retainer
12	Throttle Body Screw & L.W.	50	Pump Shaft Retainer
13	Air Horn Screw & L.W. (Short)	51	Fast Idle Adjusting Screw Spring
14	Air Horn Screw & L.W. (Long)	52	Curb Idle Adjusting Screw Spring
15	Throttle Plate Screw & L.W.		
16	Choke Plate Screw	53	Idle Needle Adjusting Screw Spring
17	E.C.S. Hinge Retainer Screw		
18	E.C.S. Vent Cover Screw	54	Fuel Inlet Filter Spring
19	Vent Valve Adjusting Screw	55	Pump Drive Spring
20	Choke Housing Screw & L.W.	56	E.C.S. Vent Spring
21	Choke Housing Gasket	57	Fast Idle Lever Nut
22	Thermostat Housing Gasket	58	Pump Lever Nut
23	Fuel Inlet Filter Gasket	59	Fast Idle Cam
24	Fuel Inlet Fitting Gasket	60	Fast Idle Rod
25	Fuel Inlet Seat Gasket	61	Choke Rod
26	Main Body Gasket	62	Pump Lever L.W.
27	Throttle Body Gasket	63	Fast Idle Lever L.W.
28	Flange Gasket	64	Washer
29	Throttle Plate	65	Pump Drive Spring Washer
30	Throttle Body & Shaft Assembly	66	Thermostat & Cap Assembly
		67	Rod Retainer
31	Idle Adjusting Needle	68	Choke Vacuum Hose
32	Float & Hinge Assembly	69	Fuel Bowl Baffle
33	Fuel Inlet Needle & Seat Assembly	70	Fuel Inlet Filter
		71	Vent Lever
34	Fuel Inlet Fitting	72	Pump Lever
35	Main Jet	73	Pump Lever Shaft
36	Pump Discharge Needle Valve	74	Choke Thermostat Lever & Piston Assembly
37	Air Vent Valve		
38	Power Valve Assembly		

Parts list for models 2210 and 2211 exploded view

Disassembly

7G.1 After removing the carburetor and mounting it on a stand (see Chapters 6 and 7A), remove the two screws holding the choke vacuum diaphragm to the carburetor cover and remove the diaphragm.

7G.2 Turn the vacuum diaphragm as it is being removed so the rod (arrow) can come out the slot on the choke lever.

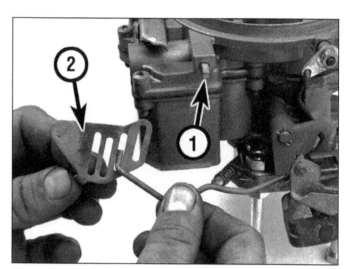

7G.3 Remove the nut from the accelerator pump lever shaft (1) in the cover and remove the accelerator pump lever (2). Note: *Write down which slot the linkage comes out of so you can reassemble it correctly later.*

7G.4 Remove the clip from the shaft for the vent valve.

7G.5 Be careful not to lose the spring (arrow) when sliding the parts off the shaft.

7G.6 Remove the two screws from the vent cover.

7G.7 Remove the vent cover and the valve with the spring (arrow). On the 2211 model, the spring is held in the cover and won't come out with the valve.

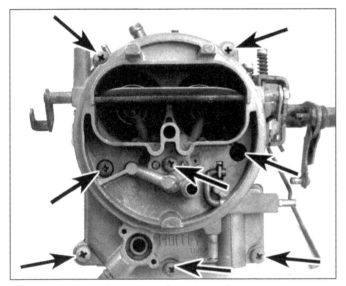

7G.8 Remove the eight screws from the carburetor cover.

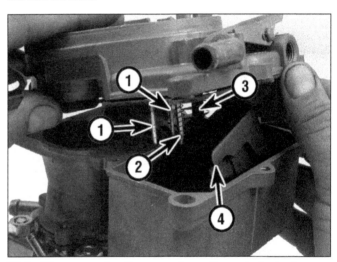

7G.9 When removing the cover from the main body, be very careful not to bend or break items attached to the cover, such as:

1	Main well metering tubes	3	Accelerator pump
2	Power valve piston	4	Float

7G.10 If the choke rod won't unhook from the plastic fast-idle cam, remove the choke shaft nut (arrow) and separate the lever from the shaft.

7G.11 With the cover removed, push up from underneath on the accelerator pump (arrow) and slide the arm out of the hole. Remove the pump assembly and set it aside - it will get a new rubber cup from the overhaul kit.

7G.12 Remove the screw (arrow) and pull off the needle-and-seat baffle.

7G.13 Pull the pivot pin from the cover and remove the float. Note: *Some types of float material become saturated with fuel and become much heavier than normal. This will cause the float to sink and raise the fuel level in the bowl, causing leakage or an over-rich condition. Not all floats can become saturated, but pinholes or breaks in the solder can develop over time, even on metal or plastic floats. If in doubt about the condition of your float(s), it's recommended that the float be replaced at the time of the overhaul.*

7G.14 Remove the needle from the seat.

7G.15 Unscrew the seat and remove the seat and gasket from the cover.

7G.16 The remaining parts can stay in the cover assembly during cleaning:

1   Main well tubes
2   Power valve piston
3   Choke plate and shaft

Tip:
If the power valve piston is sticking and won't move up or down, try cleaning it first with carburetor cleaner spray while moving the piston up and down to free it up. Replacement requires removal of the "staked" material to get the valve out. To re-stake, gently use a hammer and a small punch or screwdriver to crush a little of the cover material around the piston.

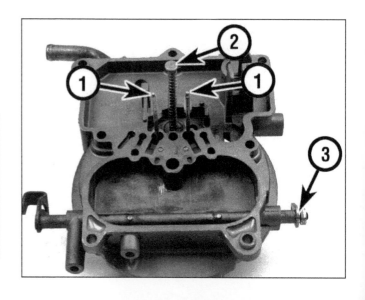

7G.17 Remove the two idle mixture screws from the throttle body. Check the tips of the needles to be sure they're smooth. Replace them if they have become pitted or severely grooved - when reinstalling the needles after cleaning, turn them clockwise until they seat lightly (DO NOT TIGHTEN THEM), then turn them counterclockwise 1-1/2 to 2 turns. This will provide an initial setting so the engine will run after overhaul until final adjustments are made.

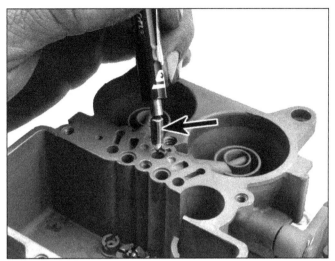

7G.18 Remove the check valve for the accelerator pump discharge nozzle.

7G.19 Remove the five screws holding the throttle body to the main body.

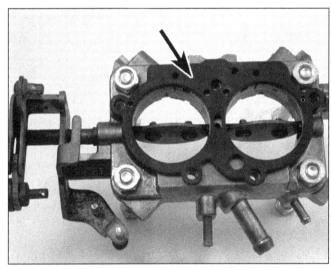

7G.20 Before removing the gasket material from the throttle body or main body, inspect, as best you can, the gasket to see if it sealed properly. Remove the gasket carefully so you can match it to the correct new gasket from the kit (the kit will provide several different gaskets to fit all applications). Then set the throttle body aside to be cleaned.

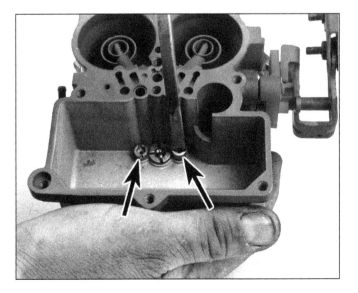

7G.21 Remove the main jets.

Tip:
The jets are usually different sizes. Write down the numbers stamped on the side of each jet and which side of the carburetor it came from (the choke housing or accelerator pump side of the bowl.) The larger numbered jet will go on the accelerator pump side.

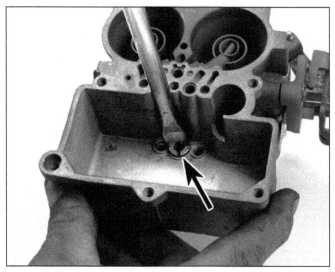

7G.22 Remove the power valve assembly from the float bowl.

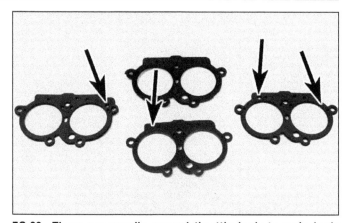

7G.23 There are usually several throttle body-to-main body gaskets in the overhaul kit; be sure to use the correct one. The gasket removed from our carburetor (upper center) matches the one on the right. The arrows indicate areas where the correct gasket differs from the other two.

Cleaning

Refer to Chapter 7A for information on "dips" and solvents used in cleaning. Remember not to place the fast-idle cam, float or any other plastic or rubber parts into a dip tank, since the chemicals will damage the parts. After cleaning (which usually requires an overnight soak in a dip tank), blow out all passages in the carburetor body with compressed air. **Warning:** *Wear eye protection!* Inspect the cleaned parts to be sure there's no gasket material remaining on any surfaces and the surfaces are not damaged or warped. Make sure all parts are completely dry before beginning reassembly.

Reassembly

Several assembly checks, tips and adjustments are shown in the following illustration sequence, but, unless covered here, reassembly is in the reverse order of disassembly. Read the special instructions in the carburetor kit and follow the illustration sequence in making adjustments.

When installing the idle mixture screws, refer to illustration 7G.17.

7G.24 Reinstall a new needle-and-seat with the new gasket - be sure the needle tip (arrow) and the area where it seats are completely clean.

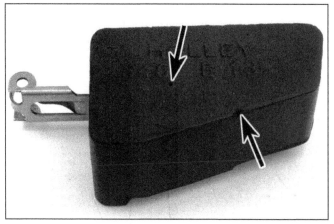

7G.25 Again check the float for signs that indicate it is saturated with fuel. A nick or gouge, as shown by the arrows, mean fuel can get through the outer coating. This float should be replaced.

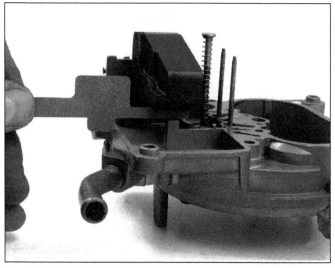

7G.26 After reinstalling the float, set the level of the float with the gauge provided in the overhaul kit (shown) or use a small ruler. Check your measurement against the specifications provided in the kit. Adjust the level by gently bending the tang that is in contact with the needle.

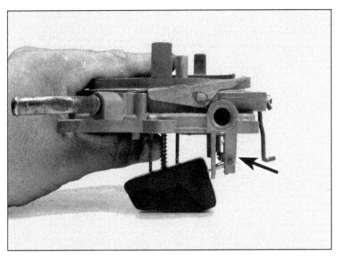

7G.27 The float on this model has a setting for float drop as well as height. With the cover held parallel to a flat table, the bottom edge of the float should also be parallel with the table. The tang to adjust drop is located at the arrow.

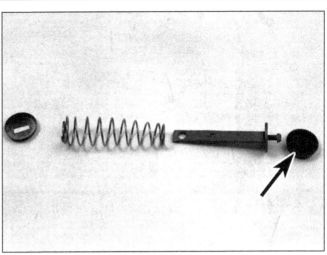

7G.28 Be sure to put a new pump cup (arrow) on the stem before reassembly. Lubricate the tips of the cup with a light film of clean engine oil.

7G.29 Place the drive spring washer in the cover first . . .

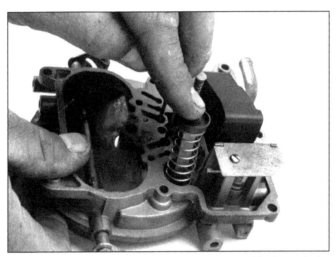

7G.30 . . . then put the accelerator pump assembly through the slot in the cover and place the pump lever shaft through the slot on the top side of the cover (refer to illustration 7G.11).

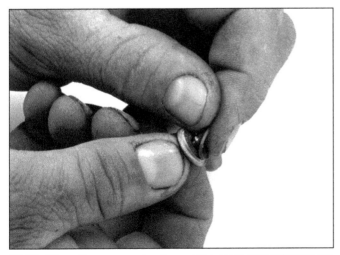

7G.31 Place the new vent valve on the stem and reinstall the washer and spring into the cover (refer to illustration 7G.7).

7G.32 To reinstall the vent valve lever, place a small screwdriver through the lever AND spring . . .

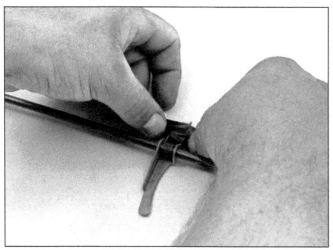

7G.33 . . . use a small amount of tape to keep the end of the spring up against the lever so it will clear the cover when sliding it on . . .

7G.34 . . . place the tip of the screwdriver against the pivot stud (arrow) and push the lever and spring onto the stud.

7G.35 Reinsert the clip in the groove on the shaft (1) and be sure the end of the spring is as shown here (2). Then remove the tape so the spring can operate freely.

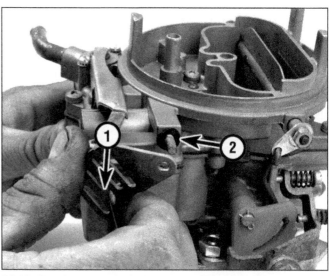

7G.36 Check your notes and place the linkage into the same slot on the lever it came out of during disassembly (1). Place the wave washer on the shaft (2), then reinstall the lever on the shaft.

Accelerator pump adjustment

7G.37 With the pump linkage in the correct slot in the lever (1) and the curb idle speed screw (2) backed out enough so the throttle plates are closed, measure the distance from the top of the pump lever to the top of the cover by the slot (3). Make the same measurement at wide open throttle. The difference between the two should be the specified travel given in the kit instructions. Bend the linkage rod (4) to make necessary adjustments.

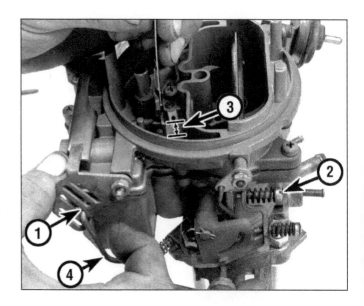

Automatic choke setting

Most of these models use a "divorced" choke with a choke coil that's mounted to the intake manifold. Refer to the procedure at the end of Chapter 7B to set the choke.

On-vehicle adjustments

The procedures for adjusting the idle speed and idle mixture on the vehicle are the same as those described in Chapter 7B. Refer to the procedures in Chapter 7B, but note that the curb idle speed screw and idle mixture needles (screws) are in different locations **(see illustrations 7G.17 and 7G.37)**.

Vacuum kick adjustment

7G.39 Open the throttle to clear the fast-idle cam and apply 10 inches of vacuum to the diaphragm (1) to pull the inner stem (2) to it's seat. Lightly close the choke plate, without pulling the diaphragm stem off its seat, and measure the choke plate opening at the air horn, which should be as specified in the overhaul kit instructions. Bend the rod (3) to make any adjustments.

Fast idle setting

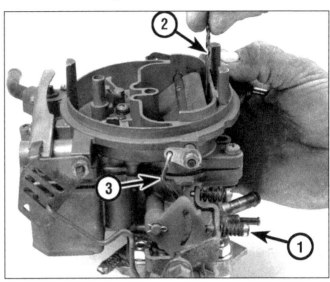

7G.38 Before installing the carburetor on the vehicle, open the throttle and position the fast-idle adjustment screw (1) on the first step of the fast-idle cam, then close the throttle. Hold the choke plate closed as far as it will go, and, using a drill bit, measure the clearance between the rear of the throttle plate and the air horn (2), which should be as specified in the overhaul kit instructions. If not, bend the rod (3) to adjust the clearance. With the screw still on the first step of the fast-idle cam, measure the clearance between the front of the primary throttle plate and the throttle body bore, which should be approximately 1/32-inch. If the throttle is not open the correct amount, adjust it with the screw. After the carburetor has been installed, all other adjustments have been made and the engine is running at normal operating temperature, open the throttle and place the adjustment screw back on the first step of the fast-idle cam. Check the idle speed, which should be approximately 1700 rpm. If the speed is not correct, adjust it with the screw. If you notice the engine idling slowly and "chugging" when the engine is cold and the choke is fully applied, try turning the screw clockwise a bit to increase the fast-idle speed. If the idle speed seems too fast with the choke applied, turn the screw counterclockwise slightly.

Unloader adjustment

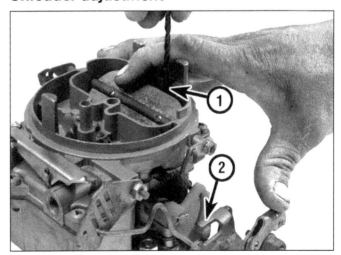

7G.40 Hold the throttle wide open, lightly close the choke plate and measure the opening in the air horn with a drill bit (1). If the clearance is not as specified in the overhaul kit instructions, bend the tang (2) to adjust it.

7 Part H
Overhaul and adjustments
Two-barrel models - 2100 and 2110

The carburetors in this group are all based on a similar design. The models covered in this group are the two-barrel models 2100 and 2110.

The overhaul procedure is covered through a sequence of photographs laid out in order from disassembly to reassembly and finally installation and adjustment on the vehicle. The captions presented with the photographs will walk you through the entire procedure, one component at a time.

The following illustration sequence depicts a model 2110 overhaul. Specific details will vary with carburetor part numbers because each carburetor is made to fit a specific application. Refer to the instruction sheet that comes with each overhaul kit. **Note:** *Because of the different models and variations of each model, it is not always possible to cover all the differences in components during the overhaul procedure. Overhaul Chapters only deal with disassembly, reassembly and basic adjustments of a typical carburetor in that model group. If you need to know more about specific carburetor variations, refer to the exploded views or go to Chapter 3.*

Warnings:

Gasoline

Gasoline is extremely flammable, so take extra precautions when you work on any part of the fuel system. Don't smoke or allow open flames or bare light bulbs near the work area, and don't work in a garage where a natural gas-type appliance (such as a water heater or clothes dryer) with a pilot light is present. A spark caused by an electrical short circuit, by two metal surfaces striking each other, or even by static electricity built up in your body, under certain conditions, can ignite gasoline vapors. Also, never risk spilling fuel on a hot engine or exhaust component. If you spill any fuel on your skin, rinse it off immediately with soap and water.

Battery

Always disconnect the battery ground (-) cable at the battery before working on any part of the fuel or electrical system.

Fire extinguisher

We strongly recommend that a fire extinguisher suitable for use on fuel and electrical fires be kept handy in the garage or workshop at all times. Never try to extinguish a fuel or electrical fire with water. Post the phone number for the nearest fire department in a conspicuous location near the phone.

Compressed air

When cleaning carburetor parts, especially when using compressed air, be very careful to spray away from yourself. Eye protection should be worn to avoid the possibility of getting any chemicals or debris into your eyes.

Lead poisoning

Avoid the possibility of lead poisoning. Never use your mouth to blow directly into any carburetor component. Small amounts of Tetraethyl lead (a lead compound) become deposited on the carburetor over a period of time and could lead to serious lead poisoning. Check passages with compressed air and a fine-tipped blow gun or place a small-diameter tube to the component and blow through the tube to be sure all necessary passages are open.

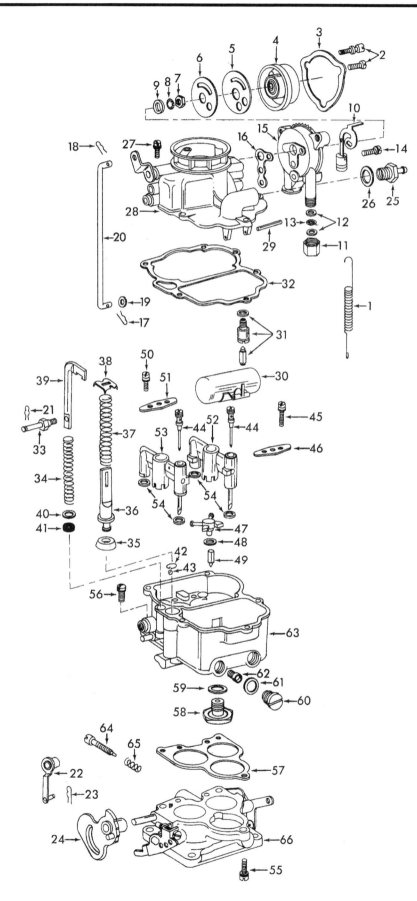

REF.
NO.

1. SPRING-THROTTLE RETURN
2. SCREW(3)-STAT COVER RETAINER
3. RETAINER-ELEC. STAT COVER
4. ELEC. STAT COVER
5. GASKET-STAT COVER
6. PLATE-CHOKE BAFFLE
7. NUT-CHOKE SHAFT
8. LOCKWASHER-CHOKE SHAFT NUT
9. SPACER-CHOKE LEVER
10. LINK & PISTON ASSY.-CHOKE
11. FITTING-SCREW
12. WASHER(2)-SCREEN
13. SCREEN-CHOKE FILTER
14. SCREW & LOCKWASHER(2)-CHOKE
    HOUSING
15. HOUSING ASSY.-CHOKE
16. GASKET-CHOKE HOUSING
17. RETAINER-CHOKE ROD
18. RETAINER-CHOKE ROD
19. WASHER-CHOKE ROD
20. ROD-CHOKE
21. RETAINER-PUMP LINK
22. LINK-PUMP
23. RETAINER-FAST IDLE CAM
24. CAM-FAST IDLE
25. FITTING-FUEL INLET
26. GASKET-FUEL INLET FITTING
27. SCREW & LOCKWASHER(5)-BOWL
    COVER
28. BOWL COVER ASSY.
29. PIN-FLOAT LEVER
30. FLOAT & LEVER ASSY.
31. NEEDLE SEAT & GASKET ASSY.
32. GASKET-BOWL COVER
33. STUD-PUMP LINK
34. SPRING-PUMP RETURN
35. CUP-PUMP PISTON
36. PISTON-PUMP
37. SPRING-PUMP
38. RETAINER-PUMP SPRING
39. ROD-PUMP OPERATING
40. WASHER-PUMP ROD LUBRICATOR
41. LUBRICATOR-PUMP ROD
42. RETAINER-CHECK BALL
43. BALL-PUMP INTAKE CHECK
44. JET(2)-IDLE
45. SCREW & LOCKWASHER(2)-NOZZLE
    BAR & PUMP JET RETAINER
46. BAR-CLAMP
47. NOZZLE-PUMP DISCHARGE
48. GASKET-PUMP DISC. NOZZLE
49. NEEDLE VALVE-PUMP DISC.
50. SCREW & LOCKWASHER(2)-NOZZLE
    BAR
51. BAR-CLAMP
52. NOZZLE ASSY. RH
53. NOZZLE ASSY. LH
54. GASKET(4)-NOZZLE ASSY.
55. SCREW & LOCKWASHER-THROTTLE
    BODY
56. SCREW & LOCKWASHER(2)-THROTTLE
    BODY
57. GASKET-THROTTLE BODY
58. PLUG-BOWL
59. GASKET-BOWL PLUG
60. PLUG(2)-MAIN JET
61. GASKET(2)-MAIN JET PLUG
62. JET(2)-MAIN
63. BOWL ASSY.
64. NEEDLE(2)-IDLE ADJ.
65. SPRING(2)-IDLE ADJ. NEEDLE
66. THROTTLE BODY ASSY.

Model 2110 exploded view (Model 2100 similar)

Disassembly

Make notes or diagrams of levers or linkage locations, slots or holes that the linkage connects with and anything you may feel could be difficult to remember later on during re-assembly. **Note:** *Remember that, on extremely dirty carburetors, overnight cleaning may be required and that reassembly may not be possible until the next day.*

7H.1 After removing the carburetor, mount it on a stand - long 5/16-inch bolts, nuts and washers make an inexpensive stand (some models have three mounting bolt holes, others have four).

7H.3 Take out the link and piston assembly (arrow) from the choke housing.

7H.5 Remove the five carburetor bowl cover screws.

7H.2 Before removing the thermostatic choke cover from the housing, make alignment marks on the housing and cover to be used in readjustment later. Remove the three retaining screws and the cover from the housing. Remove the nut from the choke plate shaft.

7H.4 Using small needle-nose pliers or a pick of some type, remove the clip (arrow) that attaches the choke rod to the lever.

7H.6 The screws for both the choke and throttle plates are "staked" (arrows), that is, the end of the threads are destroyed so the screws can't loosen and fall into the engine or let the throttle plates slip and bind in the wide open position.

The choke plate and the throttle plates don't have to be removed for cleaning. In fact, it's recommended that they are not removed for any reason.

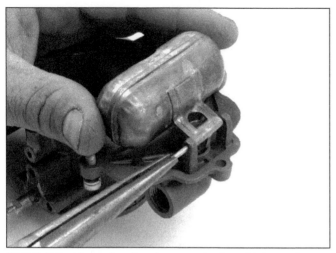

7H.7   Pull the hinge pin out and remove the float. Shake the float to see is fuel has leaked into it. If so, replace it.

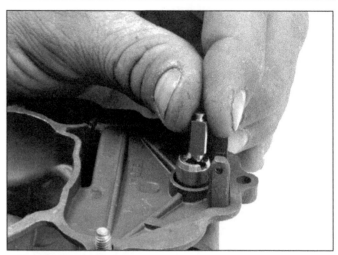

7H.8   Remove the inlet needle. . .

7H.9   . . . and, using the proper size screwdriver, remove the seat from the cover. Be sure to remove the gasket also

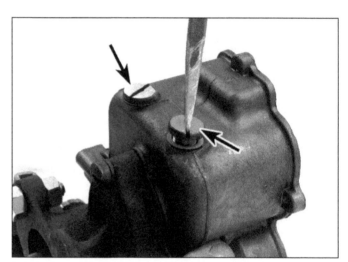

7H.10   Remove the two access plugs and gaskets from the float bowl.

7H.11   Using a screwdriver with a head narrow enough to fit through the plug hole but wide enough to completely engage the slot on each side of the jet, remove the two main metering jets

7H.12   Unscrew and remove the two idle jets

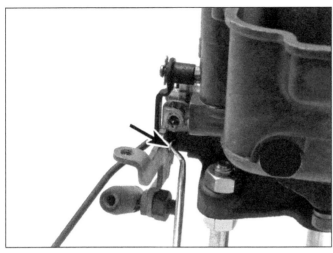

7H.13 Remove the clip (arrow) from the lower end of the choke rod

7H.14 Remove the clip from the accelerator pump stud and remove the pump link

7H.15 Unscrew the stud from the accelerator pump operating rod

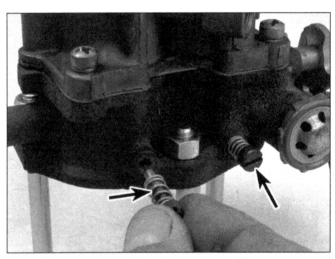

7H.16 Remove the two idle mixture screws. Check for signs of damage or extreme wear. Replace them if necessary.

7H.17 Remove the spark valve using a 1-inch wrench. Some models are equipped with a plug instead of a valve

7H.18 Remove the three screws holding the main body to the throttle body. Two of them can be removed from the top . . .

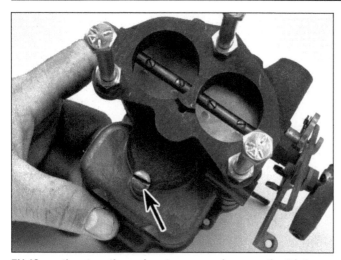

7H.19 . . . then turn the carburetor over and remove the third

7H.20 Push down on the accelerator pump spring retainer and detach the operating rod from the accelerator pump

7H.21 Remove the accelerator pump spring and plunger (1) and the operating rod (2)

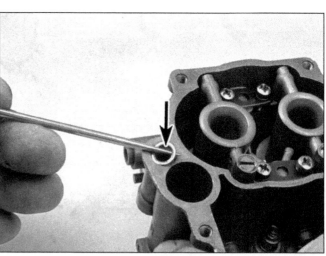

7H.22 Lift out the washer and remove the lightly lubricated packing material underneath (it's used to lubricate the pump operating rod)

7H.23 Remove the wire retainer and the inlet check ball from the bottom of the accelerator pump well

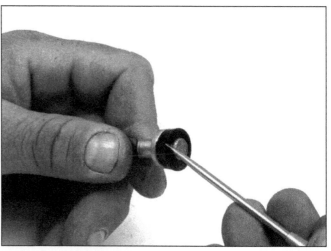

7H.24 Remove the accelerator pump cup from the end of the pump shaft. Normally they become hard enough over time that they can be cracked or split and easily removed

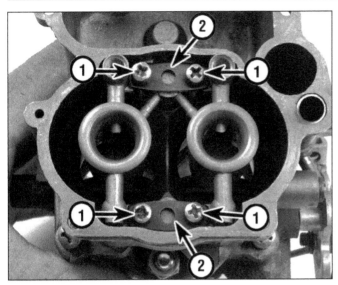

7H.25 Remove the four nozzle bar screws (1) and remove the two clamps (2)

7H.27 Remove two discharge nozzle bar assemblies and gaskets

7H.29 The bushing in the bottom of the bore for the accelerator pump operating rod should almost never have to be replaced. But if it does, pry the old bushing out and use a punch to gently drive the new one into place.

7H.26 Lift out the pump discharge nozzle and be sure to remove the gasket, then turn the carburetor over and catch the discharge check needle and set aside. Note: *Some models have a check ball and weight while others have only a check needle. Both serve the same purpose.*

7H.28 Using a 1-inch wrench, remove the power valve from the main body. Note: *Your carburetor may be equipped with a plug rather than a power valve. In either case, remove it for thorough cleaning.*

Cleaning

Refer to Chapter 7A for information on "dips" and solvents used in cleaning. Remember not to place the float or any other plastic or rubber parts into a dip tank, since the chemicals will damage the parts. After cleaning (which usually requires an overnight soak in a dip tank), blow out all passages in the carburetor body with compressed air. **Warning:** *Wear eye protection!* Inspect the cleaned parts to be sure there's no gasket material remaining on any surfaces and the surfaces are not damaged or warped. Inspect the float, as shown in illustration 7A.14. Make sure all parts are completely dry before beginning reassembly.

Reassembly

Reassembly is basically the reverse of disassembly, but several assembly checks, tips and adjustments are shown in the following illustrations to help with your areas of special concern. Be sure to follow any special instructions that may be included in the carburetor kit and follow the illustration sequence in making adjustments.

Tip:
When reassembling the carburetor, lightly spray gaskets with silicone spray lubricant. This will help eliminate the kind of sticking you saw as you disassembled your carburetor. There are also special rubber gaskets available through Holley for those who expect frequent disassembly of their carburetor, especially for performance modifications and adjustments.

7H.31 Install the nozzle bar assemblies, using new gaskets. Put the discharge check needle (arrow), or check ball and weight, back into the discharge fuel well.

7H.30 "Stake" the bushing in place with a center-punch or a small chisel. Make two or three stake marks to secure the bushing in place.

7H.32 Reinstall the pump discharge nozzle. Looking at it here from the bottom side, be sure the gasket is in place (arrow) before installation. Install the clamps and screws, tightening them securely.

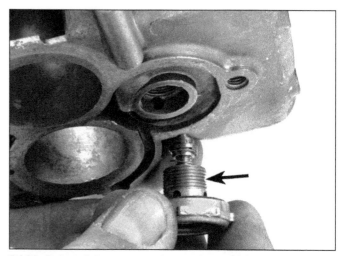

7H.33 Reinstall the power valve or plug, whichever your model has. Place the gasket on the valve or plug and install the valve up from the bottom (arrow). This will keep the gasket centered around the valve and make a proper seal.

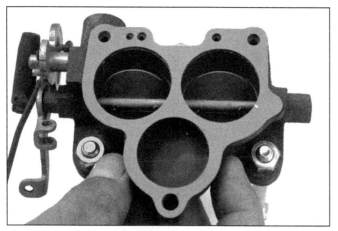

7H.34 Select the correct gasket from the assortment the kit provides and place it on the throttle body. Caution: *The gaskets provided in the overhaul kit have slight differences in hole locations and shapes. Be sure to match the old gasket with the new one selected or performance problems can result.*

7H.35 Reassemble the main body to the throttle body and install the screws, tightening them securely

7H.36 Reinstall the spark valve or plug into the throttle body. Be sure to use a new gasket.

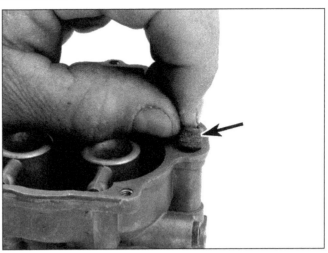

7H.37 Put the new felt packing into the opening for the accelerator pump operating rod

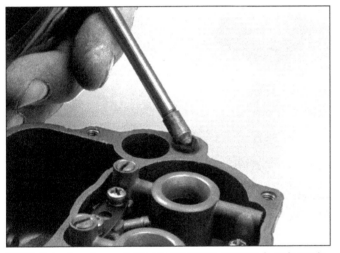

7H.38 Put a few drops of oil onto the felt to saturate it so it can lubricate the rod, then set the thin flat washer you removed earlier back onto the packing for the operating rod spring to seat against.

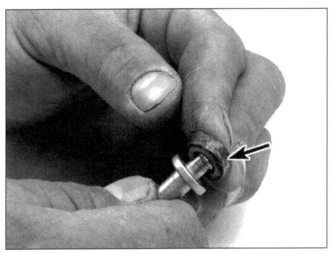

7H.39 Push the new accelerator pump cup into place on the pump shaft. Drop the new inlet check valve into its hole in the accelerator pump well, then install the wire retainer.

7H.40 Apply a light film of clean engine oil to the pump cup. Install the accelerator pump, spring and retainer, and the operating rod and spring back into their respective openings as shown here, making sure they're engaged properly.

7H.41 Hold the operating pump down with your finger while reinstalling the pump stud

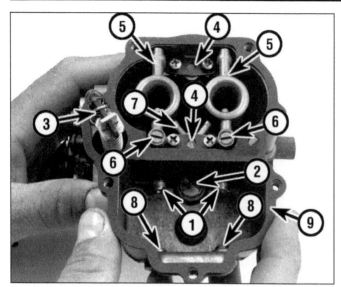

7H.42 Be sure all main body components are installed before moving on to reassembling the cover

1 Main metering jets	5 Discharge boosters
2 Power valve or plug	6 Idle jets
3 Accelerator pump	7 Discharge nozzle
assembly	8 Access plugs
4 Bar clamps	9 Gasket

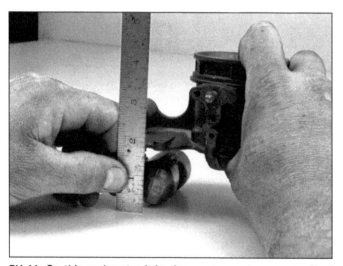

7H.44 On this carburetor it is also necessary to measure float drop. With the float hanging freely, measure from the machined surface of the cover to the bottom of the float as shown here.

Idle mixture screw initial adjustment

Make the initial idle screw settings at this time. The idle mixture screws, one for each throttle plate, are initially set between 1-1/2 to 2 turns out from the lightly seated position. This will allow the engine to idle, then the final settings need to be done while the engine is running.

Accelerator pump link setting

There are three holes (four, on some models) on the throttle shaft lever that the pump link hooks into. The center hole is the standard setting. The richer setting is the hole towards the fuel bowl end of the carburetor and the leaner setting is the hole towards the choke linkage rod. Adjust the setting after driving the vehicle to determine your needs.

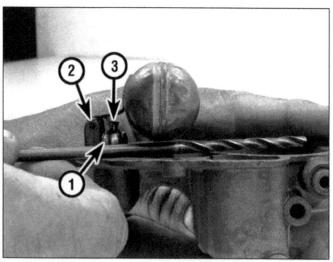

7H.43 Reinstall the new needle and seat (1) with a new gasket into the cover. Then assemble the float and hinge pin (2).

Measure and adjust the float level with the cover turned over as shown. Use the specification provided in the overhaul kit for your particular model. If adjustment is needed, bend the tang (3) to get the correct clearance.

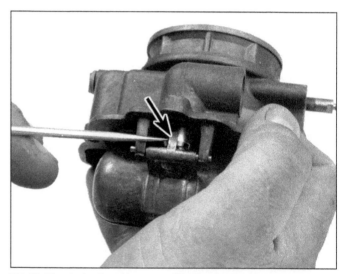

7H.45 Check the float drop specification for your particular model in the overhaul kit instructions. If adjustment is necessary, bend the tang (arrow) to get the correct float drop.

Automatic choke setting

Realign the marks you made before you removed the choke cover from the housing. You may not need as much choke as before, now that the carburetor is working properly, so readjustment may be necessary.

Rotate the thermostat cover counter-clockwise (against spring tension) until the choke plate closes completely, then rotate it clockwise until the choke plate just starts to come off from the closed position. Tighten the three retaining screws to hold the cover in place. **Note:** *The location of the mark on the cover may not line up with any marks on the housing, but it will be set correctly for your vehicle. Spring tension can deteriorate with age and use, so the location of the correct setting can change.*

Once the reassembly is complete, reinstall the carburetor on the vehicle. Refer to Chapter 6 for more details.

On-vehicle adjustments

After the carburetor is reinstalled and the fuel line and electrical connectors (if equipped) have been connected, there are several on-vehicle adjustments to be made. Hook up a tachometer in accordance with the manufacturer's instructions, start the engine and allow it to reach normal operating temperature. It may be necessary to turn the idle speed screw in to maintain sufficient rpm so the engine doesn't stall. **Warning:** *Be sure the vehicle is parked on a level surface, the parking brake is set, the wheels are blocked and the transmission is in Neutral (manual) or Park (automatic).*

Idle mixture adjustment

After the engine has reached normal operating temperature, turn the idle speed adjusting screw out as far as possible without the engine stalling, since the throttle plates must be closed when adjusting the idle mixture.

Working with one idle mixture screw at a time, turn the screw clockwise until the idle speed drops a noticeable amount. Now slowly turn the screw out until the maximum rpm is reached. Repeat the procedure for the other idle mixture screw. Keep track of the number of turns each screw is off the seat to be sure they are each out approximately the same number of turns. This will verify that the idle circuit is operating properly and balance is maintained. **Note:** *If it was necessary to turn the idle speed screw in (to keep the engine running) before setting the idle mixture, turn the idle speed screw out as far as possible without the engine stalling, then perform the mixture adjustment procedure again. This will ensure that the idle mixture is set with the throttle plates fully closed, so the engine is drawing the fuel mixture only from the idle circuit.*

Idle speed adjustment

With the idle mixture properly adjusted, the curb idle speed can now be set. Make sure the idle speed adjusting screw is set on the lowest portion of the fast idle cam **(see illustration)**. Turn the screw clockwise to increase the idle speed and counterclockwise to decrease it. Set the idle speed to approximately 750 rpm in Neutral or Park.

Fast idle speed adjustment

At this point, with the idle speed and mixture correctly set, chances are that the fast idle speed will also be correct, but the adjustment should be checked. With the engine Off, open the throttle and close the choke plate, then allow the throttle to close while holding the choke in the closed position. The idle speed adjusting screw should come to rest on the highest step of the fast idle cam **(see illustration)**. If the screw is resting on the base or any of the other steps on the fast idle cam, bend the rod as necessary to bring the highest step into position under the screw. Now open the choke plate all the way, pump the throttle once and verify the idle speed screw comes to rest on the lowest portion of the cam. If it doesn't, straighten the choke rod as necessary.

This setting should provide a fast idle speed of approximately 1500 to 1700 rpm.

7H.46 When adjusting the curb idle speed, make sure the adjusting screw is resting on the lowest portion of the fast idle cam

7H.47 When the choke plate is closed, the idle speed adjusting screw should be positioned on the highest step of the fast idle cam

7 Part I
Overhaul and adjustments

One-barrel models
1940/1945/1946/1949 and 6145/6146/6149

This group of carburetors includes a variety of models all based on a similar design. The models covered in this group are the one-barrel models 1940/1945/1946/1949 and feedback models 6145/6146/6149.

The overhaul procedure is covered through a sequence of photographs laid out in order from disassembly to reassembly and finally installation and adjustment on the vehicle. The captions presented with the photographs will walk you through the entire procedure, one component at a time.

The following illustration sequence depicts a model 1920 overhaul. Specific details will vary with carburetor part numbers because each carburetor is made to fit a specific application. Refer to the instruction sheet that comes with each overhaul kit. **Note:** *Because of the large number of different models and variations of each model it is not always possible to cover all the differences in components during the overhaul procedure. Overhaul sections only deal with disassembly, reassembly and basic adjustments of a typical carburetor in that model group. If explanation about individual differences of parts are needed to clearly understand or choose the correct part for your application, refer to the exploded views or go to Chapter 3, Carburetor Fundamentals, for more details on components not discussed in the photograph sequence.*

Warnings:
Gasoline

Gasoline is extremely flammable, so take extra precautions when you work on any part of the fuel system. Don't smoke or allow open flames or bare light bulbs near the work area, and don't work in a garage where a natural gas-type appliance (such as a water heater or clothes dryer) with a pilot light is present. A spark caused by an electrical short circuit, by two metal surfaces striking each other, or even by static electricity built up in your body, under certain conditions, can ignite gasoline vapors. Also, never risk spilling fuel on a hot engine or exhaust component. If you spill any fuel on your skin, rinse it off immediately with soap and water.

Battery

Always disconnect the battery ground (-) cable at the battery before working on any part of the fuel or electrical system.

Fire extinguisher

We strongly recommend that a fire extinguisher suitable for use on fuel and electrical fires be kept handy in the garage or workshop at all times. Never try to extinguish a fuel or electrical fire with water. Post the phone number for the nearest fire department in a conspicuous location near the phone.

Compressed air

When cleaning carburetor parts, especially when using compressed air, be very careful to spray away from yourself. Eye protection should be worn to avoid the possibility of getting any chemicals or debris into your eyes.

Lead poisoning

Avoid the possibility of lead poisoning. Never use your mouth to blow directly into any carburetor component. Small amounts of Tetraethyl lead (a lead compound) become deposited on the carburetor over a period of time and could lead to serious lead poisoning. Check passages with compressed air and a fine-tipped blow gun or place a small-diameter tube to the component and blow through the tube to be sure all necessary passages are open.

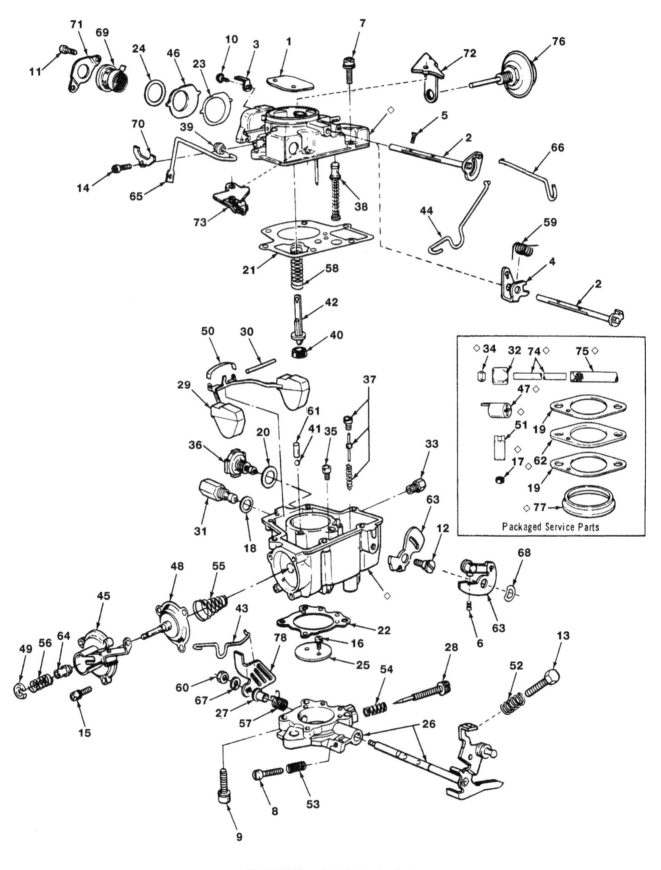

Model 1940 exploded view (typical)

Index Number	Part Name	Index Number	Part Name
◆	Air Horn & Plugs Assembly	45	Choke Diaphragm Cover Assembly
1	Choke Plate	46	Thermostat Housing Plate
2	Choke Shaft & Lever Assembly	48	Choke Diaphragm Assembly
3	Choke Thermostat Lever	49	Choke Modulator Spring Retainer
4	Choke Control Lever	50	Float Shaft Retainer
5	Choke Plate Screw	52	Fast Idle Adjusting Screw Spring
6	Fast Idle Cam Swivel Screw		
7	Air Horn To Main Body Screw & L.W.	53	Throttle Adjusting Screw Spring
8	Throttle Adjusting Screw	54	Idle Adjusting Needle Spring
9	Throttle Body to Main Body Screw & L.W.	55	Choke Diaphragm Spring
10	Choke Thermostat Lever Screw	56	Choke Modulator Spring
11	Choke Thermostat Cover Clamp Screw	57	Throttle Return Spring
		58	Pump Drive Spring
12	Fast Idle Cam Screw	59	Choke Lever Spring
13	Fast Idle Adjusting Screw	60	Pump Operating Lever Nut
14	Pump Rod Clamp Screw	61	Pump Discharge Valve Weight
15	Choke Diaphragm Cover Screw & L.W.	63	Fast Idle Cam & Swivel Assembly
16	Throttle Plate Screw	64	Choke Modulator Sleeve
◆	Main Body & Plugs Assembly	65	Pump Rod
18	Fuel Inlet Seat Gasket	66	Fast Idle Rod
20	Spark Valve Gasket	67	Pump Operating Lever Nut L.W.
21	Main Body Gasket		
22	Throttle Body Gasket	68	Spring Washer
23	Thermostat Housing Gasket	69	Thermostat & Cover Assembly
24	Thermostat Cover Gasket	70	Pump Rod Clamp
25	Throttle Plate	71	Thermostat Cover Clamp
26	Throttle Body & Shaft Assembly	72	Dashpot Bracket
27	Throttle Return Spring Bushing	73	Choke Bracket Assembly
		76	Dashpot Assembly
28	Idle Adjusting Needle	78	Pump Operating Lever
29	Float & Hinge Assembly		
30	Float Hinge Shaft	◆17	Rubber Plug
31	Fuel Inlet & Needle Seat Assembly	19	Flange Gasket
		32	Compression Nut
33	Spark Fitting	◆34	Ferrule
35	Main Jet	◆47	Idle Adjusting Needle Limiter Cap
36	Spark Valve Assembly		
		◆51	Fuel Line Clamp
37	Power Valve Assembly	◆62	Throttle Body Spacer
38	Power Valve Piston Assembly	◆74	Choke Heat Tube
39	Pump Rod Seal	◆75	Choke Heat Tube Sock
40	Pump Piston Cup	◆77	Air Cleaner Adaptor
41	Pump Discharge Valve	* *	Fuel Inlet Fitting
42	Pump Piston Stem	79	Fast Idle Cam Retainer
43	Pump Operating Link		
44	Choke Diaphragm Link		

Parts list for model 1940 exploded view

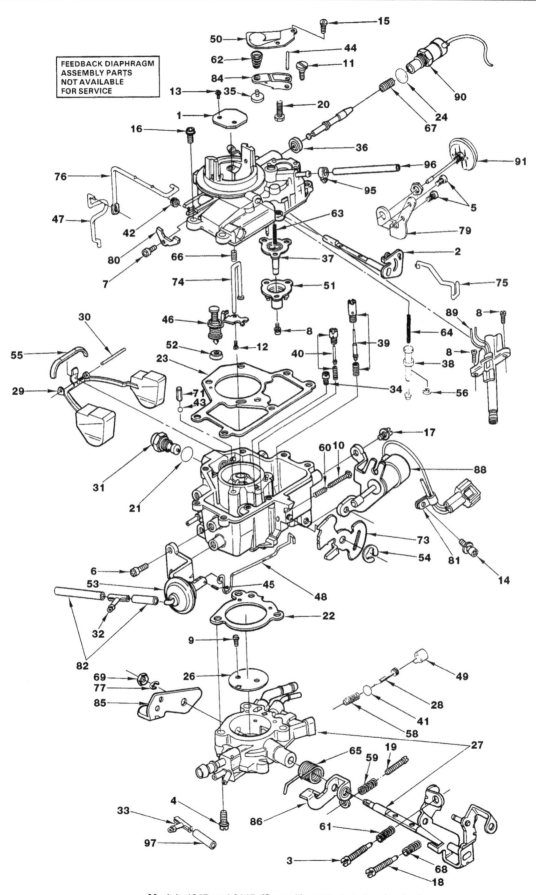

FEEDBACK DIAPHRAGM
ASSEMBLY PARTS
NOT AVAILABLE
FOR SERVICE

Models 1945 and 6145, (Stage II), exploded view (typical)

Index Number	Part Name	Index Number	Part Name
1	Choke Plate	69	Pump Operating Lever Nut
2	Choke Shaft & Lever Assembly	70	Dashpot Lock Nut
3	Fast Idle Adjusting Screw	71	Pump Discharge Valve Weight
4	Throttle Body Screw & L.W.	72	Power Valve Diaphragm Spacer*
5	Dashpot Bracket Screw	73	Fast Idle Cam
6	Choke Diaphragm Bracket Screw	74	Power Valve Rod
7	Pump Rod Clamp Screw	75	Fast Idle Rod
8	Power Valve Diaphragm Body	76	Pump Operating Rod
	Screw & L.W. (Duty Cycle Solenoid)	77	Pump Operating Lever Nut L.W.
9	Throttle Plate Screw	78	Insulator Washer*
10	Curb Idle Adjusting Screw	79	Dashpot Bracket
11	ECS Hinge Retaining Screw	80	Pump Rod Clamp
12	Retaining Plate Screw	81	Wire Hold-Down Clamp
13	Choke Plate Screw	82	Choke Diaphragm Vacuum Hose
14	Wire Hold-Down Clamp Screw	83	Idle Screw Insulator*
15	ECS Vent Cover Screw	84	ECS Vent Lever
16	Air Horn Screw & L.W.	85	Pump Operating Lever
17	Idle Solenoid Bracket Screw & L.W	86	Power Valve Lever
18	Solenoid Adjusting Screw	87	Idle Stop Switch*
19	Power Valve Lever Adjusting Screw	88	Idle Stop Solenoid
20	ECS Vent Valve Adjusting Screw	89	Duty Cycle Solenoid
21	Fuel Inlet Fitting (Needle & Seat	90	Switch Vent Solenoid & Armature
	Assy.) Gasket		Assy.
22	Throttle Body Gasket	91	Dashpot Assy.
23	Main Body Gasket	92	A/C Speed-Up Solenoid*
24	Switch Vent Solenoid Gasket	93	T.P.T. Assy.*
25	Power Valve Spacer Gasket	94	T.P.T. Bracket*
26	Throttle Plate	95	Power Valve Vacuum Hose Clamp
27	Throttle Body & Shaft Assy.	96	Power Valve Vacuum Hose
28	Idle Adjusting Needle	97	Spark Tube Vacuum Hose
29	Float & Hinge Assy.		
30	Float Hinge Shaft		
31	Fuel Inlet Needle & Seat Assy.		
32	Choke Diaphragm Hose Connector		
33	Spark Tube Hose Connector		
34	Main Jet		
35	ECS Vent Valve		
36	Switch Vent Seal		
37	Power Valve Diaphragm & Stem		
	Assy.		
38	Power Valve Piston		
39	Power Valve Assembly (Vacuum)		
40	Power Valve Assembly (Mech.)		
41	Idle Adj. Needle "0" Ring		
42	Pump-Rod Seal		
43	Pump Discharge Valve Checkball		
44	ECS Lever Hinge Pin		
45	Choke Diaphragm Link Pin		
46	Pump Stem & Retainer Plate Assy.		
47	Pump Operating Link		
48	Choke Diaphragm Link		
49	Idle Needle Limiter Cap or Plug		
50	ECS Vent Cover		
51	Power Valve Diaphragm Body		
52	Accelerator Pump Cup		
53	Choke Diaphragm Assy.		
54	Fast Idle Cam Retainer		
55	Float Shaft Retainer		
56	Power Valve Piston Retainer		
57	Wire Tie-Down Strap*		
58	Idle Adj. Needle Spring		
59	Power Valve Lever Adj. Screw		
	Spring		
60	Curb Idle Adj. Screw Spring		
61	Fast Idle Adj. Screw Spring		
62	ECS Vent Spring		
63	Power Valve Diaphragm Spring		
64	Power Valve Piston Spring		
65	Power Valve Lever Return Spring		
66	Power Valve Rod Drive Spring		*Parts not shown on
67	Switch Vent Solenoid Spring		T.V. 44-2.
68	Idle Solenoid Adj. Screw Spring		

Parts list for models 1945 and 6145 exploded view

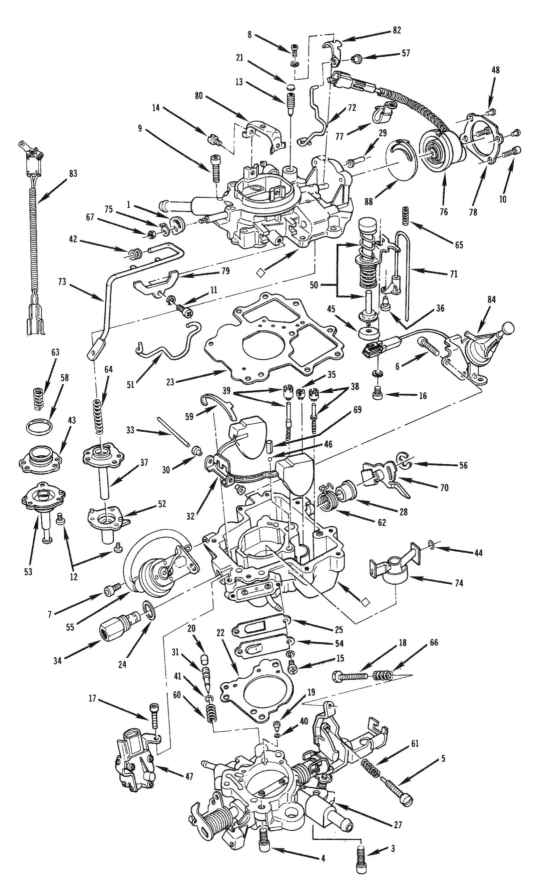

Models 1949 and 6149 exploded view (typical)

Index Number	Part Name	Index Number	Part Name
*	Choke Plate	60	Idle Mixture Screw Spring
*	Choke Shaft and Lever Assembly	*	Solenoid Adjusting Screw Spring
1	Choke Pulldown Lever	61	Fast Idle Speed Adjusting Screw Spring
3	Throttle Body to Main Body Screw	62	Fast Idle Cam Anti-Entrapment Spring
4	Throttle Body to Main Body Screw	63	Main Feedback Diaphragm Spring
*	Solenoid Adjusting Screw	64	Power Enrichment Diaphragm Spring
5	Fast Idle Adjusting Screw	65	Main Jet/Pullover Valve Actuating Rod Spring
6	Solekicker Screw	66	Idle Speed Adjusting Screw Spring
7	Choke Diaphragm Assembly Screw	*	Mech. P.V. Lever Return Spring
8	Choke Control Lever Screw & L.W.	*	Adjusting Screw Spring
9	Airhorn Screw	67	Choke Shaft Nut
10	Choke Bimetal Retaining Screw	*	Dashpot Lock Nut
11	Pump Operating Rod. Ret. Screw L.W.	68	Choke Index Plate
12	Power Enrichment/Main Feedback Diaphragm Screw	69	Pump Check Ball Weight
13	Main Feedback Diaphragm Adusting Screw (6145)	70	Fast Idle Cam
		71	Main Jet/Pullover Valve Actuating Rod
14	Air Cleaner Bail Screw	72	Fast Idle Cam Link
15	Hot Idle Compensate Cover Screw	73	Pump Operating Rod
16	Accelerator Pump Screw & L.W.	74	Booster Venturi
17	TPS Bracket Mounting Screw	75	Choke Shaft Lockwasher
*	Choke Plate Screw & L.W.	*	A/C Cut-Out Bracket Washer
18	Idle Speed Adjusting Screw	*	Throttle Shaft Nut Lockwasher
*	Adjusting Screw	76	Choke Bimetal Assembly
*	Throttle Plate Screw	77	Choke Wire Clip
*	T.P.S. Cover Screw	78	Choke Bimetal Retaining Ring
19	Idle Channel Restriction	79	Pump Operating Rod Retainer
*	Idle Mixture Screw Plug	80	Air Cleaner Bail
21	Main Feedback Diaphragm Adjusting Seal (6149)	*	T.P.S. Mounting Bracket
		82	Choke Control Lever
22	Throttle Body Gasket	*	Mech. Power Valve Lever
23	Airhorn Gasket	*	Mech. Power Valve Lever & Screw Assembly
24	Fuel Idle Needle & Seat Gasket		
25	Hot Idle Compensator Gasket	83	W.O.T. A/C Cut-Off Switch (1949 only)
*	T.P.S. Cover Gasket	84	Solekicker Assembly
*	Throttle Plate (Brass)		
*	Throttle Shaft & Pump Lever Assembly		
*	Throttle Lever & Ball Assembly		
27	Throttle Body Assembly		
28	Fast Idle Cam Bushing		
29	Bimetal Lever Bushing		
30	Float Shaft Bushing		
31	Idle Mixture Screw		
32	Float Assembly		
33	Float Shaft		
34	Fuel Inlet Needle & Seat Assembly		
35	Main Metering Jet		
36	Main Jet/Pullover Valve Seal Puck		
37	Power Enrichment Diaphragm Assembly (1949)		
38	W.O.T. Enrichment Pullover Valve		
39	Main System Feedback Valve (6149)		
40	Idle Channel Restriction O-Ring		
41	Idle Mixture Screw O-Ring		
42	Pump Operating Rod Grommet		
43	Main Feedback Diaphragm Seat		
44	Booster Venturi O-Ring		
45	Accelerator Pump Cup		
46	Pump Check Ball		
47	Throttle Position Sensor Assembly		
48	Choke Bimetal Retaining Rivet		
*	Sensor Actuating Pin		
50	Accelerator Pump Assembly		
51	Accelerator Pump Operating Link		
52	Power Enrichment Diaphragm Cover Casting		
53	Main Feedback Diaphragm Assembly (6149)		
54	Hot Idle Compensater Cover		
55	Choke Diaphragm Assembly		
56	Fast Idle Cam Retainer		
57	Fast Idle Cam Link Retainer		
58	Main Feedback Diaphragm Seat O-Ring		
59	Float Shaft Retainer		

Parts list for models 1949 and 6149 exploded view

Disassembly

One-barrel carburetors are a little awkward to handle since they only have two bolt holes in the throttle body, so bolts won't work as a stand as they do on the models with a four bolt throttle plate. If you have a vise, you can clamp a punch in the jaws and set one of the mounting holes over it. But however you handle the carburetor during overhaul, be careful not to damage the throttle plates and linkage.

Make notes or diagrams of levers or linkage locations, slots or holes that the linkage connects with and anything you may feel could be difficult to remember later on during reassembly. **Note:** *Remember that on extremely dirty carburetors overnight cleaning may be required and that reassembly may not be possible until the next day.*

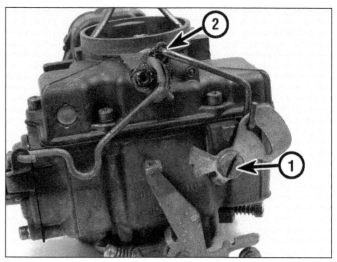

7I.1 Remove the screw (1) and fast idle cam, as well as the fast idle rod (2)

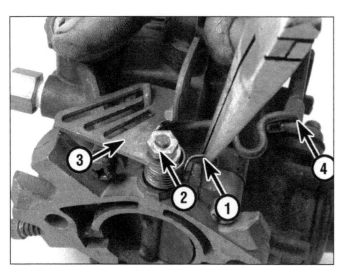

7I.2 On the other side of the carburetor lift the spring off of the throttle body casting (1). Remove the nut (2) from the throttle shaft and remove the accelerator pump operating lever (3). Note: *Make a note of which slot the linkage is removed from in order to reassemble it correctly later.* **Finally remove the linkage (4).**

7I.3 Remove the three screws attaching the throttle body to the main body. Separate the two parts carefully to try and keep the gasket intact as much as possible. You'll need to match the old gasket to one of several provided in the kit. They have slight differences, so compare them closely and get an exact match.

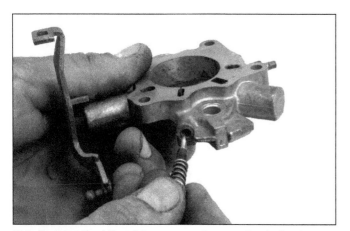

7I.4 Remove the idle mixture screw. Inspect the tip for damage (see Chapter 7, Part A) and replace the screw if necessary. Note: *Some models will have a plastic limiter cap over the mixture screw. This will have to be broken off in order to remove the screw. You won't need to replace this cap after the overhaul. If the mixture screw on your carburetor is hidden behind a concealment plug, refer to Chapter 7, Part F for the plug removal procedure.*

7I.5 Remove the six screws holding the bowl cover to the main body

7I.6 Lift the bowl cover from the main body. As you lift the cover off, be careful not to bend the metering tubes, accelerator pump or power valve. Once all the parts have cleared the body of the carburetor, turn the cover in order to disconnect the linkage from the choke diaphragm lever (arrow).

7I.7 To remove the accelerator pump assembly remove the following parts:

1   Pump rod clamp screw and clamp.
2   Pump rod
3   Accelerator pump assembly
4   Pump rod seal

7I.8 Remove the two screws and lift out the choke thermostat coil from the housing

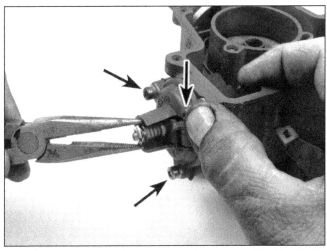

7I.9 Push down on the spring retainer and slide it off the diaphragm post, then remove the choke modulator spring. Next, remove the screws (arrows) and detach the diaphragm cover.

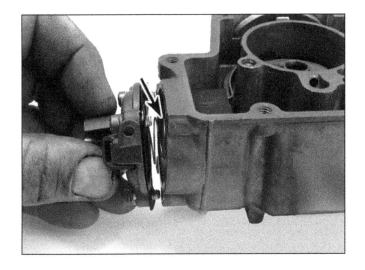

7I.10 As you remove the cover and diaphragm, note the position of the spring

7I.11 If your carburetor has a spark valve, remove the valve and the gasket

7I.12 Remove the float pin retainer

7I.13 Carefully lift out the float assembly. Be careful not to lose the hinge pin (arrow)

7I.14 Remove the accelerator pump discharge weight . . .

7I.15 . . . then the accelerator pump discharge check ball. Note: *You can tip the main body over and catch both pieces as they fall out, but be careful not to lose them.*

7I.16 Remove the power valve assembly by first unscrewing the power valve assembly from the fuel bowl (note the notch filed into the blade of the screwdriver to clear the stem) . . .

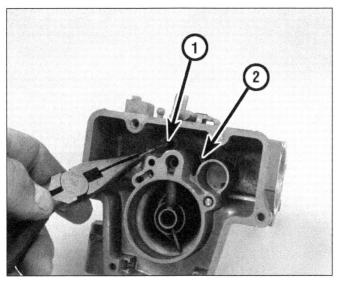

7I.17 . . . then pull the power valve stem and spring from the main body (1). Next, remove the main jet (2) from the fuel bowl.

7I.18 Unscrew and remove the needle and seat assembly.

Cleaning

Refer to Chapter 7A for information on "dips" and solvents used in cleaning. Remember not to place the float or any other plastic or rubber parts into a dip tank, since the chemicals will damage the parts. After cleaning (which usually requires an overnight soak in a dip tank), blow out all passages in the carburetor body with compressed air. **Warning:** *Wear eye protection!* Inspect the cleaned parts to be sure there's no gasket material remaining on any surfaces and the surfaces are not damaged or warped. Inspect the float, as shown in illustration 7A.14. Make sure all parts are completely dry before beginning reassembly.

Reassembly

Reassembly is basically the reverse of disassembly, but several assembly checks, tips and adjustments are shown in the following illustrations to help you with areas of special concern. Be sure to follow any special instructions that may be included in the carburetor kit and follow the illustration sequence in making adjustments.

7I.19 Put the gasket over the tang on the thermostat coil, then place it against the cover . . .

7I.20 . . . likewise, place the tang through the hole of the thermostat housing plate first when reassembling these two pieces.

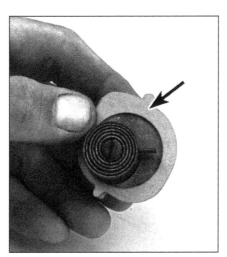

7I.21 Place a new gasket on the thermostat housing plate

7I.22 Install the thermostat assembly into the cover, place the cover clamp over it and install the two screws

7I.23 Place the pump rod seal (1) in the cover first then slide the pump rod (2) through the seal as shown. Tip: *Lightly grease either the rod or the inner surface of the seal to help reassembly.*

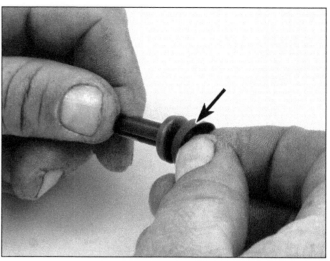

7I.24 Replace the old accelerator pump cup with the new one that comes in the overhaul kit

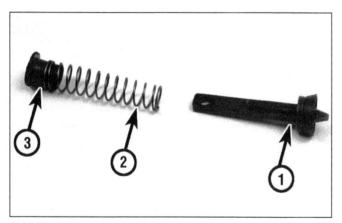

7I.25 Reassemble the accelerator pump parts:

1   *Piston stem and cup*
2   *Spring*
3   *Retainer (some models may not have this as a separate piece)*

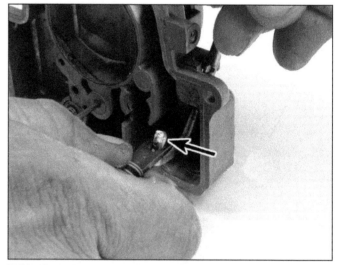

7I.26 Hook the pump rod to the shaft and install the accelerator pump parts (refer to illustration 7I.7). Apply a light film of clean oil to the pump cup.

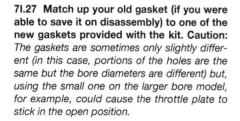

7I.27 Match up your old gasket (if you were able to save it on disassembly) to one of the new gaskets provided with the kit. Caution: *The gaskets are sometimes only slightly different (in this case, portions of the holes are the same but the bore diameters are different) but, using the small one on the larger bore model, for example, could cause the throttle plate to stick in the open position.*

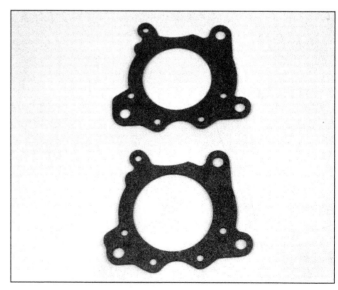

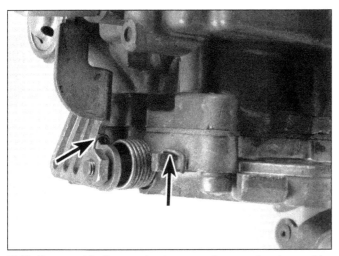

71.28 Reassemble the accelerator pump lever and spring and be sure the ends of the spring are in place as shown.

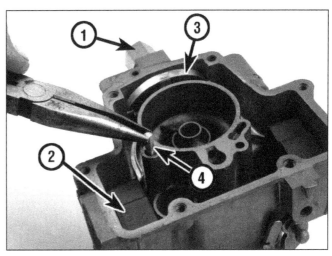

71.29 Reassemble the main body components:

1   Needle and seat assembly, with new gasket
2   Float and float pin
3   Float pin retainer
4   Check ball and weight

71.30 Match the new main body gasket with the old one to be sure all the necessary passages are open.

71.31 When reassembling the carburetor be sure to hook the choke diaphragm linkage in the proper slots as the parts are being put together. It can't be installed after the after the components are assembled.

Preliminary adjustments

Once the reassembly is complete, perform the following preliminary adjustments, then reinstall the carburetor (see Chapter 6) and make the on-vehicle adjustments. **Note:** *There may be other adjustments that apply to your particular carburetor model. Refer to the instruction sheet furnished with the overhaul kit.*

Fast idle adjustment

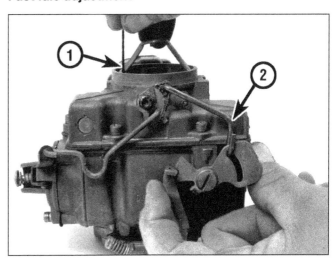

71.32 Hold the throttle lever against the second step of the fast idle cam and, using a drill bit of the specified size, measure the opening between the upper edge of the choke plate and the air horn (1). If your carburetor has a 1-7/16 inch throttle bore, the clearance should be 1/16-inch (use that size drill bit). If your carburetor has a 1-11/16 inch throttle bore, the clearance should be 5/64-inch (check the specifications sheet furnished with the overhaul kit for the exact setting on your model). Bend the linkage (2) to provide the correct clearance, if necessary.

Vacuum pulldown adjustment

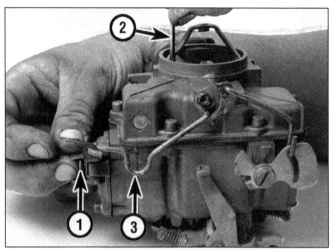

7I.33 Press on the choke modulator diaphragm spring (1) until it seats. Measure the opening between the upper edge of the choke plate and the air horn (2) using a drill bit of the specified size. If your carburetor has a 1-7/16 inch throttle bore, the clearance should be 1/8-inch. If your carburetor has a 1-11/16 inch throttle bore, the clearance should be 5/32-inch (check specifications sheet furnished with the overhaul kit for the exact setting on your model). Bend the linkage (3) if adjustment is necessary.

Idle mixture screw initial adjustment

7I.35 Make the initial idle mixture screw setting at this time. The idle mixture screw (arrow) has an initial setting between 1-1/2 to 2 turns out from the lightly seated position. This will allow the engine to idle, then the final setting needs to done while the engine is running at normal operating temperature.

On-vehicle adjustments

After the carburetor is reinstalled and the fuel line, vacuum hoses and electrical connectors (if equipped) have been connected, there are three on-vehicle adjustments to be made. Hook up a tachometer in accordance with the manufacturer's instructions, start the engine and allow it to reach normal operating temperature. It may be necessary to turn the idle speed screw in to maintain sufficient rpm so the engine doesn't stall.

Accelerator pump adjustment

7I.34 Hold the throttle in the closed position. Measure the distance from the casting (1) to the center of the hole in the pump rod (2). Bend the linkage (3) by squeezing or spreading the linkage to get the proper pump stroke (refer to the specifications furnished with the overhaul kit for the exact setting, but 27/32-inch is a good approximate setting)

Warning: *Be sure the vehicle is parked on a level surface, the parking brake is set, the wheels are blocked and the transmission is in Neutral (manual) or Park (automatic).*

Idle mixture adjustment

After the engine has reached normal operating temperature, turn the idle speed adjusting screw out as far as possible without the engine stalling, since the throttle plate must be closed when adjusting the idle mixture.

Turn the mixture adjusting screw clockwise until the idle speed drops a noticeable amount. Now slowly turn the screw out until the maximum rpm is reached. **Note:** *If it was necessary to turn the idle speed screw in (to keep the engine running) before setting the idle mixture, turn the idle speed screw out as far as possible without the engine stalling, then perform the mixture adjustment procedure again. This will ensure that the idle mixture is set with the throttle plate fully closed, so the engine is drawing the fuel mixture only from the idle circuit.*

Idle speed adjustment

With the idle mixture properly adjusted, the curb idle speed can now be set. Make sure the idle speed adjusting screw is set on the lowest portion of the fast idle cam. Turn the screw clockwise to increase the idle speed and counterclockwise to decrease it. Set the idle speed to the curb idle speed listed on the sheet furnished with the overhaul kit. If the specification isn't available, set the idle speed to approximately 700 rpm in Neutral or Park.

Fast idle speed adjustment

With the transmission in Park or Neutral, and the parking brake set, place the throttle lever on the second step of the fast idle cam **(see illustration 7I.32)**. Start the engine and check the engine rpm. If adjustment is necessary, insert a screwdriver in the slot on the throttle arm and bend it, as necessary, to increase or decrease the fast idle rpm.

7 Part J
Overhaul and adjustments

One-barrel models 1920/1904/1908/1960

This group of carburetors includes a variety of models all based on a similar design. The models included in this group are the one-barrel models 1920, 1904, 1908 and 1960.

The overhaul procedure is covered through a sequence of photographs laid out in order from disassembly to reassembly and finally installation and adjustment on the vehicle. The captions presented with the photographs will walk you through the entire procedure, one component at a time.

The following illustration sequence will use model 1920. Specific details will vary with carburetor part numbers because each carburetor is made to fit a specific application. Refer to the instruction sheet that comes with each overhaul kit. **Note:** *Because of the large number of different models and variations of each model, it is not always possible to cover all the differences in components during the overhaul procedure. Overhaul Chapters only deal with disassembly, reassembly and basic adjustments of a typical carburetor in that model group. If explanation about individual differences is necessary, refer to the exploded views at the beginning of this Chapter or refer to Chapter 3.*

Warnings:

Gasoline

Gasoline is extremely flammable, so take extra precautions when you work on any part of the fuel system. Don't smoke or allow open flames or bare light bulbs near the work area, and don't work in a garage where a natural gas-type appliance (such as a water heater or clothes dryer) with a pilot light is present. A spark caused by an electrical short circuit, by two metal surfaces striking each other, or even by static electricity built up in your body, under certain conditions, can ignite gasoline vapors. Also, never risk spilling fuel on a hot engine or exhaust component. If you spill any fuel on your skin, rinse it off immediately with soap and water.

Battery

Always disconnect the battery ground (-) cable at the battery before working on any part of the fuel or electrical system.

Fire extinguisher

We strongly recommend that a fire extinguisher suitable for use on fuel and electrical fires be kept handy in the garage or workshop at all times. Never try to extinguish a fuel or electrical fire with water. Post the phone number for the nearest fire department in a conspicuous location near the phone.

Compressed air

When cleaning carburetor parts, especially when using compressed air, be very careful to spray away from yourself. Eye protection should be worn to avoid the possibility of getting any chemicals or debris into your eyes.

Lead poisoning

Avoid the possibility of lead poisoning. Never use your mouth to blow directly into any carburetor component. Small amounts of Tetraethyl lead (a lead compound) become deposited on the carburetor over a period of time and could lead to serious lead poisoning. Check passages with compressed air and a fine-tipped blow gun or place a small-diameter tube to the component and blow through the tube to be sure all necessary passages are open.

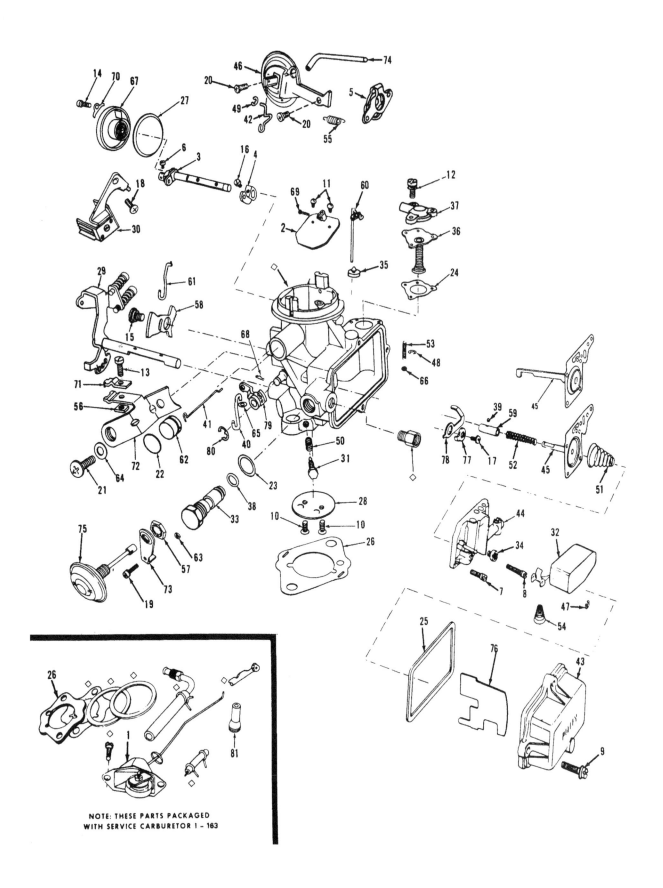

Model 1920 exploded view (typical)

Index Number	Part Name	Index Number	Part Name
1	Automatic Choke Assembly	41	Choke Piston Link
2	Choke Plate Assembly	42	Choke Diaphragm Link
3	Choke Shaft & Lever Assy.	43	Float Bowl
4	Choke Lever	44	Main Well & Econ. Body &
5	Choke Control Lever		Plugs Assy.
6	Choke Control Lev. Swivel Scr.	45	Pump Diaphragm & Rod Assy.
7	Main Well & Economizer Body	46	Choke Diaphragm Assy.
	Scr. & L.W. (Short)		Complete
8	Main Well & Economizer Body	47	Float Spring Retainer
	Scr. & L.W. (Long)	48	Mechanical Vent Spring Retainer
9	Fuel Bowl Scr. & L.W.	49	Choke Diaphragm Link Retainer
10	Throt. Plate Scr. & L.W.	50	Idle Adjusting Needle Spring
11	Choke Plate Scr. & L.W.	51	Pump Return Spring
12	Econ. Body Scr. & L.W.	52	Pump Operating Spring
13	Wire Clamp Scr. & L.W.	53	Mechanical Vent Spring
14	Therm. Cover Clamp Scr. & L.W.	54	Float Spring
15	Fast Idle Cam Screw	55	Choke Lever Spring
16	Choke Lever Screw	56	Wire Clamp Screw Nut
		57	Dashpot Nut
		58	Fast Idle Cam
17	Mechanical Vent Scr. & L.W.	59	Pump Push Rod Sleeve
18	Throt. Adapter Scr. & L.W.	60	Mechanical Vent Rod
19	Dashpot Bracket Screw	61	Fast Idle Rod
20	Choke Brkt. Retainer Screw	62	Choke Piston Assy.
21	Wire Bracket Screw	63	Dashpot Bracket Lock Washer
22	Choke Piston Plug	64	Wire Brkt. Scr. L.W.
23	Fuel Inlet Seat Gasket	65	Pump Operating Link Washer
24	Economizer Body Gasket	66	Mechanical Vent Spring Washer
25	Float Bowl Gasket	67	Choke Therm. & Cap Assy.
26	Flange Gasket	68	Pump Operating Link Retainer
27	Therm. Housing Gasket	69	Choke Piston Link Retainer Ping
28	Throttle Plate	70	Choke Therm. Cover Clamp
29	Throt. Lever & Shaft Assy.	71	Wire Clamp
30	Throt. Lever Drive Assy.	72	Wire Bracket
31	Idle Adjusting Needle	73	Dashpot Bracket
32	Float & Lever Assy.	74	Choke Vacuum Hose
33	Fuel Inlet Needle & Seat Assy.	75	Dashpot Assembly
34	Main Jet	76	Fuel Inlet Baffle Assy.
35	Mechanical Vent Valve	77	Mechanical Vent Control Lever
36	Economizer Stem Assembly	78	Mechanical Vent Operating Lever
37	Economizer Body Cover Assy.	79	Pump Operating Lever
38	Fuel Inlet Seat "O" Ring Seal	80	Pump Operating Lever Retainer
39	Diaphragm Push Rod Sleeve Ball	81	Throttle Lever Bushing
40	Pump Operating Link		

Parts list for model 1920 exploded view

Disassembly

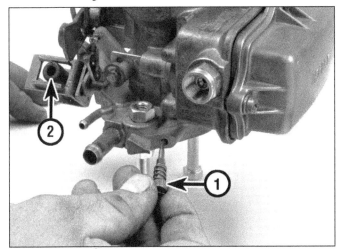

7J.1 Remove the idle mixture screw (1) and the rubber insert for the throttle linkage (2), if equipped. Inspect the idle mixture screw. The tip should be evenly tapered and smooth. If not, replace it with a new idle mixture screw.

7J.2 Remove the four screws holding the fuel bowl to the main body of the carburetor

7J.3 Remove the fuel bowl and gasket. Then lift out the baffle (arrow) from the bowl.

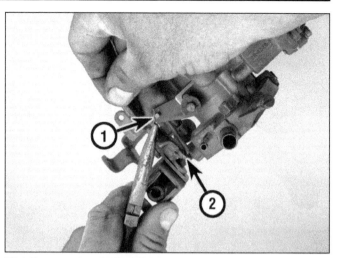

7J.4 Remove the clip from the accelerator pump lever (1). Slip the linkage rod from the operating lever. Before removing the linkage rod from the other connection (2), make a note of which hole the rod is installed in (there are three holes).

7J.5 Remove the needle-and-seat assembly (arrow)

7J.6 Remove the C-clip on the float assembly pivot pin.

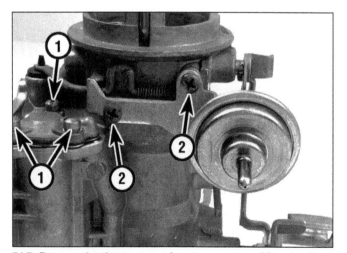

7J.7 Remove the three economizer cover screws (1) and pull out the economizer valve assembly. Note: *The economizer valve is sometimes called a power valve.* Next, take out the two screws for the choke vacuum diaphragm (2) and remove the diaphragm and mounting bracket. Note: *Some models have an on-carburetor choke housing. Before removing any of the parts, mark the location of the cover to the housing so you can set the choke correctly during assembly.*

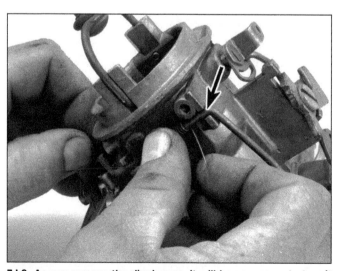

7J.8 As you remove the diaphragm, it will be necessary to turn it slightly to unhook the linkage from the choke operating lever. The bowl vent rod and spring sit behind the mounting plate for the choke and can now be removed.

7J.9 Using the proper-size screwdriver, remove the main metering jet from the economizer body assembly

7J.10 Remove the five screws . . .

7J.11 . . . and, while holding the accelerator pump in it's fully extended position (left hand), remove the body from the fuel bowl. Be sure to catch the pump return spring between the diaphragm and the metering body as it's being disassembled. **Note:** *There is also a small spring on the pump diaphragm rod that needs to be placed on the new pump diaphragm before reassembly.*

7J.12 After the accelerator pump is removed, remove the clip and the operating lever (if equipped) from the pivot pin (early style shown, later models are similar).

Cleaning

Refer to Chapter 7A for information on "dips" and solvents used in cleaning. Remember not to place the float or any other plastic or rubber parts into a dip tank, since the chemicals will damage the parts. After cleaning (which usually requires an overnight soak in a dip tank), blow out all passages in the carburetor body with compressed air. **Warning:** *Wear eye protection!* Inspect the cleaned parts to be sure there's no gasket material remaining on any surfaces and the surfaces are not damaged or warped. Inspect the float, as shown in illustration 7A.14. Make sure all parts are completely dry before beginning reassembly.

7J.13 These two screws (arrows) secure the throttle plate to the throttle shaft. Under normal circumstances, you **DON'T** need to remove them during overhaul. The screws are staked in place and must be re-staked on installation; if this is not done correctly, the screws could vibrate loose and fall into the engine! For this reason, if the throttle shaft or throttle body is worn or damaged, we recommend obtaining a new throttle body assembly or taking the throttle body to a reputable carburetor rebuilder for repair.

Reassembly

Several assembly checks, tips and adjustments are shown in the following illustration sequence, but, unless specifically detailed here, reassembly is in the reverse order of disassembly. Read the special instructions in the carburetor kit and follow the illustration sequence in making adjustments.

7J.14 Place the pump operating spring on the pump rod and reinstall the diaphragm and gasket (arrow) into the main body of the carburetor

7J.16 Install the economizer valve parts into the float bowl . . .

7J.18 Reinstall the float and spring in the direction shown here. Install the C-clip (arrow) on the pivot pin.

7J.15 If your model has an operating lever like this, make sure the accelerator pump arm engages in the operating lever pocket (arrow)

7J.17 . . . and tighten down the three screws

Tip:
Use a little white grease (arrow) on the threads of the screws during reassembly; this will keep the screws from binding on the rubber gasket (and possibly tearing the gasket) as they are tightened.

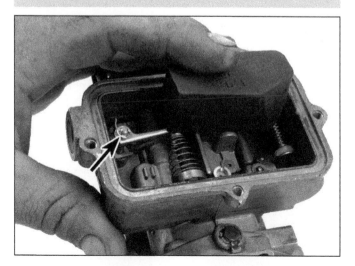

7J.19 When installing the needle-and-seat assembly into the bowl, rotate the carburetor so the fuel inlet side is facing down. If you don't do this, there's a chance the gasket (1) can fall out of place and not sit evenly around the opening. Also, note the positions of the needle tip (2) and float adjustment tang (3). With the seat assembly tightened, these two will be in contact with each other.

Float Adjustment

7J.20 Turn the carburetor upside down to adjust the float level. Measure the distance between the float and the fuel bowl (a drill bit of the correct size works well). If adjustment is necessary, gently bend the tang to get the correct specification (listed on the specification sheet that comes with the overhaul kit).

Idle mixture screw initial adjustment

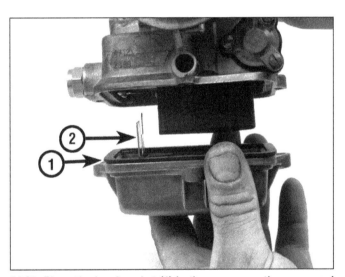

7J.21 Place the bowl gasket (1) in the groove on the cover and install the baffle (2), then install the fuel bowl on the carburetor body. **Note:** *Holding the carburetor with the fuel bowl down will make this operation easier.*

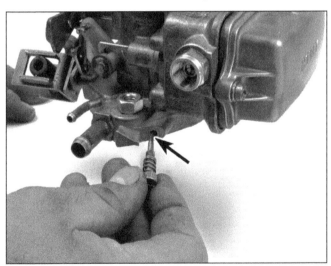

7J.22 Reinstall the idle mixture screw by turning it in until it seats lightly, then backing it out 1-1/2 to 2 turns. The vehicle will run on this setting, but final adjustment will need to be made while the engine is running.

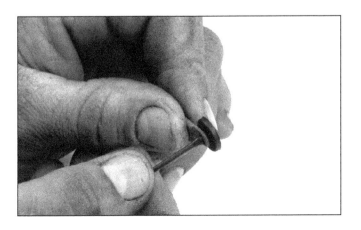

7J.23 Install a new vent seal from the overhaul kit

7J.24 Reinstall the vent rod and spring (1) and reattach the diaphragm assembly (2). Note: *Be certain the hook on the end of the spring is over the rod, as shown here, so the vent will close properly*

Vacuum kick adjustment

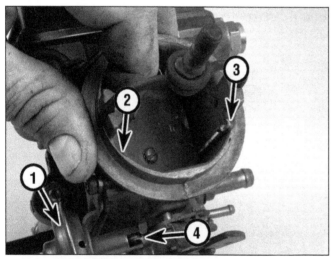

7J.25 Apply vacuum to the diaphragm (1); this will pull the choke plate towards the closed position. Using a drill bit of the correct size (see the specifications in the overhaul kit), measure the gap between the plate and the air horn, either at the front edge (2) or the back edge (3) of the plate. If the measurement is not correct, bend the rod (4) to adjust the gap.

Automatic choke setting

Most of these models use a "divorced" choke with a choke coil that's mounted to the intake manifold; however, some later models are equipped with a choke housing on the carburetor. Refer to the procedure at the end of Chapter 7B to set the choke.

Fast idle setting

7J.26 Before installing the carburetor on the vehicle, open the throttle and position the fast-idle adjustment screw on the first step of the fast-idle cam (1), then close the throttle. Using a 1/32-inch drill bit, measure the clearance between the front of the throttle plate and the throttle body bore. If the throttle is not open about 1/32-inch, adjust the opening with the fast-idle screw. After the carburetor has been installed, all other adjustments have been made and the engine is running at normal operating temperature, open the throttle and place the adjustment screw back on the first step of the fast-idle cam. Check the idle speed, which should be approximately 1700 rpm. If the speed is not correct, adjust it with the screw. If you notice the engine idling slowly and "chugging" when the engine is cold and the choke is fully applied, try turning the screw clockwise a bit to increase the fast-idle speed. If the idle speed seems too fast with the choke applied, turn the screw counterclockwise slightly. Note the location of the curb idle speed screw (2) that you'll be using later to adjust the curb idle speed.

On-vehicle adjustments

The procedures for adjusting the idle speed and idle mixture on the vehicle are the same as those described in Chapter 7B. Refer to the procedures in Chapter 7B, but note that the curb idle speed screw and idle mixture screw are in different locations **(see illustrations 7J.22 and 7J.26).**

8 Selection and modification

Finding the right carburetor

Carburetor selection

Matching the carburetor to the engine and application is critical for performance and/or economy. Many hot rodders like the look of a huge carburetor, or carburetors, on their engine. They fall into the trap of "bigger is better" or "if a little is good, more is better". This might hold true for cubic inches, but it almost never should apply to carburetion.

If an engine is over-carbureted, it will have poor throttle response, bog and hesitate at low speeds and won't start to run well until very high rpm. Fuel economy and emissions will be poor.

Larger displacement engines and engines that run at high rpm need larger capacity carburetors than smaller engines running at lower speeds. The most important factors in carburetor size selection are engine displacement, maximum rpm and volumetric efficiency.

Volumetric efficiency

Volumetric efficiency is a measure of the engine's ability to fill the cylinder completely, and is given as a percentage. For example, a 100 cubic inch engine that gets 80 cubic inches of air/fuel mix into the combustion chamber on each intake stroke has a volumetric efficiency of 80-percent.

A volumetric efficiency of about 80 to 85-percent is generally what an average, well built, four-barrel equipped high-performance street engine will provide. Stock engines achieve approximately 70 to 75-percent volumetric efficiency. You must decide what rpm range you want your engine to run best at.

When selecting a carburetor, use the following formula or chart to estimate the required cfm rating **(see illustration)**. Round off all results to the nearest carburetor size. Be realistic; you're only hurting the end result of all your hard work by overestimating.

Put in your specific numbers for cubic inches and RPM range of the engine. The remaining number in the equation

		Engine RPM					
		4000	4500	5000	5500	6000	6500
Engine displacement (cu. in.)	**250**	245	260	290	320	350	380
	275	255	290	320	350	380	420
	300	280	315	350	380	420	450
	325	300	340	380	415	450	490
	350	325	365	405	445	490	525
	375	350	390	435	480	520	565
	400	370	420	465	510	555	600
	425	400	450	500	550	600	650
	450	420	470	520	580	625	700

8.1 As a general rule, high performance small-block engines will need 400 to 600 cfm carburetors, and big-block engines will need 600 to 800 cfm carburetors (exact size will depend upon the actual engine modifications that have been made). Note: *The table shown here is for stock street engines. For high performance models increase the CFM rating about ten percent.*

8.2 When a carburetor is subjected to heat over a period of time the heat has a tendency to warp the individual components. This can create a vacuum or fuel leak that just tightening the bolts and screws cannot correct. The arrows indicate the areas that should be milled in order to make flat surfaces

are predetermined except for the volumetric efficiency percentage (VE). For the chart shown, a figure of 80% VE is used but for an engine that is more highly modified a figure of 85% will be closer to your actual cfm needs.

The formula to use in figuring your own requirement (cfm) is:

(cubic inches ÷ 2) x (maximum rpm ÷ 1728) x VE = cfm

Example:

You have a 350 cubic inch engine that will reach a maximum of 7000 rpm. Let's also say you've built an engine that's well thought out with the right combination of parts, so you're VE is roughly 85%, compared to the 80% used for milder engines. Plug these numbers into the formula:

(350 ÷ 2) x (7000 rpm ÷ 1728) x 85% = 602 cfm

So you would need approximately a 600 cfm carburetor to handle the needs of the engine. **Note:** *As a general rule, high-performance small-block engines will need 400 to 600 cfm carburetors and big-block engines will need between 600 to 800 cfm, depending on actual displacement and level of modification. Smaller carburetors generally give better throttle response but fall off slightly in power at top end.*

Carburetor modifications

Most carburetors have seven basic operating systems:

> *Fuel inlet/float circuit*
> *Idle circuit*
> *Transformer/off idle circuit*
> *High speed/main metering circuit*
> *Power enrichment circuit*
> *Accelerator pump circuit*
> *Choke circuit*

The main purpose for modification of one or more of these systems is to increase engine performance.

When engine and other vehicle modifications for performance applications are performed, some carburetor operating

8.3 When you see the base idle screw adjusted all the way in you probably have a good indication of an idle adjustment problem . . .

characteristics change. These changes necessitate "dialing-in" the carburetor to achieve peak engine performance.

Anyone having a good basic working knowledge of carburetor systems can "dial-in" or modify a carburetor. The purpose of this section is to explain and illustrate some of the available parts and methods for modifying carburetor operating characteristics to achieve the best possible performance. Complete system operations are covered in Chapter 3. Description, disassembly, reassembly, and adjustment procedures are covered in Chapter 7. Specifications are listed in Chapter 9.

All carburetor tuning begins with the disassembly and inspection of the carburetor. This serves to familiarize yourself with the systems and components of the unit. **Note:** *Modifying or tuning a carburetor is usually a trial and error process. If you anticipate disassembling your carburetor frequently, it's a good idea to use special rubber or teflon gaskets, which can be reused many times before they are replaced. Check with your local speed shop or auto parts store regarding the availability of these gaskets.*

Main body

If the carburetor has been in service for any length of time, you should inspect the flat surfaces for warpage. Anything in excess of 0.010-inch warpage in the areas indicated **(see illustration)** will require resurfacing to insure a proper gasket seal. We recommend draw filing using a medium-fine flat file to true the surface. Except in extreme cases, it is usually not necessary to have the surface machined.

Throttle body

Many engines that have been modified for higher performance seem to have a problem maintaining a low, smooth idle. The main reason for this condition is the selection of the camshaft. The more "extreme" the camshaft profile, the larger the lift and duration, the harder it is for the engine to produce vacuum. This makes it necessary to raise the idle speed to compensate. Sometimes you'll find the base idle screw turned in as far as it can go **(see illustration)** and still the car has a hard time idling.

Because the vacuum is lower and the idle rpm requirement is higher, the throttle plates must be opened for addi-

8.4 . . . the gap shown here by the arrow is what happens to the throttle plates when the screw is turned as shown in the previous illustration. The gap between the throttle plates and the bore . . .

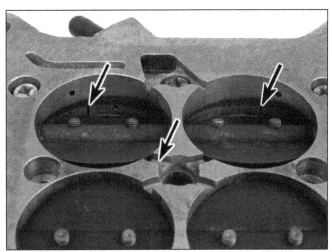

8.5 . . . exposes too much of the transfer slots (arrows) - when that happens you loose the ability to get adjustment from the idle mixture screws

8.6 The holes (upper arrows) drilled in the primary throttle plates will allow you to back the base idle screw out to a normal position and you will regain control of the idle mixture. The lower arrow points to the secondary throttle stop screw

8.7 There are many different power valves available to fit the range of manifold vacuum readings that you may get from your specific engine. Choose the power valve that is approximately 1 to 1-1/2 inches of vacuum lower than the reading on your vacuum gauge. (The one on the right isn't a power valve, but a plug to take the place of a power valve in some applications)

tional airflow (see illustration). Unfortunately, since transfer slot exposure is increased, fuel flow is also increased, which usually makes the engine run too rich and makes adjustment difficult. The idle mixture screws are no longer useful because as the throttle plates are opened and you lose vacuum at the idle ports (see illustration) and the ability of the idle mixture screws to meter fuel is lost. One way to close the throttle plates back down to partially cover the slots and still provide the additional airflow that is required, is to drill a 3/32-inch hole in each primary throttle plate (see illustration). Note: If the engine is radically modified, it may be necessary to drill the holes in the secondaries as well.

In some cases, idle airflow can be increased without drilling holes in the throttle plates by simply increasing the secondary airflow. To do this, turn the secondary throttle stop screw concealed in the throttle body casting (see illustration 8.6) clockwise about a turn and a half. This opens up the secondary throttle plates, resulting in increased airflow without a significant increase in fuel flow. Essentially, it will spread the airflow and give a better balance between the primary and sec-

ondary throttle bores. As a result, the primary throttle plates can then be lowered down into the bore, putting them closer to the proper relationship with the transfer slot, which improves idle mixture screw adjustability.

In some applications where the vacuum is unusually low, drilling holes in the primary throttle plates may still be necessary. The proper idle setting should expose 0.045 to 0.060-inch of the idle transfer slot below the bottom edge of the plate.

Metering block

Proper idling characteristics also require fuel metering matched to the vacuum values your engine produces. Correct power valve timing aids in obtaining a "clean" idle and part throttle. The opening point of the power valve (see illustration) should be 1 to 1-1/2 inches less than manifold vacuum at idle. With the camshafts presently in use, a number 65 (or 6.5 inch) power valve is usually sufficient to maintain proper control.

8.8 There is a wide range of main metering jets available for all Holley models. One of the easiest ways to keep all of them organized and undamaged is to get a pre-drilled and threaded holder from your local auto parts store or speed shop. The tool on the left is a special driver just for jet removal and installation. It captures the jet in the end of the tool so you can get it into or out of tight places without damaging the jet

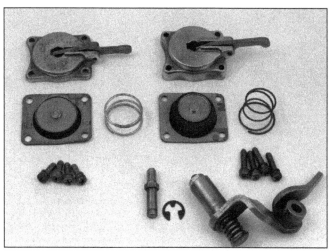

8.10 Here is an example of the two different capacity accelerator pumps. The one on the left is a standard 30 cc pump while the one on the right is the 50 cc setup, including pump cover and screws, diaphragm and spring, lever assembly and mounting stud. Note: *Some applications will require a 1/4-inch spacer to get enough clearance to operate properly*

Note: *If you have an extremely high-performance camshaft it may be necessary to go all the way to a 3.5 inch power valve (first you'll have to accurately measure the manifold vacuum at idle on your engine).*

The power valve is designed to supplement the fuel flow through the main jets. During acceleration, low manifold vacuum allows the power valve to open. Fuel flow through the power valve and channel restrictions effectively increases fuel discharge in the main well approximately 6 to 10 jet numbers during the time it's operating.

The main system in the metering block incorporates the main jet, main well, and the booster venturi. Performance carburetors are designed with more than adequate capacity in the main system passages and therefore do not require modifica-

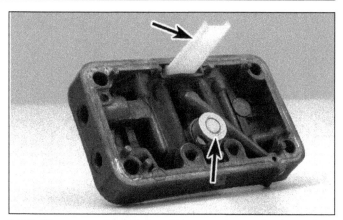

8.9 Baffles are used mostly in racing or marine applications where the fuel sloshing around from cornering or rough water can cause a flooding over condition or a lean condition from fuel moving away from the power valve and opening to nothing but air

tion. The main jets that come with the carburetors are generally in the ballpark for most applications. Of course, this varies with individual applications. There is usually no need to modify bleeds, since this changes the system balance and booster venturi pull-over point, which can cause driveability problems. Consequently, any air/fuel ratio changes in the main system should be made with the jets **(see illustration)**.

Another modification that will improve performance, even though it doesn't directly control fuel flow, is through the use of two different types of baffles **(see illustration)**. As fuel sloshes around in the fuel bowl during acceleration, braking and cornering, fuel can either flood over at vent areas or cavitate at the power valve. Neither of these is a desirable situation. Installation of these baffles can offset both of these conditions.

Float bowl

There are no modifications to make to the actual float or bowl components. But fuel pressure will affect how the whole system operates.

The inlet pressure should be between 7 and 8 psi at idle. Anything in excess of that might have a tendency to create flooding problems by overpowering the needle and seat assembly. Another precaution when initially setting up the carburetor on the vehicle is to set the float level with the engine running. It's recommend that you use Viton needles and seats unless the type of fuel being used is one of the modern racing fuels where octane levels are increased considerably or when octane boosters are used in proportions in excess of the manufacturer's suggested levels. In some cases, the use of these fuels will create problems by attacking the Viton on the needle.

Accelerator pump(s) and controls

In the muscle car era, cars, trucks or boats weighed much more than they do now. The more mass you have to move or the quicker you want to move it the more fuel it's going to take. There are three areas in the accelerator pump circuit that can be altered to fit your desired performance needs.

The first option you have is to use a 50 cc accelerator pump vs. the standard 30 cc pump **(see illustration)** or possibly two pumps if your application needs that much fuel on acceleration.

The second is to change the shape of the plastic cam **(see illustration)** that determines the type of accelerator pump shot that is delivered. The height and shapes of the curved upper

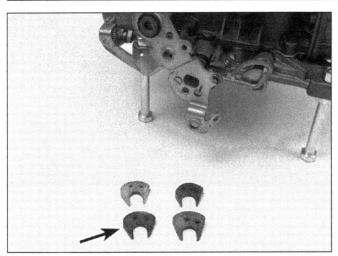

8.11 Another way to modify the way the accelerator pump delivers the fuel shot is to change the plastic cam that operates the lever. As you can see the shapes (profiles) are all slightly different and will control fuel discharge differently

8.12 There can be either one or two of these discharge nozzles in your carburetor, depending on the model. They're available in a variety of sizes and tip types. The numbering system is similar to the one used on the metering jets, a number stamped in the side indicating orifice size

edge of the cam can deliver more fuel sooner or later and for longer or shorter bursts during the stroke of the accelerator pump.

The third is to use different styles of fuel discharge nozzles, sometimes referred to as "squirters" or "shooters" **(see illustration)**. These have a numbering system, similar to main metering jets, stamped on the side of the nozzle. They also have different types of tips on the nozzles. The performance versions are longer and discharge the fuel out further into the air stream in the center of the venturi. While the standard shooters have no extending tips from the body of the shooter.

The accelerator pump shooter size has a direct affect on the initial off-line or "launch" performance. If the initial acceleration produces a hesitation and then picks up, the pump shooter size should be increased. On the other hand, if the pump shooter is too large, it may cause a bog or sluggish response from too much fuel. Another indication that the shooters are too large is a puff of black smoke on acceleration.

You can come close to the correct size during assembly by using the recommended original sizes that Holley assembled the carburetor with. But you'll probably need to fine tune your vehicle by some trial and error to get it exactly right.

> **Tip:**
> If the vehicle has an automatic transmission you should increase the size of the shooter over the original size. Also be sure that as the shooter size is increased, especially up over a 0.042-inch shooter, the hollow accelerator pump discharge nozzle screw should also be used to ensure that there'll be enough fuel flow.

8.13 On models with mechanical secondaries, the adjustment for the secondary throttle plates is made by bending the linkage in the middle (arrow). Be sure the secondary plates open fully (perpendicular to the throttle body)

8.14 The modification for the vacuum secondary opening speed is very simple, since the entire vacuum secondary control assembly can be removed or just the top can be taken off to change the diaphragm control spring

Mechanical and vacuum secondaries

The two choices you get for secondary throttle plate operation are mechanical and vacuum.

All "double-pumpers" are equipped with mechanically actuated secondaries. There isn't much to modify with this type of secondary operation, except to check for the full opening of the secondary throttle plates at wide-open throttle. If the plates

don't open completely, slightly bend the secondary link **(see illustration)** in the center of the linkage to allow the throttle plates to fully open.

Vacuum secondary operation on the other hand can be modified for a wide range of performance. The vacuum secondary diaphragm **(see illustration)** is the best bet for all

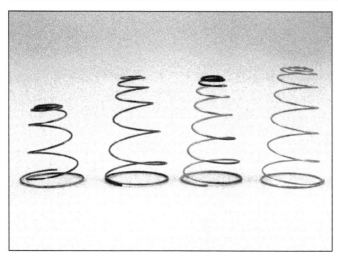

8.15 Here are four different springs for the vacuum secondary diaphragm. There are actually seven different spring rates available. Each spring is designated by color. The colors for the spring rating and there relative load are as follows:

White - lightest   *Plain (no color) - medium*
Pale Yellow - lighter   *Brown - medium heavy*
Darker yellow - light   *Black - heavy*
Purple - medium light

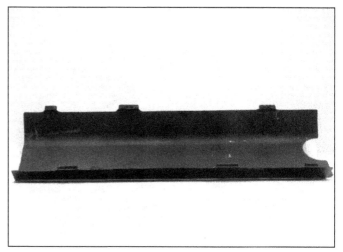

8.16 Some types of manifolds don't allow the installation of a windage tray, but most do. The advantage to these is that they keep hot oil from contacting the bottom of the manifold. The cooler you can keep the fuel mixture, the more power your engine will make

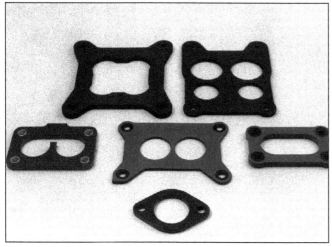

8.17 These are a few of the different types of insulator gaskets found in the overhaul kits used on all models of Holley carburetors

8.18 Some of the insulating spacers now available are made of injection-molded plastic and come in varying thicknesses. You can find insulating spacers for both square bore and spread bore applications

around driveability and versatility. It can be adjusted for stock vs. hi-performance, manual vs. automatic transmission and a heavy vs. light vehicle. The way to accommodate all these different variables is by the use of a spring which controls the timing and speed at which the secondary throttle plates are allowed to open. A kit is available from Holley that will give you several choices, from a very lightweight, low resistance spring to a stiffer, slower opening spring **(see illustration)**. Here again, as in previous modifications, trial and error is still the best way to fine tune your vehicle to reach its maximum performance.

Heat

While heat isn't considered a modification, if it gets out of control it will certainly modify the way the carburetor performs. The ability to cool the fuel mixture has a direct impact on the way the vehicle will run. The cooler the fuel mixture, the more

power that can be made from that mixture, since it will be more dense.

There are several ways to keep the carburetor and fuel isolated or insulated from the heat of the engine.

First, if you had the intake manifold off, before reinstalling the intake manifold check to see if the manifold will allow the installation of a windage tray between the heads **(see illustration)** in order to keep hot oil from splashing against the bottom of the manifold. This will keep the air/fuel mixture cooler just before it enters the combustion chamber.

Second, there are a number of different carburetor base gaskets and spacers available to be used when installing the carburetor on the intake manifold **(see illustrations)**. These are made of different types of heat insulating materials, from paper to a phenolic composite. If space allows the installation of any of these, by all means use them to keep the carburetor (and ultimately the fuel mixture) cooler.

Third, there are heat shields available from the manufac-

8.19 This type of insulator acts more as a cooling "fin" rather than an insulator. It's made from aluminum so it absorbs heat quickly and spreads it over the large surface area where the air passing over the exposed surface can dissipate the heat before it can all get up to the carburetor

8.20 One of the most complete sources of parts for improving the performance of your Holley carburetor comes in the Holley overhaul "Trick Kit." Even though these parts can be purchased separately, this overhaul kit will give you more modification possibilities than you may ever need

turer or from many aftermarket sources that are installed between the manifold and carburetor **(see illustration)** that are made of aluminum and act as a heat sink. The heat from the manifold enters the shield before it reaches the carburetor and, because of the large surface area, dissipates the heat quickly.

Last, but one of the most overlooked causes of unnecessarily preheating the fuel is the routing of the fuel lines and filters. Many people go to great lengths to buy all the right parts, and do all the "trick" stuff to their engine and overlook the detail work. Running fuel lines too close to headers or touching the engine block at some point creates a "hot spot," which if severe enough can even cause vapor lock.

Many of the parts used for modifications and performance tuning can be found in individual kit form or in one complete overhaul package from Holley called the "Trick Kit" **(see illustration)**. You can find these kits at most local parts stores. So do it yourself and learn why things really do or don't work, and in the process save yourself some money.

Remember, nothing is so small or simple that it should be overlooked or taken for granted. The details are everything - they distinguish a professional approach to doing a job versus the backyard "hack" repair that always seems to have to be done over and over again because of impatience and/or poor planning. Do it right and do it once.

Engine modifications

In these times of ever-increasing fuel costs, the beleaguered enthusiast is constantly reminded of the trade-offs required to drive a performance vehicle. It takes fuel to produce horsepower, and the faster you go, the more you need.

But performance and economy don't have to be mutually exclusive. By properly matching components and careful tuning, you can improve the power and efficiency of the drivetrain to obtain the best of both worlds.

Automotive designs are, by necessity, fraught with compromises. Factory engineers must allow for wide variations in production line tolerances, driving techniques, low octane fuel, carbon build-up, wear, emissions certification, neglect and

lack of maintenance while keeping costs down.

Stock production passenger cars and light trucks are designed for a balance between everyday stop-and-go driving around town and cruising on the highway. The engines and drivelines are optimized for low and mid-range power rather than high rpm peak horsepower.

Engines are basically air pumps that mix fuel and air and produce power from the combustion. Anything you can do to increase the flow of air (assuming the fuel system is capable of delivering sufficient fuel in the correct proportions) through the engine will increase power. Other ways to increase power and/or economy are to reduce weight, friction and drag.

Every engine is designed to operate most efficiently in a certain speed range (measured in revolutions per minute [rpm]). The length and diameter of the intake and exhaust ports and the intake and exhaust manifolds (or headers) help determine the power band of the engine. Long, small-diameter intake and exhaust runners improve low-rpm torque and decrease high-rpm power. Conversely, short, large cross-section passages favor high rpm power.

The type and rating of intake and exhaust systems, the camshaft, valve springs and lifters, ignition, cylinder heads, valve diameters and bore/stroke relationship are matched at the factory to ensure a good combination of economy, power, driveability and low emissions. Additionally, the transmission characteristics, differential gear ratio and tire diameter must all work in harmony with the engine.

For street driving, low and mid-range torque is much more useful (and economical) than theoretical ultimate horsepower at extremely high rpm. A street-driven engine that produces high torque over a wide range of rpm will deliver more average horsepower during acceleration through the gears than an engine that delivers higher peak power in a narrow range of rpm.

Heavy vehicles with relatively small engines should have lower gearing (higher numerically) than light vehicles with relatively large engines. Also, the engine in a heavy vehicle should be optimized for maximum torque in the low and middle rpm range, since it takes more torque to get a heavy vehicle moving and to accelerate it.

New cars and trucks have low numerical axle ratios, lock-up torque converters, overdrives and more ratios in the trans-

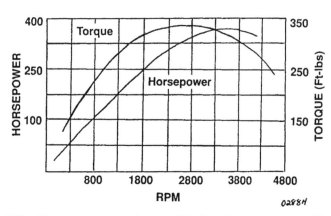

8.21 Note how torque drops off before horsepower on this typical engine

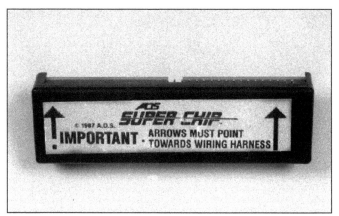

8.22 Special chips are available for computer equipped vehicles

missions to get good mileage and acceleration. One of the best ways to improve performance and economy at the same time on an older vehicle is to install a transmission with more forward gears and a differential with better ratios than stock. Frequently, these parts are available from late-model vehicles at wrecking yards for reasonable prices.

Most racing engines run in a narrow range at high rpm and don't need the flexibility low-speed torque provides. Many hot rodders succumb to the temptation of putting a radical racing camshaft or a huge carburetor on an otherwise stock engine. This increases the theoretical air flow capacity in one part of the engine without changing the flow characteristics of other components. Since the components aren't matched, intake air charge velocity slows down and fuel doesn't mix properly with the air. The engine no longer has an optimum rpm band. This results in a gas-guzzling slug that isn't as fast as stock.

Torque, usually measured in foot-pounds (ft-lbs) in the USA, is a measure of the twisting force produced by the engine. Horsepower is a measure of work done by the engine. Torque (in ft-lbs) multiplied by engine speed (rpm) divided by 5,250 equals horsepower.

Engines produce the most power from a given amount of fuel at their peak torque. This is the rpm the factory engine design is optimized for. Peak horsepower is achieved by spinning the engine faster than this most- efficient speed. The torque peak always occurs at a lower rpm than the peak horsepower **(see illustration)**. Horsepower peaks when the gains made by running faster are balanced by the losses caused by running above the optimum speed the engine components are tuned for.

You can tell a lot about an engine from its power specifications. On a high-performance engine, maximum horsepower will usually be higher than maximum torque, and peak power will occur at a relatively high rpm. Also, as a general rule, high-performance street engines put out approximately one horsepower per cubic inch or better. For example, a hypothetical standard engine might have 300 cubic inches of displacement, maximum torque of 275 ft-lbs @ 3,000 rpm and 200 hp @ 4,200 rpm. The high-performance version with the same displacement might have 245 ft-lbs @ 3,800 rpm and 325 hp @ 5,600 rpm. **Note**: *Late-model (about 1972 and later) engine horsepower ratings are net (with accessories connected) whereas earlier models are rated for gross or brake horsepower (without accessories). Net ratings tend to be lower, but more*

realistic than the gross ratings.

Before you select components to modify your vehicle, you should plan out realistically what you want it to do. Your engine must be in excellent condition to start with, otherwise it will quickly self-destruct. Check the condition of the engine thoroughly. If necessary, rebuild it now; you can include modifications during the overhaul and it will cost less than if you did the work separately.

Find out what the factory rated horsepower and torque are, and at what rpm the peaks occur. Then determine what rpm the engine is turning at your usual highway cruising speed and what the gearing is. If your vehicle isn't equipped with a tachometer, temporarily connect a test meter with long wires run into the interior. To determine the axle ratio, check the tag on the differential.

Once you know these things, you can decide which way to go. Generally, if you modify the vehicle so it has to run at higher rpm and/or you want a large increase in power, expect to sacrifice a considerable amount of fuel economy and reliability.

Some of the more popular ways to dramatically raise power output are supercharging, turbocharging, nitrous oxide injection or by swapping a much larger displacement engine into the vehicle. There are many entire books devoted to each of these methods, and these subjects are beyond the scope of this Chapter.

Depending on the year and model of your vehicle, you may be able to make substantial improvements in mileage through careful tuning, by changing axle ratios, tires, intake and exhaust modifications, camshaft replacement and ignition improvements. Vehicles from the 1950s through the 1970s are most responsive to these changes, which must be carefully planned and coordinated.

Newer computer-controlled models have many of these changes incorporated in them already and get better mileage than their predecessors. They are so sensitive to modifications that even a change in tire diameter can affect the driveability of the vehicle.

There are very few modifications that can be done to computer-controlled vehicles without making them violate emission laws. Several aftermarket manufacturers produce intake manifolds, camshafts, exhaust systems and computer chips **(see illustration)** that can increase the performance of late-model vehicles. Shop carefully and read the fine print to determine computer compatibility before you purchase any components.

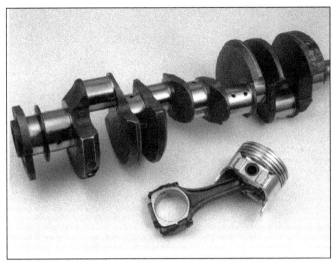

8.23 Long-stroke crankshafts and oversize pistons can boost torque and horsepower by increasing displacement

8.24 The camshaft plays a major role in determining the operating characteristics of the engine

If you are planning to rebuild your engine, you may want to make many modifications during the overhaul. While the engine is apart, you can easily change the cylinder heads, pistons, connecting rods, crankshaft and camshaft. Modified cylinder heads can provide substantial gains in high rpm power. For a mild street engine, a good three-angle valve job and matching the intake ports to the intake manifold will make it run better without sacrificing driveability. Older engines can benefit from the addition of hardened valve seats and special valves to enable them to run on low-lead or unleaded gasoline.

High-compression pistons improve power and efficiency at all speeds, but if you exceed about 9:1, premium fuel is necessary. Flat-top pistons produce a better flame front in the combustion chamber than dished (concave) ones. Forged pistons are stronger than cast pistons; however, cast pistons work fine on the street.

Longer stroke crankshafts with matching connecting rods and large-bore pistons **(see illustration)** can increase horsepower without sacrificing driveability or low-end torque. However, if you intend to build a high rpm engine, don't get carried away with this: long-stroke engines can limit high-rpm potential.

Before you assemble the engine, have an automotive machine shop blueprint and balance the parts; it's a way of finding extra horsepower that doesn't require more fuel.

In this Section we will discuss the pros and cons of various component changes and how they affect other parts of the vehicle. Usually, if you change one part, you'll have to modify or replace others that work in conjunction with it. Check the fuel mileage of your vehicle and measure performance carefully with a stop watch before and after each modification to determine its effect. Test before and after under the same conditions on the same roads to ensure accuracy.

Many books and articles have been written on how to turn your street vehicle into a race car by spending a fortune. Of course, that leaves you back at square one, because you don't have a vehicle to drive to work anymore. We will attempt to limit the discussion to modifications that can be done at home for a reasonable cost and that will not prevent the vehicle from being used as daily transportation.

Make sure any modifications are legal according to the latest federal, state and local laws. Many states conduct "smog"

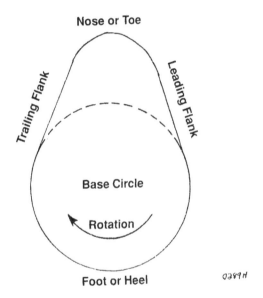

8.25 Each portion of the camshaft profile has a specific name and function

inspections for street-driven vehicles. Removing or modifying emission control systems is often illegal, and usually results in little or no improvements in power or economy. Besides, we all have to breathe the air. Many jurisdictions forbid loud exhausts, too. Be sure to check with your state department of motor vehicles and state or local police regarding current regulations.

Camshaft selection

The camshaft is the mechanical "brain" of the engine **(see illustration)**. It determines when and how fast the valves will open and close, and how long they stay open by pushing up valve lifters with elliptical (egg-shaped) lobes as it turns **(see illustration)**.

The camshaft, more than any other single part, determines the running characteristics (or "personality") of the engine. A single camshaft design can not provide maximum power from idle to redline. Like everything else in a motor vehicle, camshaft design is a compromise. If a camshaft is designed to produce gobs of low-end torque, good driveability and fuel economy, it must give up some peak horsepower at high rpm. Conversely,

8.26 Hydraulic lifters can be identified by the retainer clips in the top (arrow)

8.27 Aftermarket roller lifters

camshafts designed for high rpm run poorly at idle and lower engine speeds.

Before you can make an informed choice on camshaft selection, you need to know some basic design parameters and terminology. Camshafts must be designed to work with a certain type of valve lifter and must not be used with any other type. There are three basic types of valve lifters: mechanical, hydraulic and roller. Mechanical lifters, also known as solid or flat tappets, are the oldest, simplest and least expensive type. Because of their light weight, mechanical lifters allow an engine to rev slightly higher before the valves float. The main disadvantages of mechanical lifters are the necessity of frequent valve adjustments and the noise they produce.

Hydraulic lifters **(see illustration)** are the most common type used in V8 engines. They have a small internal chamber where engine oil collects and a check valve to prevent backflow. This feature allows the lifter to automatically compensate for differences in valve lash, or clearance. Standard hydraulic lifters are relatively inexpensive and are maintenance-free; however, at high rpm, they tend to pump up and float the valves. Special high-performance lifters are available which extend the rpm range high enough for virtually any street ma-

chine. Hydraulic lifters are the most popular type of lifter used in performance street engines, and work well in most applications.

Roller valve lifters **(see illustration)** are the best and most expensive type available. They increase horsepower and improve fuel economy by reducing friction. Roller lifters are available in both solid and hydraulic versions. If you can afford it, buy a roller camshaft and lifters. Hydraulics are next best and mechanical are least desirable for a street engine.

Every engine has a certain speed it runs best at, which is a result of the "tuning" of components to achieve an optimum flow velocity of air/fuel mixture.

The main reason engines don't run at maximum efficiency throughout the power band is because air has mass, and therefore inertia. As engine speed increases, the amount of time available for the gases to enter and exit the combustion chamber becomes less. Camshaft designers compensate for this by opening the valves earlier and holding them open longer. But the valve timing that works well at low speed is inefficient at high speed.

Several quantifiable factors determine camshaft characteristics. The most important and widely advertised items are LIFT, DURATION and OVERLAP **(see illustration)**.

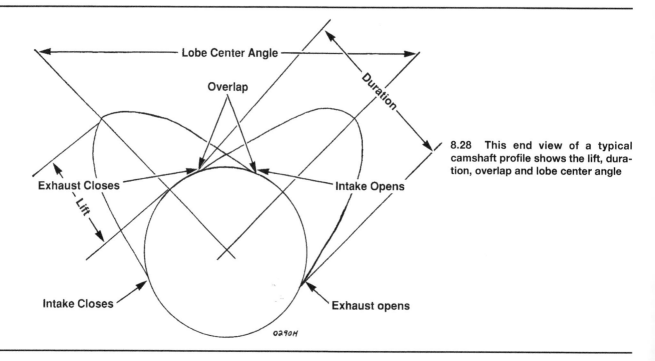

8.28 This end view of a typical camshaft profile shows the lift, duration, overlap and lobe center angle

ENGINE SPECIFICATION QUESTIONNAIRE

NOTE To be filled out only when customer desires the recommendations of our Technical Assistance Department

ENGINE MAKE/MODEL: _____ Year: _____ Original Cubic Inches: _____

Cylinder Bore Size: _____ Stroke: _____ Present Total Displacement: _____

NUMBER OF CYLINDERS: *(Circle)*   1   2   4   V4   Flat 4   6   V6   Straight 6   8   V8

ENGINE TYPE: *(Circle)*   OHV   OHC   DOHC   L-HEAD

ROCKER ARMS: Make: _____ Year: _____ Rocker Arm Ratio: _____ ☐ Stock   ☐ Hi-Lift

VALVE SIZE: Intake: _____ Exhaust: _____ Stock Valve Size: _____ Valve stem dia.: 3/8"-11/32"-5/16"

CARBURETORS: Make: _____ Model: _____ How Many: _____ Jet Size: _____

INTAKE MANIFOLD: Type: _____ Model: _____ Fuel Injection: _____

CYLINDER HEADS: Year/Type: _____ Ported? _____ Milled? _____

PISTONS: Stock ☐   Replacement ☐   Brand: _____ Compression Ratio: _____

SUPERCHARGED ☐   Ratio: _____ TURBOCHARGED ☐   Turbo Make/Type: _____

CHASSIS: Make: _____ Year: _____ Weight: _____

TYPE OF USE: ☐ Street Use   ☐ Street & Strip   ☐ Dragstrip   ☐ Oval Track   ☐ Banked Oval   ☐ Marine

FOR STREET APPLICATIONS: ☐ Max. Low Speed Torque & Economy   ☐ Fuel Economy + Performance
☐ Passenger Car   ☐ Van   ☐ Truck   ☐ Camper   ☐ Motorhome   ☐ Towing
☐ Mostly City Driving   ☐ City & Highway   ☐ Mostly Highway Driving

Gear Ratio: _____ Tire Size: _____ Tire Diameter: _____ Size of Track: _____

RPM Range During Competition: _____ to _____ Type of Transmission: _____ Auto Stall Speed: ___

Is Idling Speed Important? _____ Type of Fuel: _____ Competition Class: _____

ANY OTHER PERTINENT INFORMATION? _____

8.29 This questionnaire is typical of the information camshaft manufacturers need to choose the right camshaft for a specific application

Lift is simply the amount of movement imparted by the camshaft lobe. Lift specifications can be confusing because the rocker arms multiply the actual lift at the camshaft lobe by a ratio of approximately 1:1.5 to 1:1.7. Most camshaft manufacturers provide "net" lift specifications which is the maximum amount of lift (or movement) that occurs at the valve. Actual lobe lift measured at the camshaft is considerably less than net lift. Up to about 0.50 inch net lift, more lift produces more power. Beyond this point, there are diminishing returns. High lifts also result in greater wear rates and premature failure of valve train components.

Duration indicates how long a valve stays open and is measured in degrees of crankshaft rotation (remember that camshafts turn at one half of crankshaft speed). Long duration increases high rpm power at the expense of economy, exhaust emissions and low-end power.

Comparing duration of different camshafts is difficult because not all manufacturers use the same method of measuring. Some companies measure duration from the exact point where the valve lifts off the seat. This produces a higher number, but in practice, the fuel/air mixture does not begin to flow significantly until the valve has lifted a certain amount.

Most camshaft experts have agreed to measure duration starting and ending at 0.050-inch lift. This method produces a smaller number that is more representative of flow characteristics. For street use, cams having about 230-degrees of duration (measured at 0.050-inch lift) work well. Be sure you know which method of measurement is being used when you compare camshafts from different manufacturers.

Overlap is a measurement of crankshaft rotation in degrees during which both the intake and exhaust valves are open. Like duration, long overlap also increases high rpm power at the expense of economy, exhaust emissions and low-end power.

Two factors influence valve overlap specifications. First and easiest to understand is the amount of valve duration. Second is the lobe centerline angle or lobe displacement of the camshaft.

The lobe center angle varies indirectly with the overlap. That is, when overlap increases, lobe center angle decreases, and vice versa. Increasing the angle generally increases low-end torque and reducing the angle improves high-rpm horsepower.

Another area of design that affects camshaft characteristics is the lobe profile. The rate of lift, acceleration and rate of valve closing are determined by the shape of the lobes and affect the way the engine runs. By opening and closing the valves more quickly, cam designers can get more flow from a given amount of duration.

Camshafts and valve train components must be matched to work together properly. In addition, you must carefully match the camshaft/valve train to the other components used on the engine and vehicle, especially the intake and exhaust, gearing and transmission.

Most camshaft manufacturers have technical departments that help customers determine the best camshaft and components for their specific application. In order make accurate recommendations, they need complete information on the vehicle **(see illustration)**. If you have a heavy vehicle with a relatively small engine, you must be conservative with the camshaft and other components. Follow the recommendations of the camshaft manufacturers; they have done a lot of testing and research.

Modified camshaft timing can result in valve-to-piston interference and severe engine damage. Be sure to check the following when you install a camshaft with different specifications than the original one:

8.30 Use a feeler gauge to check for coil bind

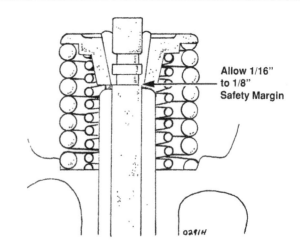

8.31 Check for interference between the valve guide and the spring retainer

8.32 Press the modeling clay onto a piston where the valves come close to the piston crown

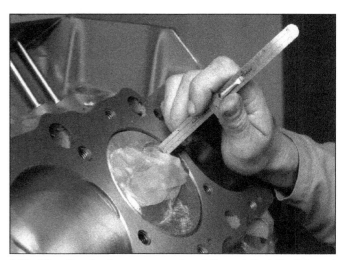

8.33 After the clay is compressed, cut through it at the thinnest point and measure it's thickness at that part

Valve spring coil bind

Whenever a camshaft with higher lift than stock is in-stalled, the springs should be checked for coil bind. Due to the increased travel, the valve spring coils may hit together (bind), causing considerable damage.

Perform this check with the new camshaft and lifters in place and the valve covers off. The cylinder heads, rocker arms and push rods must be in place and adjusted. Using a socket and breaker bar on the front crankshaft bolt inside the lower pulley, carefully turn the crankshaft through at least two com-plete revolutions (720-degrees). When the valve is completely open (spring compressed), try to slip a 0.010-inch feeler gauge between each of the coils **(see illustration)**. It should slip through at least two or three of the coils. If any spring binds, stop immediately and back up slightly. Then correct the prob-lem before continuing. Usually, the valve springs must be re-placed with special ones compatible with the camshaft.

Spring retainer-to-valve guide clearance

Sometimes high-lift camshafts will cause the valve spring retainers to hit the valve guide. To check for this, rotate the crankshaft as described above and check for guide-to-retainer

interference **(see illustration)**. There should be at least 1/16-inch clearance.

Piston-to-valve clearance

Remove a cylinder head and press modeling clay onto the top of a piston **(see illustration)**. Temporarily reinstall the cylinder head with the old gasket and tighten the bolts. Install and adjust the rocker arms and pushrods for that cylinder. Ro-tate the crankshaft through at least two complete revolutions (720-degrees). Remove the cylinder head and slice through the clay at the thinnest point **(see illustration)**. Measure the thick-ness of the clay at that point. It should be at least 0.080-inch thick on the intake side and 0.10-inch thick on the exhaust side. If clearance is close to the minimum, check each cylinder to be sure a variation in tolerances doesn't cause valve-to-pis-ton contact.

Degreeing the camshaft

Every manufactured part has a design tolerance, and when several parts are put together, these tolerances can combine to create a significant error, called tolerance stacking. A lot of power can be lost by assembling an engine without

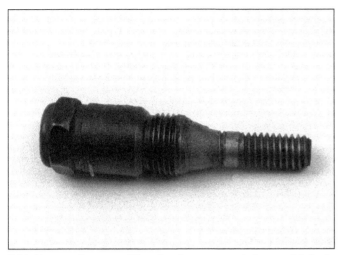

8.34 You can buy a positive stop from a speed shop or fabricate one yourself

8.35 Mount a degree wheel on the front of the crankshaft and install a pointer as shown

checking and, if necessary, correcting the camshaft timing.

The first step is to find true Top Dead Center (TDC); original factory marks can be off by several degrees. Remove the number one spark plug. Using a socket and breaker bar, rotate the crankshaft slowly in a clockwise direction until air starts to blow out of the spark plug hole. Stop rotating the crankshaft at this point.

Obtain a positive stop tool **(see illustration)**. Screw it into the number one cylinder spark plug hole; it prevents the piston from going all the way to the top of the bore during this procedure.

Mount a degree wheel securely to the front of the crankshaft and fabricate a pointer out of coat hanger wire **(see illustration)**. Rotate the crankshaft very slowly in the same direction (clockwise when viewed from the front) until the piston touches the stop. **Note**: *Use a large screwdriver on the teeth of the flywheel/driveplate to turn the crankshaft; otherwise you may disturb the degree wheel. Note what number the pointer aligns with and mark the degree wheel with a pencil.*

Now rotate the crankshaft slowly in the opposite direction until it touches the stop. Again note what number the pointer aligns with and mark the degree wheel with a pencil. True TDC is exactly between the two marks you made.

Double-check your work, remove the piston stop, then rotate the crankshaft to true TDC and check the factory mark. If it is off by a degree or more, correct the factory timing marks.

Now that you've found true TDC, you can check camshaft timing in relation to your crankshaft. Advancing the camshaft in relation to the crankshaft improves low-end performance and retarding the camshaft benefits high-rpm performance at the expense of low-end performance. Whenever cam timing is changed, ignition timing must be reset, and valve-to-piston clearance should be checked if a large change is made.

Read the specification tag included with the camshaft, it should have the opening and closing points (in degrees) of the intake and exhaust valves at 0.050-inch lift. **Note**: *This is the industry standard; if the cam manufacturer uses a different lift, follow their recommendations.*

Begin this check with the engine on true TDC and the degree wheel set at zero. Mount a 0-to-1-inch dial indicator on the intake pushrod of the number one cylinder as shown **(see illustration)**. The dial indicator shaft must be in exact align-

8.36 Mount a dial indicator in line with the pushrod, as shown

ment with the centerline of the pushrod. Preload the indicator stem about 0.100-inch and then reset the dial to zero. **Note**: *If the cylinder heads are not installed, mount the indicator so it presses against the lifter.*

Rotate the crankshaft clockwise until 0.050-inch movement occurs at the dial indicator. Note the position of the pointer in relation to the degree wheel. Record your reading and continue rotating the crankshaft until the lifter again returns to 0.050-inch lift. Record this reading.

Repeat the above procedure on the adjacent exhaust valve lifter and record the results. Compare the readings to the camshaft timing specification tag. If you want to be really thorough, check every cylinder - camshaft machining can be imperfect in some cases.

The readings you have taken may be two to four crankshaft degrees off of the timing tag figures. If this is the case, there are probably slight errors in the dowel pin hole location on the camshaft sprocket. These variations can be corrected by using an offset cam sprocket bushing. These are readily available in speed shops and from camshaft manufacturers. Follow the instructions provided with the kits. **Note**: *Some cam manufacturers build in one or two degrees of advance to compensate for timing chain wear.*

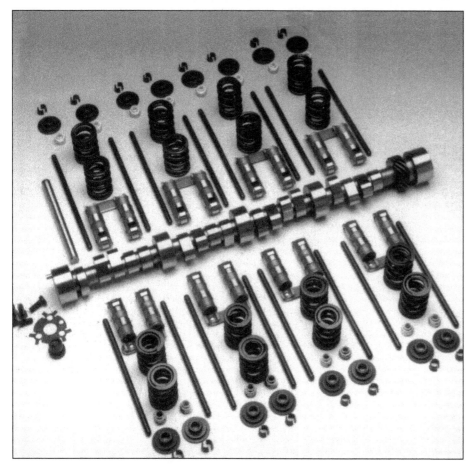

8.37 Buying a complete camshaft kit ensures compatibility of components

8.38 Roller rocker arms reduce friction, add horsepower and ensure more stable valve lash

Matching valve train components

Most camshaft manufacturers sell camshaft kits **(see illustration)** that include the matching lifters, valve springs, pushrods, retainers and sometimes even the rocker arms. Purchasing all of the components from the same source, ensures compatibility of the parts. Also, if you have a problem, it's easier to deal with only one manufacturer.

Precautions

On hydraulic and mechanical lifter camshafts, don't install used lifters on a new camshaft. Used roller lifters may be installed on a new camshaft designed for roller lifters if the lifters are in good condition; however, it's always best to replace all components at the same time.

Apply a generous coating of assembly lube to all components prior to installation. Follow the manufacturer's instructions and don't take any short cuts. Let the cam and lifters break in, then change the oil frequently.

Rocker arms

Installing roller rocker arms **(see illustration)** is an easy way to gain a few horsepower and improve mileage by reducing friction. Roller rockers are available from many manufacturers and are compatible with all types of lifters. Buy a known quality brand and use quality hardware and pushrods to complete the installation.

Exhaust modifications

Stock manifolds

The original stock cast iron exhaust manifolds are quiet,

sturdy, compact and you probably already have them. If you intend to modify your vehicle only slightly, they'll probably work fine.

Smog regulations in some areas prohibit the use of aftermarket tubular headers with catalytic converters. If these regulations apply to your vehicle, you can still pick up some additional power by adding dual exhaust while still retaining the original type manifolds.

Several good cast iron manifolds were used on high-performance and police models, especially in the 1960s. On small-block engines, the "ram's horn" design used in the 50's and 60's flows very well. Manifolds designed for Corvette are usually better than others. These may still be available from your dealer or a salvage yard. If you intend to use these on a different body style than they were originally designed for, be sure there is enough clearance and the outlet is in the right place.

Aftermarket headers

Tubular steel exhaust headers generally provide increases in power without significant changes in fuel economy. They do this by allowing the exhaust gases to flow out of the engine more easily, thereby reducing pumping losses. Headers also "tune" or optimize the engine to run most efficiently in a certain rpm range. They do this by using the inertia of exhaust gases to produce a low-pressure area at the exhaust port just as the valve is opening. This phenomenon, known as scavenging, can be exploited by choosing headers of the correct type, length and diameter to match your engine type, driving style and vehicle.

Header manufacturers produce a large variety of headers designed for different purposes. Most of them provide detailed

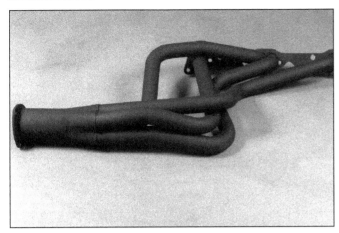

8.39 Four-into-one headers are the most common type

information, recommendations and specifications on their products to assist you in selecting the right headers for your application. Some companies also have technical hot lines to answer customer questions.

Types

Most headers come in a conventional four-into-one design **(see illustration)**. This means that all four pipes on one side of the engine come together inside a large diameter collector pipe. These four-into-one designs tend to produce more peak power at high rpm at the expense of low and mid-range torque.

Another, less common approach to header design is commonly known as "Tri-Y" because of their appearance. The initial four pipes are paired into two pipes and then these are paired again into one pipe at the collector. This design provides more low and mid-range power, but sacrifices a slight amount of horsepower over about 6,000 rpm compared to the four-into-one design.

One of the more recent developments in exhaust system technology is the Anti-reversion (or AR) header. They have a small cone inside the pipe near the cylinder head that prevents reflected pressure waves from creating extra backpressure. The AR design also allows the use of larger diameter tubing without sacrificing a lot of low-end torque. Anti-reversion headers broaden and extend the torque curve and improve throttle response.

Length and diameter

Selection is not an exact science, but rather a series of compromises. Generally, long tubing and collector lengths favor low-end torque and shorter lengths produce their power in the higher rpm range. Larger diameter tubing should be used on large displacement engines and smaller cross-section tubing should be used on smaller displacement models.

Most street headers have primary tubes that are 30 to 40 inches long. Primary tube diameter should be about the same as exhaust port diameter. For small block engines this is about 1-5/8 inch. On big block engines, 1-7/8 inch is typical. The collectors are usually about 10 to 20-inches long and three to four inches in diameter.

Component matching

It's also important to match the characteristics of the intake and exhaust systems. Keep in mind that heavier vehicles with relatively small engines need more low-end torque and

higher numerical gearing to launch them than lighter vehicles with larger engines. Follow the manufacturer's recommendations to ensure that you obtain the correct model for the application.

Open-plenum, single-plane intake manifolds with large, high-CFM carburetors should only be used with four-into-one headers designed for high rpm. This combination should also include a high-performance camshaft matched to the other components and a fairly high numerical gear ratio.

If you want a more tractable street engine, you may want to go for a 180-degree dual-plane manifold with a somewhat smaller carburetor and "Tri-Y" type headers or four-into-one headers with relatively long tubes. A mid-range camshaft would help complete the package.

Buying tips

Be sure to check the manufacturer's literature carefully to determine if the headers will fit your vehicle and allow you to retain accessories such as power steering, air conditioning, power brakes, smog pumps, etc. Note if there is sufficient clearance around the mounting bolts for a wrench. Also, some headers interfere with the clutch linkage on manual transmission models, so check for this, if applicable.

Some applications require the engine to be lifted slightly or even removed to allow header installation. Find out how the mufflers and tailpipes connect to the headers, also. Most installations require some cutting and modifications to the existing exhaust system, which may require a trip to the muffler shop. Check the instructions, ask questions and know what you are getting into before you purchase any components. Don't let poor planning get you stuck with a ticket for driving to the muffler shop with open headers!

Most headers don't come with heat shields like stock manifolds do and tend to melt original spark plug wires and boots. Plan on purchasing a set of heat-resistant silicone spark plug wires and boots, along with the necessary wire holders or looms to keep them away from the headers.

Also, look for header kits that have adapters for heat riser valves and automatic choke heat tubes. Some models with air pumps need threaded fittings in the headers to mount the air injection rails. If your vehicle has an exhaust gas oxygen sensor, be sure the replacement headers have a provision for mounting the sensor.

After the headers are installed, the fuel mixture may be too lean. The carburetor will usually require rejetting and may need some other adjustments. Test the vehicle on an engine analyzer after installation and tune the engine as necessary; failure to do so may result in driveability problems and/or burned valves.

Disadvantages

Headers have several disadvantages which should be considered before you run out and buy a set. Because of their thin tubing construction, headers emit more noise than cast-iron exhaust manifolds. Most headers produce a "tinny" sound in and around the engine compartment and some also produce a resonance inside the vehicle at certain engine speeds.

Tubular headers also have more surface area than cast-iron manifolds, so they give off more heat to the engine compartment. Engines with stock manifolds transmit this heat into the exhaust system instead.

Another drawback is port flange warpage. Some lower-quality headers have thin cylinder head port flanges which are more prone to warpage and exhaust leaks. Compare the thick-

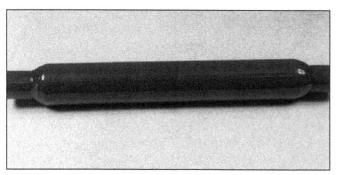

8.40 Glass pack mufflers use a "straight through" design - the larger diameter portion is packed with fiberglass or a similar material to help deaden noise

8.41 Turbo mufflers have large passages and high flow capacity

ness of the flanges on various brands and also check the gaskets and hardware supplied with the kits. The gaskets should be fairly thick and well made and the hardware should be of a high grade and fit your application. Look for the best design and quality control. Before you install headers, check the flanges for warpage with a straightedge, and if warpage is excessive, return them before they are used.

If you live in the "Rust Belt," be prepared for corrosion. Regular paint will quickly burn off and rusting will begin. Before installation, thoroughly coat the headers with heat-resistant paint, porcelainizing or aluminizing spray.

Aftermarket tubular exhaust headers are not approved for use on catalytic converter-equipped vehicles in California and some other jurisdictions (engines equipped with tubular headers from the factory are approved). Additionally, vehicles originally equipped with air injection tubes in the manifolds must retain this system. Check the regulations in your area before you remove the original manifolds.

Exhaust systems

Substantial gains in both performance and economy throughout the rpm range can be obtained by properly modifying the stock exhaust system. The less restriction, the more efficient the engine will be. By increasing the flow capacity, you can unleash the potential horsepower in your engine.

The original mufflers, pipes and catalytic converters on most vehicles have small passages and tight bends. Perhaps the most difficult task when choosing exhaust components for a performance street vehicle is to make the system reasonably quiet while still having low restriction. Also, vehicles originally equipped with catalytic converters won't pass emissions inspections if the catalysts are removed.

Before you purchase a new exhaust system or individual components, you need to know what is available and the pros and cons of various items. Determine what you want your vehicle to be like, what your budget is, then design a balanced system that meets your requirements.

Many special parts and even complete high-performance systems are available for some popular models. If you have a less-common vehicle, you may have to devise a system composed of "universal" parts or use items originally designed for other vehicles.

If you intend to do the modifications on your vehicle yourself, find a level concrete area to work. Raise the vehicle with a floor jack and support it securely with sturdy jackstands. **Warning:** *Never work under a vehicle supported solely by a jack!*

Most exhaust system work requires only a small number of common hand tools. Sometimes, special equipment such as an acetylene torch, a pipe expander or pipe bender will be necessary. Muffler shops usually have this equipment and can modify, fabricate and install custom exhaust systems. But remember, the more work you do yourself, the more money you can save.

One of the easiest ways to reduce backpressure and increase flow capability is by adding dual exhaust. This effectively doubles the capacity of the exhaust without making the vehicle appreciably louder. Sometimes you can reuse most of the original parts on one side and install new parts on the other. If your vehicle is required to have a catalytic converter, you can install one on each side of the system to stay within the law.

As a general rule, mufflers and pipes with large inlet and outlet openings can handle more flow volume than smaller ones, all other factors being equal. Also, shorter mufflers are usually louder and have less restriction than longer ones. So keep this in mind when choosing mufflers.

Conventional or original equipment mufflers

This is the type your vehicle came with from the factory. They are usually of reverse flow construction, and are quite heavy and restrictive. The only time you should consider using these is if you are adding dual exhaust and want to match the original muffler to save money.

Glasspack mufflers

Glasspack mufflers **(see illustration)** are usually less expensive than conventional or "turbo" mufflers, and have less restriction than original equipment types. Unfortunately, they're also too loud to meet most noise regulations and they vary greatly in quality and performance.

Turbo mufflers

Originally, factory engineers designed a large, high flow muffler for use on the turbo Chevrolet Corvair. Hot rodders found out and started using them for all sorts of vehicles. The name "turbo muffler" was first applied to these Corvair parts and gradually became generic, covering all high-flow oval shaped mufflers.

Turbo mufflers **(see illustration)** offer the best features of stock and glasspack mufflers. They are simply high-capacity mufflers that have low restriction without making much more

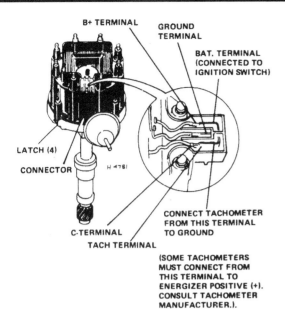

8.42 A typical Chevrolet electronic ignition distributor

8.43 A typical points-type ignition distributor

noise than the original units. Today, virtually every muffler manufacturer sells a "turbo" muffler. They have expanded coverage to include many models; there should be one to fit your application.

Exhaust crossover pipes

Every dual exhaust system should have an exhaust crossover pipe located upstream of the mufflers. This tube balances the pulses between the two sides of the engine and allows excess pressure from one side to bleed off into the opposite muffler. This reduces noise and increases capacity. Be sure that any system you purchase incorporates this design, which improves power throughout the rpm range.

Stainless steel

Stainless steel mufflers and pipes are preferable to regular steel ones if you plan to keep your vehicle for a long time and/or do a lot of short-trip driving. Several specialty manufacturers produce ready-made and custom-built exhaust components from stainless steel. These parts are extremely resistant to corrosion and most are warranteed for the life of the vehicle. They usually cost several times more than standard components, so consider that when you make a purchase decision.

Pipe diameter

Exhaust pipe diameter is measured inside the pipe. Small increases in diameter result in large increases in flow capacity. As a general rule, small block engines displacing 5.0 to 5.8 liters run well with 1-7/8 to 2 inch diameter exhaust pipes, with the more powerful ones using 2-1/4 inch. Engines over 6.5 liters almost always use 2-1/4 to 2-1/2 inch pipes.

Ignition systems

Pre-1975 model engines were equipped with point-type distributors. Standard models came with single-point distributors, while high-performance models often have dual-point versions. If your engine has a single-point distributor, it will probably benefit from an aftermarket performance-oriented replacement unit.

Breakerless electronic ignitions available in the aftermarket and those installed by the factory since 1975 (see illustration) make conventional point-type distributors outmoded (see illustration). However, all non-computer controlled ignitions can benefit from being checked and the advance curve prepped by an ignition shop on a distributor tester. If your engine has a high-performance dual-point distributor, you may want to retain it to keep the engine original. Stock distributors use an advance curve that is a compromise between economy, low-octane fuel and emissions. For most applications, total spark advance should not exceed 38-degrees. So if the centrifugal advance in the distributor produces 28-degrees advance at the crankshaft, initial timing should be about 10-degrees BTDC.

Be very careful when you vary significantly from original specifications. Too little advance results in lost power, overheating and reduced fuel economy. Too much advance can cause internal damage to the engine and high exhaust emissions. Be sure the timing is not advanced to the point where the engine pings continuously under load. A slight rattle when you accelerate is normal, but continued pinging will damage pistons, rods and bearings.

If you have a late-model vehicle with computer-controlled electronic ignition, about the only thing you can do is install an aftermarket control module. Some of these will cause a vehicle to fail an emissions test; check with the manufacturer of the product before you buy.

Some of the more recent model vehicles have knock sensors mounted in the block that detect any pinging and signal the ignition control module to retard the spark timing until the noise goes away. This feature also allows the ignition timing to advance for higher octane fuel and to retard if pinging is detected due to lower octane. Don't tamper with this system; it works well as it is.

Every ignition upgrade should include a new, high-quality distributor cap, rotor, spark plugs and ignition wires (and new points and condenser, if equipped). For street machines, use radio interference suppression wires (TVRS); the solid non-suppression wires will wipe out radio reception (for you and cars and houses near you) and won't provide a measurable gain in performance. Use looms to keep the wires apart. Long parallel runs can cause induced crossfire.

A quality high-performance ignition coil is a nice extra. Be sure to get one that is compatible with the rest of the ignition system.

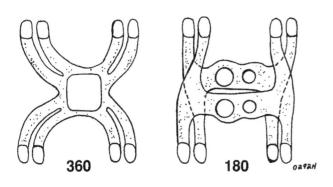

8.44 On 360-degree designs, all of the runners feed from a single chamber (or plenum) - on 180 degree designs, half the runners connect to one plenum and half to the other

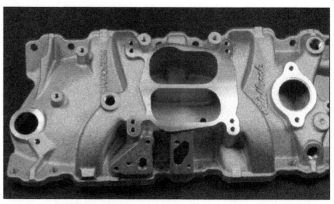

8.45a A typical dual-plane manifold

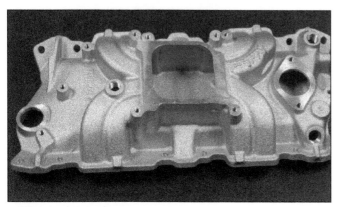

8.45b A typical single-plane manifold

Intake manifolds

A new intake manifold can unlock a considerable amount of power and improve economy at the same time if it is selected properly. Another benefit of using a high-performance manifold is the low weight of aluminum compared to the original cast-iron units.

Although exotic multi-carburetor high-rise and cross-ram manifolds look great, for street use, simple is better. Almost all current high-performance street manifolds are topped off with a single four-barrel carburetor. These provide power with economy and reliability at relatively low cost. Multiple carburetor setups are expensive to buy and difficult to tune and maintain. A single, properly selected carburetor will provide all the flow necessary for a street engine.

Intake manifolds, like many other parts of the engine, are "tuned" to perform best in a certain rpm range. Generally, manifolds with long runners produce more low rpm torque and manifolds with relatively short runners increase high rpm horsepower.

There are two basic intake manifold designs available for street machines, single plane and dual plane, also known as 360-degree and 180-degree designs **(see illustration 8.44)**.

Virtually all stock V8 intake manifolds use the dual-plane design because they enhance low and mid-range power, economy, driveability and produce low emissions. Dual-plane manifolds **(see illustration)** are divided so every other cylinder in the firing order is fed from one side of the carburetor and the remaining cylinders are fed from the other side. This effectively

8.45c A typical tunnel-ram manifold, which is the combination of two single plane manifolds on extended length runners. Each carburetor feeds its four cylinders from a common plenum.

improves intake velocity and throttle response at low and midrange rpm at the expense of some top-end power.

On single-plane (360-degree) intake manifolds **(see illustration)**, all of the intake runners are on the same level and approximately the same length. This helps improve mixture distribution, which is a problem with some dual-plane manifolds. Recently, single-plane manifolds have been designed with better low and mid-range performance. If you want increased mid and top-end power instead of low and mid-range torque, a single-plane manifold may be the way to go. Keep in mind, though, that the most usable power produced by an engine is in the low and mid rpm range.

Intake manifolds are available in low, mid and high-rise versions. The low-rise type is designed to fit under low hood lines and generally sacrifices some power relative to the higher rise type. For most applications, if there is enough hood clearance, go with the high-rise type.

There are two other common designs that you should know about, which are used primarily for racing. Unfortunately, these manifolds decrease low end performance, driveability and fuel economy and increase exhaust emissions.

The cross-ram manifold mounts two four-barrel carburetors side-by-side, instead of one in front of the other. Each carburetor feeds the cylinders on the opposite bank and provides a "ram-tuning" effect.

Tunnel-ram manifolds **(see illustration)** also mount two four-barrel carburetors, but they are installed in-line instead of

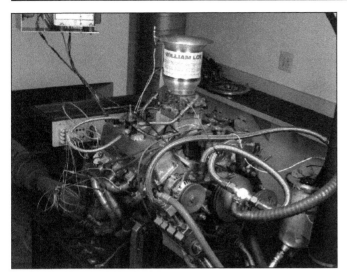

8.46 Carburetor manufacturers, as well as other engine builders, constantly test their products on dynamometers. Every part change or modification as well as flow improvements are tested and recorded while loading the engine to simulate actual driving conditions.

8.47 Some companies make kits with compatible camshafts and intake manifold combinations

side-by-side. They look wild on "Pro-street" cars with the carburetors sticking through the hood, but don't work well on street-driven vehicles.

There are a bewildering number of high-performance factory and aftermarket intake manifolds available for carbureted engines. Manufacturers provide a wealth of information in their catalogs to assist the buyer in making an informed purchase. However, the engines must be identical to the engines listed in the tests to produce the same amount of power. All of the components must work together for best results.

Once you decide on the basic type (single-plane or dual-plane) and rpm range, you must get the right manifold to fit your engine and carburetor. If the carburetor currently on the engine is in good condition and meets your needs, you can retain it and save money. If the carburetor is defective or doesn't suit the application, now is the time to change it.

The intake manifold mounting flanges for various carburetors are different, so you must decide what carburetor to use before you buy a manifold. For example, spread bore and regular bore carburetors have different bore sizes and bolt pattern spacing.

Many performance enthusiasts over-carburate their engines, which hurts throttle response, economy and emissions. The carburetor must be matched to the manifold and the engine for best results. .See the beginning of this chapter for further information.

Be sure to match the manifold characteristics with the overall theme and purpose of your vehicle. Carburetor manufacturers spend countless hours testing their products on dynamometers **(see illustration)** testing flow (CFM) and performance variations with every modification and adjustment change. Some companies sell matched camshaft and manifold kits **(see illustration)**. Before you plunk down your hard-earned cash, find out from the manufacturer what the manifold is designed to do and what it will be like on your engine.

Emission-controlled models may be required to have a functional Exhaust Gas Recirculation (EGR) valve. Some aftermarket manifolds don't have a provision for EGR; this could cause you to fail an emissions test. Also, many engines ping

under load when the EGR is disconnected.

Make sure the manifold has exhaust crossover passages to improve warm-up driveability. Without these, the engine will hesitate and stumble until it is fully warmed up.

Sometimes replacement manifolds require different throttle linkage and mounting hardware. Some engines have a choke stove and many have numerous vacuum fittings. Be sure the parts are readily available in kit form or separately before you purchase a manifold.

Be sure to check local regulations before modifying any emission-controlled vehicle.

Installation tips

Always use new gaskets and seals during installation. Follow the instructions provided with the gasket set and the intake manifold and use quality, name-brand products.

If the cylinder heads have been resurfaced, be sure to have the machine shop mill the manifold surfaces to prevent leaks.

Aluminum manifolds are slightly more prone to warpage than their cast-iron counterparts. Follow the recommended tightening procedure, which is usually an alternating sequence from the center toward the ends, working from side-to-side. Use a torque wrench to tighten the fasteners to the torque specified by the manifold manufacturer

Install a new thermostat and gasket whenever you replace the intake manifold. Use the correct heat range thermostat for the vehicle. Most emission-controlled models use a 195-degree F. thermostat. Earlier models generally use a 180-degree unit.

After the installation is complete, carefully adjust the accelerator linkage to allow full throttle opening and check for binding and sticking before starting the engine.

Be sure there is sufficient fresh oil and coolant in the engine. Run the engine, set the ignition timing, adjust the carburetor and check for oil, fuel and coolant leaks.

8.48 This Holley high-performance carburetor is certified for use on many Chevrolets through the mid 1970's - certified replacement carburetors like this are available for many earlier models that don't have a computer

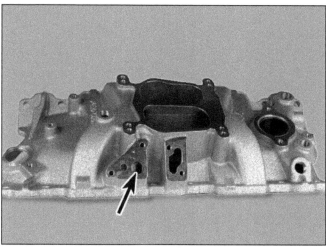

8.49 If your vehicle is equipped with EGR, make sure the certified aftermarket intake manifold you select has a mounting point for the EGR valve (arrow)

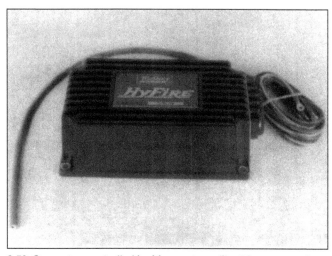

8.50 Computer-controlled ignition systems like this one can give a slight increase in performance and fuel economy to vehicles that don't already have a computer - check to be sure they're certified for use on your vehicle

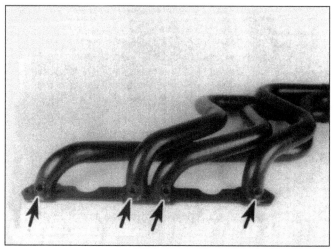

8.51 If you have an air injection system on your vehicle and the injection tubes are threaded into the exhaust manifold, make sure the certified exhaust headers you select have provisions for mounting the tubes in the same place (arrows), and . . .

Emission considerations

Can you modify an emissions-controlled vehicle?

Basically, yes!, so long as you leave all emissions systems intact and the EPA (or, if you live in California, the California Air Resources Board [CARB]) has certified all the components you're planning to install. **Note:** *The components must be certified for use on your particular vehicle. Although, anything you do on the vehicle can be considered "tampering".*

What is tampering?

EPA regulations and many state laws prohibit tampering with the emissions components originally installed on the vehicle. They also prohibit replacing an emissions-related component with a non-original component, unless that component has been specifically certified for use on the vehicle. "Emissions-related" components are any components that have an

effect on emissions. Although their primary function is not emissions control, authorities consider many engine, fuel and exhaust system components to be emissions related. These include camshafts, intake and exhaust manifolds, air cleaners and carburetors or fuel-injection systems. Even aftermarket replacement computer chips must be certified for use in the vehicle.

Tampering with emissions system components (for example, removing an air pump, disconnecting an EGR vacuum hose or even replacing an air cleaner with an aftermarket unit that doesn't have the original emissions provisions) is not only illegal, it can decrease your engine's performance. Vehicles designed for operation with emissions control devices often run better when the systems are hooked up and working properly. This is particularly true of newer, computer-controlled vehicles in which the computer uses information from the emissions systems to control the engine's operation.

Since most smog inspections involve a visual inspection as well as exhaust gas analysis, you'll be caught if you tamper with your emissions systems or emissions-related compo-

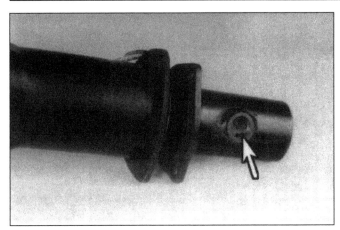

8.52 . . . if your vehicle is equipped with an oxygen sensor, be sure to use the collector on the end of the header that has a mounting hole for the sensor(arrow)

nents. And there's usually no limit on how much you have to spend to correct components that have been tampered with; you'll have to put everything back the way it was, regardless of cost, or you won't get registered.

And don't figure you can fool the inspectors. They have on-line information or books that tell what equipment must be installed on the vehicle, and inspectors are usually pretty good at spotting non-original equipment, such as a non-certified high-performance carburetor.

Don't fall into the trap of figuring you can change a camshaft without worry, since it's inside the engine and won't be found on a visual inspection. Many long-duration camshafts cause the engine to emit excessive pollutants at idle. You may pass the visual inspection, but you might fail the exhaust analysis. Considering the amount of work and expense involved in replacing a camshaft, it pays to make sure it is certified for use in your vehicle.

What modifications can you make?

Aftermarket equipment

Many aftermarket manufacturers offer equipment that is designed for use on emission-controlled vehicles. These components include carburetors, intake manifolds, electronic ignition systems, exhaust headers and camshafts **(see illustrations)**. Avoid components that are for racing use only. These often say "for off-highway use only" or "not for use in pollution-controlled vehicles" in their product literature. If you doubt any piece of equipment, check with the manufacturer to see if the component is EPA or CARB certified.

Reference and specifications

Introduction`

The following Chapter covers information on various interchangeable parts of the different Holley carburetor models. The tables and charts contained here cover such information as:

Carburetor tables with the original components of each model listed individually
Needle and seat assembly types
Float designs and materials available
Vacuum secondary spring rates
Main metering jet sizes

Secondary metering plate size combinations
Power valve (single-stage and two-stage styles) opening rates
Accelerator pump discharge nozzle styles and sizes
Original Equipment (O.E.) carburetors on muscle car era vehicles
Standard/Metric conversion table

Because of the many variations of carburetor models and applications, the individual assembly and adjustment specifications are given in either Chapter 7, *Carburetor Overhaul and Adjustments* or in the individual specifications provided in the overhaul kit purchased from your local auto parts store.

Carburetor Part No.	Carb. Model No.	CFM	Renew Kit	Trick Kit	Primary & Secondary Needle & Seat	Primary Main Jet	Secondary Main Jet or Plate	Primary Power Valve	Primary Discharge Nozzle Size
0-1848-1	4160	465	37-119	37-912	6-506	122-57	134-3	125-85	.025
0-1849	4160	550	37-119	37-912	6-506	122-62	134-5	125-85	.025
0-1850-2	4160	600	37-119	37-912	6-506	122-66	134-9	125-65	.025
0-1850-3,-4	4160	600	37-119	37-912	6-506	122-66	134-9	125-65	.025
0-2818-1	4150	600	37-333	37-905	6-503	122-65	122-76	125-65	.025
*0-3310-1	4150	780	37-1084	37-915	6-504	122-72	122-76	Notes 12,13	.025
*0-3310-2	4160	750	37-754	37-921	6-504	122-72	134-21	125-65	.025
*0-3310-3,-4	4160	750	37-754	37-921	6-504	122-72	134-21	125-65	.025
0-3418-1	4150	855	37-1273	N/A	6-504	122-76T/122/82C	122-80T/122-82C	Notes 15,21	.028
*0-4118	4150	725	37-1184	37-906	6-504	122-68	122-78	125-85	.025
0-4224	4160	660	37-424	N/A	6-508	122-76	134-12	N/R	.025
0-4412	2300	500	37-474	37-901	6-504	122-73	N/R	125-50	.028
0-4412-1 -2	2300	500	37-474	37-901	6-504	122-73	N/R	125-50	.028
0-4452-1	4160	600	37-119	37-912	6-506	122-63	134-39	125-85	.031
0-4548	4160	450	37-119	N/A	6-506	122-57	134-30	25R-219-85	.031
**0-4575	4500	1050	37-487	37-920	6-504	122-84	122-84	125-65 (15)	.035
0-4691-2	2110	300	37-496	N/A	6-509	122-63	N/R	N/R	.021
0-4742	4150	600	37-433	N/A	6-504	122-63	122-72	25R-237-85	.031
**0-4776	4150	600	37-485	37-910	6-504	122-69	122-71	125-65	.025
**0-4776-1	4150	600	37-485	37-910	6-504	122-66	122-76	125-65	.028
**0-4776-2	4150	600	37-485	37-910	6-504	122-66	122-76	125-65	.028
**0-4776-3,-4	4150	600	37-485	37-910	6-504	122-66	122-73	125-65	.028
**0-4777	4150	650	37-485	37-910	6-504	122-71	122-76	125-65	.025
**0-4777-1	4150	650	37-485	37-910	6-504	122-67	122-76	125-65	.028
**0-4777-2	4150	650	37-485	37-910	6-504	122-67	122-76	125-65	.028
**0-4777-3,-4	4150	650	37-485	37-910	6-504	122-67	122-73	125-65	.028
**0-4778	4150	700	37-485	37-910	6-504	122-66	122-71	125-65	.025
**0-4778-1	4150	700	37-485	37-910	6-504	122-66	122-76	125-65	.028
**0-4778-2	4150	700	37-485	37-910	6-504	122-66	122-76	125-65	.028
**0-4778-3,-4	4150	700	37-485	37-910	6-504	122-69	122-78	125-65	.028
**0-4779	4150	750	37-485	37-910	6-504	122-75	122-76	125-85	.025
**0-4779-1	4150	750	37-485	37-910	6-504	122-70	122-80	125-85	.028
**0-4779-2	4150	750	37-485	37-910	6-504	122-70	122-80	125-65	.028
**0-4779-3	4150	750	37-485	37-910	6-504	122-70	122-73	125-65	.028
*0-4779-4,-5	4150	750	37-485	37-910	6-504	122-70	122-80	125-65	.028
**0-4780	4150	800	37-485	37-910	6-504	122-72	122-76	Notes 12,21	.031
**0-4780-1	4150	800	37-485	37-910	6-504	122-70	122-76	Notes 12,21	.031
**0-4780-2	4150	800	37-485	37-910	6-504	122-70	122-85	125-65	.031
**0-4780-3,-4	4150	800	37-485	37-910	6-504	122-71	122-85	125-65	.031
**0-4781	4150	850	37-485	37-916	6-504	122-80	122-80	125-65 (15)	.035
**0-4781-1	4150	850	37-485	37-916	6-504	122-80	122-80	125-65 (15)	.031
**0-4781-2	4150	850	37-485	37-916	6-504	122-80	122-80	125-65 (15)	.031
**0-4781-3,-4	4150	850	37-485	37-916	6-504	122-80	122-78	125-65 (15)	.031
0-4782	2300	355	37-396	N/A	6-504	122-64	N/R	125-65	.031
0-4783	2300	500	37-396	N/A	6-504	122-82	N/R	N/R	.028
**0-4788	4150	830	37-485	37-916	6-504	122-80	122-80	125-65 (B)	.031
**0-4788-1	4150	830	37-485	37-916	6-504	122-80	122-80	125-65 (B)	.031
0-6109	4150	750	37-485	37-910	6-504	122-75	122-76	125-85	.025
0-6210-1	4165	650	37-605	37-918	Notes 16,17	122-602	122-632	Notes 14,15	.025
0-6210-2	4165	650	37-605	37-918	Notes 16,17	122-602	122-83	125-85	.025
0-6210-3	4165	650	37-605	37-918	Notes 16,17	122-602	122-83	125-85	.025
0-6211	4165	800	37-605	37-918	Notes 16,17	122-62	122-85	Notes 14,15	.025
0-6211-1	4165	800	37-605	37-918	Notes 16,17	122-602	122-85	Notes 14,15	.025
**0-6212	4165	800	37-606	37-919	6-504	122-63	122-86	Notes 14,15	.025
**0-6213	4165	800	37-606	37-919	6-504	122-62	122-85	Notes 14,15	.025
**0-6214	4500	1150	37-608	37-914	6-504	122-95	122-95	N/R	.026
0-6244-1	2110	200	37-496	N/A	6-509	122-47	N/R	N/R	.021
0-6262	4165	800	37-605	37-918	Notes 16,17	122-62	122-85	Notes 14,15	.025
0-6270-1	4160	600	37-397	N/A	18R-203	122-64	34R-6153-39	25R-237-85	.032
0-6291	4160	600	37-119	N/A	6-506	122-62	134-39	125-85	.031
0-6299-1	4160	390	37-654	37-911	6-506	122-50	134-34	N/A	.025
0-6425	2300	650	37-656	37-903	6-504	122-82	N/R	125-65	.031
**0-6464	4500	1050	37-487	37-920	6-504	122-88	122-88	N/R	.035
0-6468-1	4165	650	37-605	37-918	Notes 16, 17	122-60	122-83	125-85	.025
0-6468-2	4165	650	37-605	37-918	Notes 16, 17	122-602	122-83	125-85	.025
0-6497	4165	650	37-605	37-918	Notes 16, 17	122-582	122-602	Notes 14, 15	.025
0-6498	4165	650	37-605	37-918	Notes 16, 17	122-592	122-602	Notes 14, 15	.025
**0-6499	4165	650	37-606	37-919	6-504	122-60	122-63	Notes 14, 15	.025
0-6512	4165	650	37-605	37-918	Notes 16, 17	122-60	122-60	Notes 14, 15	.025
0-6520	4160	600	37-119	N/A	6-506	122-62	134-39	125-85	.031
0-6528	4165	650	37-605	37-918	Notes 16, 17	122-61	122-60	Notes 14, 15	.025
0-6619-1	4160	600	37-720	37-912	6-506	122-642	134-39	125-65	.031
**0-6708	4150	650	37-1272	32-925	6-504	122-552	122-752	Notes 21, 22	.025
**0-6708-1	4150	650	37-1272	32-925	6-504	122-542	122-85	125-65	.025
**0-6709	4150	750	37-1272	32-925	6-504	122-652	122-652	Notes 21, 22	.025

*Indicates Center Inlet Dual Feed Bowls
**Indicates Dual Accel. Pump & Dual Feed Bowls

5. Main Body Gasket
12. 125-85 Secondary

13. 125-105 Primary
14. 125-85 Primary

15. 125-65 Secondary
16. 6-511 Primary

17. 6-510 Secondary

9.1a Carburetor specifications (by list number)

Secondary Nozzle Size or Spring Color	Primary Bowl Gasket	Primary Metering Block Gasket	Secondary Bowl Gasket	Secondary Metering Block Gasket	Secondary Metering Plate Gasket	Venturi Diameter Primary	Venturi Diameter Secondary	Throttle Bore Diameter Primary	Throttle Bore Diameter Secondary
Green	108-33	108-29	108-30	108-30	108-27	1-3/32	1-3/32	1-1/2	1-1/2
Plain	108-33	108-29	108-30	108-30	108-27	1-3/16	1-1/4	1-1/2	1-1/2
Plain	108-33	108-29	108-30	108-30	108-27	1-1/4	1-5/16	1-9/16	1-9/16
Plain	108-33	108-29	108-30	108-30	N/R	1-1/4	1-5/16	1-9/16	1-9/16
Purple	108-33	108-29	108-33	108-29	N/R	1-1/4	1-5/16	1-9/16	1-9/16
Plain	108-33	108-29	108-33	108-29	N/R	1-3/8	1-7/16	1-11/16	1-11/16
Plain	108-33	108-29	108-30	108-30	108-27	1-3/8	1-7/16	1-11/16	1-11/16
Plain	108-33	108-29	108-30	108-30	108-27	1-3/8	1-7/16	1-11/16	1-11/16
Yellow	108-33	108-29	108-33	108-29	N/R	1-9/16	1-9/16	1-3/4	1-3/4
Yellow	108-33	108-29	108-33	108-29	N/R	1-5/16	1-3/8	1-11/16	1-11/16
.025	108-33	108-29	108-30	108-30	108-27	1-1/4	1-5/16	1-11/16	1-11/16
N/R	108-33	108-29	N/R	N/R	N/R	1-3/8	1-11/16	N/R	N/R
N/R	108-33	108-29	N/R	N/R	N/R	1-3/8	1-11/16	N/R	N/R
Purple	108-33	108-29	108-30	108-30	108-27	1-1/4	1-5/16	1-9/16	1-9/16
Brown	108-33	108-29	108-30	108-30	108-27	1-3/32	1-3/32	1-1/2	1-1/2
.035	108-33	108-29	108-33	108-29	N/R	1-11/16	1-11/16	2	2
N/R	N/R	N/R	N/R	N/R	N/R	1-5/32	N/R	1-7/16	N/R
Purple	108-33	108-29	108-33	108-29	N/R	1-1/4	1-5/16	1-9/16	1-9/16
.032	108-33	108-29	108-33	108-29	N/R	1-1/4	1-5/16	1-9/16	1-9/16
.032	108-33	108-29	108-33	108-29	N/R	1-1/4	1-5/16	1-9/16	1-9/16
.032	108-33	108-29	108-33	108-29	N/R	1-1/4	1-5/16	1-9/16	1-9/16
.032	108-33	108-29	108-33	108-29	N/R	1-1/4	1-5/16	1-9/16	1-9/16
.025	108-33	108-29	108-33	108-29	N/R	1-1/4	1-5/16	1-11/16	1-11/16
.028	108-33	108-29	108-33	108-29	N/R	1-1/4	1-5/16	1-11/16	1-11/16
.028	108-33	108-29	108-33	108-29	N/R	1-1/4	1-5/16	1-11/16	1-11/16
.028	108-33	108-29	108-33	108-29	N/R	1-1/4	1-5/16	1-11/16	1-11/16
.032	108-33	108-29	108-33	108-29	N/R	1-5/16	1-3/8	1-11/16	1-11/16
.031	108-33	108-29	108-33	108-29	N/R	1-5/16	1-3/8	1-11/16	1-11/16
.031	108-33	108-29	108-33	108-29	N/R	1-5/16	1-3/8	1-11/16	1-11/16
.031	108-33	108-29	108-33	108-29	N/R	1-5/16	1-3/8	1-11/16	1-11/16
.032	108-33	108-29	108-33	108-29	N/R	1-3/8	1-3/8	1-11/16	1-11/16
.031	108-33	108-29	108-33	108-29	N/R	1-3/8	1-3/8	1-11/16	1-11/16
.031	108-33	108-29	108-33	108-29	N/R	1-3/8	1-3/8	1-11/16	1-11/16
.031	108-33	108-29	108-33	108-29	N/R	1-3/8	1-3/8	1-11/16	1-11/16
.031	108-33	108-29	108-33	108-29	N/R	1-3/8	1-3/8	1-11/16	1-11/16
.031	108-33	108-29	108-33	108-29	N/R	1-3/8	1-7/16	1-11/16	1-11/16
.031	108-33	108-29	108-33	108-29	N/R	1-3/8	1-7/16	1-11/16	1-11/16
.031	108-33	108-29	108-33	108-29	N/R	1-3/8	1-7/16	1-11/16	1-11/16
.031	108-33	108-29	108-33	108-29	N/R	1-3/8	1-7/16	1-11/16	1-11/16
.025	108-33	108-29	108-33	108-29	N/R	1-9/16	1-9/16	1-3/4	1-3/4
.031	108-33	108-29	108-33	108-29	N/R	1-9/16	1-9/16	1-3/4	1-3/4
.031	108-33	108-29	108-33	108-29	N/R	1-9/16	1-9/16	1-3/4	1-3/4
.031	108-33	108-29	108-33	108-29	N/R	1-9/16	1-9/16	1-3/4	1-3/4
N/R	108-33	108-29	N/R	N/R	N/R	1-3/16	N/R	1-1/2	N/R
N/R	108-33	108-29	N/R	N/R	N/R	1-9/16	N/R	1-3/4	N/R
.031	108-33	108-29	108-29	N/R	1-9/16	1-9/16	1-11/16	1-11/16	
.031	108-33	108-29	108-33	108-29	N/R	1-9/16	1-9/16	1-11/16	1-11/16
.032	108-33	108-29	108-33	108-29	N/R	1-3/8	1-3/8	1-11/16	1-11/16
.037	108-32	108-31†	108-32	108-31†	N/R	1-5/32	1-3/8	1-3/8	2
.037	108-32	108-31	108-32	108-31	N/R	1-5/32	1-3/8	1-3/8	2
.037	108-32	108-31	108-32	108-31	N/R	1-5/32	1-3/8	1-3/8	2
.037	108-32	108-31†	108-32	108-31†	N/R	1-5/32	1-23/32	1-3/8	2
.037	108-32	108-31†	108-32	108-31†	N/R	1-5/32	1-23/32	1-3/8	2
.037	108-32	108-31†	108-32	108-31†	N/R	1-5/32	1-23/32	1-3/8	2
.037	108-32	108-31†	108-32	108-31†	N/R	1-5/32	1-23/32	1-3/8	2
.026	108-33	108-36	108-33	108-36	N/R	1-13/16	1-13/16	2	2
N/R	N/R	N/R	N/R	N/R	N/R	1-5/16	N/R	1-7/16	N/R
.037	108-32	108-31†	108-32	108-31†	N/R	1-13/16	1-23-32	1-3/8	2
Orange	108-33	108-34	108-30	108-30	108-27	1-1/4	1-5/16	1-9/16	1-9/16
Purple	108-33	108-31	108-30	108-30	108-27	1-1/4	1-5/16	1-5/16	1-9/16
Plain	108-33	108-29	108-30	108-27	108-28	1-1/16	1-1/16	1-7/16	1-7/16
N/R	108-32	108-35	N/R	N/R	N/R	1-7/16	N/R	1-3/4	N/R
.035	108-33	108-36	108-33	108-36	N/R	1-11/16	1-11/16	2	2
.037	108-32	108-31	108-32	108-31	N/R	1-5/32	1-3/8	1-3/8	2
.037	108-32	108-31	108-32	108-31	N/R	1-5/32	1-3/8	1-3/8	2
.037	108-32	108-31	108-32	108-31	N/R	1-5/32	1-3/8	1-3/8	1-3/8
.037	108-32	108-31	108-32	108-31	N/R	1-5/32	1-3/8	1-3/8	2
.037	108-32	108-31	108-32	108-31	N/R	1-5/32	1-3/8	1-3/8	2
.037	108-32	108-31	108-32	108-31	N/R	1-5/32	1-3/8	1-3/8	2
Purple	108-33	108-31	108-30	108-30	108-27	1-1/4	1-5/16	1-9/16	1-9/16
.037	108-32	108-31	108-32	108-31	N/R	1-5/32	1-3/8	1-3/8	2
Black	108-33	108-31	108-30	108-30	108-27	1-1/4	1-5/16	1-9/16	1-9/16
.037	108-32	108-31	108-32	108-31	N/R	1-3/32	1-9/16	1-1/2	1-3/4
.037	108-32	108-31	108-32	108-31	N/R	1-3/32	1-9/16	1-1/2	1-3/4
.037	108-32	108-31	108-32	108-31	N/R	1-1/4	1-9/16	1-1/2	1-3/4

21. 125-65 Primary
22. 125-35 Secondary

24. 25R-475A-13

† Early versions must use 108-29 to seal pump passage.

9.1b Carburetor specifications (continued)

Carburetor Part No.	Carb. Model No.	CFM	Renew Kit	Trick Kit	Primary & Secondary Needle & Seat	Primary Main Jet	Secondary Main Jet or Plate	Primary Power Valve	Primary Discharge Nozzle Size
**0-6710	4165	800	37-606	37-919	6-504	122-63	122-86	Notes 21, 22	.025
0-6711	4165	650	37-605	37-918	Notes 16, 17	122-602	122-632	Notes 21, 22	.025
0-6772	4165	650	37-605	37-918	Notes 16, 17	122-592	122-602	Notes 14, 15	.025
0-6773	4165	650	37-605	37-918	Notes 16, 17	122-592	122-602	Notes 14, 15	.025
0-6774	4165	650	37-605	37-918	Notes 16, 17	122-572	122-602	Notes 14, 15	.025
0-6853	4165	650	37-605	37-918	Notes 16, 17	122-62	122-60	Notes 14, 15	.025
0-6895	4150	390	37-739	N/A	6-504	122-50	122-62	125-85	.025
0-6909	4160	600	37-119	37-912	6-506	122-622	134-39	125-65	.031
**0-6910	4165	800	37-606	37-919	6-504	122-612	122-86	Notes 14, 15	.025
0-6919	4160	600	37-1415	37-912	6-506	122-622	134-39	125-206	.031
0-6946-1	4160	600	3-1016	N/A	6-504	122-612	134-41	25R-609-13A	.025
0-6947	4160	600	3-1011	N/A	6-504	122-612	134-41	25R-475-18A	.025
0-6979	4160	600	37-747	37-912	6-506	122-642	134-39	125-85	.031
0-6979-1	4160	600	37-747	37-912	6-506	122-642	134-39	125-208	.031
0-6989	4160	600	37-1415	37-912	6-506	122-622	134-39	125-206	.031
0-7001	4165	650	37-743	37-918	Notes 16, 17	122-582	122-602	Notes 15, 24	.025
0-7002-1	4175	650	37-732	37-922	Notes 16, 17	122-582	134-21	125-85	.025
0-7004-1	4175	650	37-741	37-922	Notes 16, 17	122-562	134-45	125-212	.025
0-7004-2	4175	650	37-741	37-922	Notes 16, 17	122-542	134-50	125-211	.025
0-7005-1	4175	650	37-741	37-922	Notes 16,17	122-562	134-45	125-212	.025
0-7005-2	4175	650	37-741	37-922	Notes 16,17	122-542	134-50	25R-619A-16	.025
0-7006-1	4175	650	37-741	37-922	Notes 16,17	122-562	134-45	125-212	.025
0-7006-2	4175	650	37-741	37-922	Notes 16,17	122-542	134-50	125-211	.025
0-7009-1	4160	600	37-1415	37-912	6-506	122-622	134-39	125-206	.031
*0-7010	4160	780	37-740	37-925	6-506	122-662	134-42	125-65	.025
0-7053-1	4160	600	37-119	—	6-506	122-632	134-39	125-85	.031
0-7054	4165	650	37-605	37-918	Notes 16, 17	122-592	122-602	Notes 14, 15	.025
0-7154	4160	600	37-119	—	6-506	122-62	134-43	125-85	.031
**0-7320	4500	1150	37-487	37-920	6-504	122-95	122-95	N/A	.031
0-7343	5200	230	37-716	N/A	6-512	22R-103-132	22R-103-140	125-36	.020
0-7344	5210	255	37-687	N/A	6-512	22R-103-132	22R-103-142	125-36	.021
0-7351	4175	650	37-741	37-922	Notes 16, 17	122-592	134-21	25R-475-13	.037
0-7397	4175	650	37-741	37-922	Notes 16, 17	122-582	134-21	25R-475-13	.037
0-7410	4150	340	37-739	N/A	6-504	122-50	122-62	125-85	.025
0-7411	4150	370	37-739	N/A	6-504	122-50	122-62	125-85	.025
0-7413	4160	600	37-119	—	6-506	122-632	134-39	125-85	.031
0-7448	2300	350	37-749	37-901	6-504	122-61	N/A	125-85	.031
0-7454	4360	450	37-750	N/A	6-514	124-215	124-550	125-36	.028
0-7455	4360	450	37-750	N/A	6-514	124-215	124-537	125-36	.028
0-7456	4360	450	37-750	N/A	6-514	124-215	124-550	125-36	.028
0-7555	4360	450	37-750	N/A	6-514	124-215	124-550	125-36	.028
0-7556	4360	450	37-750	N/A	6-514	124-215	124-550	125-36	.028
0-7850	4160	600	37-830	—	6-506	122-622	134-39	125-85	.031
0-7855	4175	650	37-741	N/A	Notes 16, 17	122-562	134-45	125-212	.028
0-7955	4360	450	37-1138	N/A	6-514	124-219	124-550	125-201	.028
0-7956	4360	450	37-1138	N/A	6-514	124-239	124-550	125-201	.028
0-7957	4360	450	37-1138	N/A	6-514	124-219	124-550	125-201	.028
0-7958	4360	450	37-1138	N/A	6-514	124-219	124-550	125-201	.028
0-7985	4160	600	37-840	37-912	6-506	122-632	134-39	125-208	.031
0-7986	4160	600	37-840	37-912	6-506	125-652	134-39	125-208	.031
0-7987	4160	600	37-840	37-912	6-506	122-612	134-39	125-208	.031
0-8001	4360	450	37-1138	N/A	6-514	124-215	124-550	125-201	.028
0-8002	4360	450	37-1138	N/A	6-514	124-215	124-550	125-201	.028
0-8003	4360	450	37-1138	N/A	6-514	124-235	124-550	125-201	.028
0-8004	4160	600	37-840	37-912	6-506	122-632	134-39	125-208	.031
0-8005	4160	600	37-840	37-912	6-506	122-622	134-39	125-208	.031
0-8006	4160	600	37-840	37-912	6-506	122-622	134-39	125-208	.031
0-8007	4160	390	37-720	37-912	6-506	122-51	134-34	125-65	.025
0-8059	4175	650	37-741	N/A	Notes 16, 17	122-582	134-21	25R-475A-13	.037
0-8059-1	4175	650	37-741	N/A	Notes 16, 17	122-582	134-49	125-211	.025
0-8060	4175	650	37-741	N/A	Notes 16, 17	122-582	134-21	25R-475A-15	.037
0-8060-1	4175	650	37-741	N/A	Notes 16, 17	122-582	134-49	125-24	.025
**0-8082	4500	1050	37-487	37-920	6-504	122-84	122-84	125-65	.035
**0-8082-1	4500	1050	37-487	37-920	6-504	122-88	122-88	125-65(15)	.035
0-8149	4360	450	37-1138	N/A	6-514	124-231	124-550	125-36	.028
0-8149-1	4360	450	37-1138	N/A	6-514	124-215	124-550	125-201	.028
**0-8156	4150	750	37-485	37-910	6-504	122-70	122-83	125-65	.028
0-8158	4360	450	37-750	N/A	6-514	124-219	124-550	125-201	.028
0-8162	4150	850	37-485	37-916	6-504	122-80	122-80	125-65	.031
0-8181	4160	600	37-1415	N/A	6-504	122-80	122-80	125-65(15)	.031
0-8203	4360	450	37-1138	N/A	6-514	124-211	124-550	125-201	.028
0-8204	4360	450	37-1138	N/A	6-514	124-215	124-550	125-201	.028
0-8206	4360	450	37-1138	N/A	6-514	124-203	124-550	125-201	.028
0-8207	4160	600	37-830	—	6-506	122-622	134-39	125-85	.031
0-8276	4175	650	37-732	N/A	Notes 16, 17	122-572	134-21	125-85	.025

*Indicates Center Inlet Dual Feed Bowls
**Indicates Dual Accel. Pump & Dual Feed Bowls

5. Main Body Gasket
12. 125-85 Secondary

13. 125-105 Primary
14. 125-85 Primary

15. 125-65 Secondary
16. 6-511 Primary

17. 6-510 Secondary

9.1c Carburetor specifications (by list number)

Secondary Nozzle Size or Spring Color	Primary Bowl Gasket	Primary Metering Block Gasket	Secondary Bowl Gasket	Secondary Metering Block Gasket	Secondary Metering Plate Gasket	Venturi Diameter Primary	Venturi Diameter Secondary	Throttle Bore Diameter Primary	Throttle Bore Diameter Secondary
.037	108-32	108-31	108-32	108-31	N/R	1-5/32	1-23/32	1-3/8	2
.028	108-32	108-31	108-32	108-31	N/R	1-5/32	1-3/8	1-3/8	2
.040	108-32	108-31	108-32	108-31	N/R	1-5/32	1-3/8	1-3/8	2
.040	108-32	108-31	108-32	108-31	N/R	1-5/32	1-3/8	1-3/8	2
.037	108-32	108-31	108-32	108-31	N/R	1-5/32	1-3/8	1-3/8	2
.037	108-32	108-31	108-32	108-31	N/R	1-5/32	1-3/8	1-3/8	2
.025	108-33	108-29	108-33	108-29	N/R	1-1/16	1-1/16	1-7/16	1-7/16
Black	108-33	108-31	108-30	108-30	108-27	1-1/4	1-5/16	1-9/16	1-9/16
.037	108-32	108-31	108-32	108-31	N/R	1-5/32	1-23/32	1-3/8	2
Black	108-33	108-31	108-30	108-30	108-27	1-1/4	1-5/16	1-9/16	1-9/16
Plain	108-33	108-31	108-30	108-30	108-27	1-3/16	1-1/4	1-1/2	1-1/2
Plain	108-33	108-31	108-30	108-30	108-27	1-3/16	1-1/4	1-1/2	1-1/2
Black	108-33	108-31	108-30	108-30	108-27	1-1/4	1-5/16	1-9/16	1-9/16
Black	108-33	108-31	108-30	108-30	108-27	1-1/4	1-5/16	1-9/16	1-9/16
Black	108-33	108-31	108-30	108-30	108-27	1-1/4	1-5/16	1-9/16	1-9/16
.037	108-32	108-31	108-32	108-31	N/R	1-5/32	1-3/8	1-3/8	2
Black	108-32	108-31	108-30	108-30	108-27	1-5/32	1-3/8	1-3/8	2
Plain	108-32	108-31	108-30	108-30	108-27	1-5/32	1-3/8	1-3/8	2
Plain	108-32	108-31	108-30	108-30	108-27	1-5/32	1-3/8	1-3/8	2
Plain	108-32	108-31	108-30	108-30	108-27	1-5/32	1-3/8	1-3/8	2
Plain	108-32	108-31	108-30	108-30	108-27	1-5/32	1-3/8	1-3/8	2
Plain	108-32	108-31	108-30	108-30	108-27	1-5/32	1-3/8	1-3/8	2
Plain	108-32	108-31	108-30	108-30	108-27	1-5/32	1-3/8	1-3/8	2
Black	108-33	108-31	108-30	108-30	108-27	1-1/4	1-5/16	1-9/16	1-9/16
Black	108-32	108-31	108-30	108-30	108-27	1-1/4	1-9/16	1-1/2	1-3/4
Purple	108-33	108-31	108-30	108-30	108-27	1-1/4	1-5/16	1-9/16	1-9/16
.037	108-32	108-31	108-33	108-31	N/R	1-5/32	1-3/8	1-3/8	2
Purple	108-33	108-31	108-30	108-30	108-27	1-1/4	1-5/16	1-9/16	1-9/16
.035	108-33	108-29	108-33	108-29	N/R	1-13/16	1-13/16	2	2
N/R	N/R	N/R	N/R	N/R	N/R	1-1/25	1-1/16	1-7/25	1-7/16
N/R	N/R	N/R	N/R	N/R	N/R	1-1/33	1-11/16	1-1/4	1-7/25
Black	108-32	108-31	108-30	108-30	108-27	1-13/64	1-13/32	1-3/8	2
Black	108-32	108-31	108-30	108-30	108-27	1-13/64	1-13/32	1-3/8	2
.025	108-33	108-29	108-33	108-29	N/R	1-1/16	1-1/16	1-7/16	1-7/16
.025	108-33	108-29	108-33	108-29	N/R	1-1/16	1-1/16	1-7/16	1-7/16
Purple	108-33	108-31	108-30	108-30	108-27	1-1/4	1-5/16	1-9/16	1-9/16
N/R	108-33	108-29	N/R	N/R	N/R	1-3/16	N/R	1-1/2	N/R
N/R	108-26[5]	N/R	N/R	N/R	N/R	1-1/16	1-3/16	1-3/8	1-7/16
N/R	108-26[5]	N/R	N/R	N/R	N/R	1-1/16	1-3/16	1-3/8	1-7/16
N/R	108-26[5]	N/R	N/R	N/R	N/R	1-1/16	1-3/16	1-3/8	1-7/16
N/R	108-26[5]	N/R	N/R	N/R	N/R	1-1/16	1-3/16	1-3/8	1-7/16
Plain	108-33	108-31	108-30	108-30	108-27	1-1/4	1-5/16	1-9/16	1-9/16
Plain	108-33	105-31	108-30	108-30	108-27	1-13/32	1-13/64	1-3/8	2
N/R	108-26[5]	N/R	N/R	N/R	N/R	1-1/16	1-3/16	1-3/8	1-7/16
N/R	108-26[5]	N/R	N/R	N/R	N/R	1-1/16	1-3/16	1-3/8	1-7/16
N/R	108-26[5]	N/R	N/R	N/R	N/R	1-1/16	1-3/16	1-3/8	1-7/16
N/R	108-26[5]	N/R	N/R	N/R	N/R	1-1/16	1-3/16	1-3/8	1-7/16
Black	108-33	108-31	108-30	108-30	108-27	1-1/4	1-5/16	1-9/16	1-9/16
Black	108-33	108-31	108-30	108-30	108-27	1-1/4	1-5/16	1-9/16	1-9/16
Black	108-33	108-31	108-31	108-30	108-27	1-1/4	1-5/16	1-9/16	1-9/16
N/R	108-26[5]	N/R	N/R	N/R	N/R	1-1/16	1-3/16	1-3/8	1-7/16
N/R	108-26[5]	N/R	N/R	N/R	N/R	1-1/16	1-3/16	1-3/8	1-7/16
N/R	108-26[5]	N/R	N/R	N/R	N/R	1-1/16	1-3/16	1-3/8	1-7/16
Black	108-33	108-31	108-30	108-30	108-27	1-1/4	1-5/16	1-9/16	1-9/16
Black	108-33	108-31	108-30	108-30	108-27	1-1/4	1-5/16	1-9/16	1-9/16
Black	108-33	108-31	108-30	108-30	108-27	1-1/4	1-5/16	1-9/16	1-9/16
Plain	108-33	108-31	108-30	108-30	108-27	1-1/16	1-1/16	1-7/16	1-7/16
Black	108-32	108-31	108-30	108-30	108-27	1-13/64	1-13/32	1-3/8	2
Black	108-32	108-31	108-30	108-30	108-27	1-13/64	1-13/32	1-3/8	2
Black	108-32	108-31	108-30	108-30	108-27	1-13/64	1-13/32	1-3/8	2
Black	108-32	108-31	108-30	108-30	108-27	1-13/64	1-13/32	1-3/8	2
.035	108-33	108-29	108-33	108-29	N/R	1-11/16	1-11/16	2	2
.035	108-33	108-29	108-33	108-29	N/R	1-11/16	1-11/16	2	2
N/R	108-26[5]	N/R	N/R	N/R	N/R	1-1/16	1-3/16	1-3/8	1-7/16
N/R	108-26[5]	N/R	N/R	N/R	N/R	1-1/16	1-3/16	1-3/8	1-7/16
.031	108-33	108-29	108-33	108-29	N/R	1-3/8	1-3/8	1-11/16	1-11/16
N/R	108-26[5]	N/R	N/R	N/R	N/R	1-1/16	1-3/16	1-3/8	1-7/16
.031	108-33	108-29	108-33	108-29	N/R	1-9/16	1-9/16	1-3/4	1-3/4
.031	108-33	108-29	108-33	108-29	N/R	1-9/16	1-9/16	1-3/4	1-3/4
N/R	108-26[5]	N/R	N/R	N/R	108-27	1-1/16	1-3/16	1-3/8	1-7/16
N/R	108-26[5]	N/R	N/R	N/R	N/R	1-1/16	1-3/16	1-3/8	1-7/16
N/R	108-26[5]	N/R	N/R	N/R	N/R	1-1/16	1-3/16	1-3/8	1-7/16
Plain	108-33	108-31	108-30	108-30	108-27	1-1/4	1-5/16	1-9/16	1-9/16
Black	108-32	108-31	108-30	108-30	108-27	1-13/64	1-13/32	1-3/8	2

21. 125-65 Primary
22. 125-35 Secondary

24. 25R-475A-13

† Early versions must use 108-29 to seal pump passage.

9.1d Carburetor specifications (continued)

Carburetor Part No.	Carb. Model No.	CFM	Renew Kit	Trick Kit	Primary & Secondary Needle & Seat	Primary Main Jet	Secondary Main Jet or Plate	Primary Power Valve	Primary Discharge Nozzle Size
0-8302	4175	650	37-732	N/A	Notes 16, 17	122-582	134-21	125-85	.025
0-8479	4360	450	37-1138	N/A	6-514	124-219	124-589	125-201	.028
0-8516	4360	450	37-1138	N/A	6-514	124-167	124-423	125-203	.028
0-8517	4360	450	37-1138	N/A	6-514	124-203	124-524	125-201	.028
0-8546	4175	650	37-732	N/A	Notes 16, 17	122-582	134-21	125-85	.025
0-8642	4360	450	37-1138	N/A	6-514	124-215	124-550	125-201	.028
0-8677	4360	450	37-1138	N/A	6-514	124-219	124-524	125-200	.028
0-8679	4175	650	37-732	37-922	Notes 16, 17	122-592	134-27	125-85	.025
0-8700	4175	650	37-732	37-922	Notes 16, 17	122-582	134-21	125-85	.025
0-8771	4360	450	37-1138	N/A	6-514	124-207	124-537	125-201	.028
**0-8804	4150	830	37-485	37-916	6-504	122-80	122-80	125-65(B)	.028
0-8874	4360	450	37-1138	N/A	6-514	124-219	124-589	125-201	.028
0-8875	4360	450	37-1180	N/A	6-514	124-231	124-576	125-204	.028
0-8876	4360	450	37-1179	N/A	6-514	124-231	124-550	125-205	.028
0-8877	4360	450	37-1160	N/A	6-514	124-231	124-550	125-202	.028
0-8879	4175	650	37-732	37-922	Notes 16, 17	122-592	134-21	125-65	.025
**0-8896	4500	1050	37-487	37-920	6-504	122-88	122-88	N/R	.035
0-8914	4360	450	37-1138	N/A	6-514	124-207	124-537	125-201	.028
0-8958	4360	450	37-1138	N/A	6-514	124-195	124-550	125-200	.028
0-9002	4160	600	37-1176	N/A	6-506	122-632	134-37	125-208	.031
0-9040	4160	600	37-1177	N/A	6-85	122-661	134-23	125-211	.031
0-9088	4360	450	37-1316	N/A	6-514	124-215	124-550	125-214	.028
0-9105	4360	450	37-1160	N/A	6-514	124-195	124-550	125-200	.028
0-9112	4360	450	37-1138	N/A	6-514	124-211	124-563	125-201	.028
0-9162	4360	450	37-1138	N/A	6-514	124-203	124-537	125-201	.028
0-9185	4360	450	37-1138	N/A	6-514	124-191	124-550	125-201	.028
0-9188	4150	780	37-1184	37-915	6-504	122-72	122-76	Notes 12, 21	.025
0-9192	4360	450	37-1138	N/A	6-514	124-231	124-550	125-201	.028
0-9193	4360	450	37-1138	N/A	6-514	124-211	124-589	125-201	.028
0-9210	4160	600	37-1176	N/A	6-506	122-612	134-39	125-208	.031
0-9219	4160	600	37-1176	N/A	6-506	122-632	134-39	125-208	.031
0-9228	5200	280	37-1279	N/A	6-512	124-163	124-231	125-36	.023
0-9254	4160	600	37-1275	N/A	6-506	122-622	134-39	125-211	.031
**0-9375	4500	1050	37-487	37-920	6-504	122-92	122-92	N/R	.035
**0-9377	4500	1150	37-487	37-920	6-504	122-94	122-94	N/R	.035
**0-9379	4150	750	37-485	37-910	6-504	122-68	122-81	125-65	.028
**0-9380	4150	850	37-485	37-916	6-504	122-78	122-78	125-65 (15)	.031
**0-9381	4150	830	37-485	37-916	6-504	122-78	122-78	125-65 (15)	.028
0-9429	5200	280	37-1279	N/A	6-512	124-183	124-231	125-36	.023
0-9441	5200	280	37-1279	N/A	6-512	124-163	124-231	125-36	.023
0-9444	5200	280	37-1279	N/A	6-512	124-163	124-231	125-36	.023
0-9446	5200	280	37-1279	N/A	6-512	124-163	124-231	125-36	.023
0-9545	5200	280	37-1279	N/A	6-512	124-183	124-231	125-36	.023
0-9626	4160	600	37-1274	N/A	6-506	122-612	134-39	125-206	.031
0-9644	6520	280	37-1205	N/A	N/A	124-179	124-283	N/A	.020
0-9645	4150	750	37-1312	37-923	6-515	122-80	122-80	125-165	.045
0-9646	4150	850	37-1312	37-923	6-515	122-92	122-92	125-165	.045
0-9647	2300	500	37-1311	37-924	6-515	122-81	N/R	125-45	.040
0-9655	6520	280	37-1205	N/A	N/A	124-195	124-299	N/A	.020
0-9659	6520	280	37-1205	N/A	N/A	124-131	134-267	N/A	.020
0-9678	4360	450	37-1160	—	6-514	124-211	124-550	125-202	.028
0-9681	5200	280	37-1279	N/A	6-512	124-171	124-215	125-31	.023
0-9682	6520	280	37-1205	N/A	N/A	124-219	124-283	N/A	.020
0-9688	5200	280	37-1279	N/A	6-512	124-163	124-251	125-36	.023
0-9689	5200	280	37-1279	N/A	6-512	124-159	124-251	125-36	.023
0-9694	4360	450	37-1138	N/A	6-514	124-171	124-485	125-201	.028
0-9767	5200	280	37-1279	N/A	6-512	124-179	124-259	125-36	.023
0-9776	4160	450	37-1321	N/A	6-506	122-582	134-6	125-85	121-31
0-9777	4360	450	37-750	N/A	6-514	124-255	124-550	125-36	N/A
0-9781	5200	280	37-1279	N/A	6-512	124-159	124-251	125-36	.023
0-9810	6520	280	37-1205	N/A	N/A	124-195	124-299	N/A	.020
0-9811	6520	280	37-1205	N/A	N/A	124-155	124-271	N/A	.020
0-9834	4160	600	37-720	37-912	6-506	122-642	134-39	125-65	.031
0-9834-1	4160	600	37-720	37-912	6-506	122-661	134-39	125-65	.031
0-9834-2,-3	4160	600	37-720	37-912	6-506	122-68	134-39	125-65	.031
0-9864	5200	280	37-1279	N/A	6-512	124-159	124-219	125-36	.023
0-9875	4360	450	37-1316	N/A	6-514	124-199	124-576	125-214	.028
0-9895	4175	650	37-741	N/A	Notes 16, 17	122-592	134-21	25R-475-13	.037
0-9896	6510	280	37-1286	N/A	N/A	124-104	124-271	N/A	.020
0-9899	5200	280	37-1279	N/A	6-512	124-147	124-231	125-36	.023
0-9923	4175	650	37-741	N/A	Notes 16, 17	122-542	134-50	125-211	.025
0-9925	5200	280	37-1279	N/A	6-512	124-147	124-251	125-36	.023
0-9931	4360	450	37-1138	N/A	6-514	124-239	124-550	125-203	.028
0-9932	5200	280	37-1279	N/A	6-512	124-159	124-219	125-36	.023
0-9935	4360	450	37-1138	N/A	6-514	124-207	124-589	125-203	.028

*Indicates Center Inlet Dual Feed Bowls   5. Main Body Gasket   13. 125-105 Primary   15. 125-65 Secondary   17. 6-510 Secondary
**Indicates Dual Accel. Pump & Dual Feed Bowls   12. 125-85 Secondary   14. 125-85 Primary   16. 6-511 Primary

9.1e Carburetor specifications (by list number)

Secondary Nozzle Size or Spring Color	Primary Bowl Gasket	Primary Metering Block Gasket	Secondary Bowl Gasket	Secondary Metering Block Gasket	Secondary Metering Plate Gasket	Venturi Diameter Primary	Venturi Diameter Secondary	Throttle Bore Diameter Primary	Throttle Bore Diameter Secondary
Black	108-32	108-31	108-30	108-30	108-27	1-13/64	1-13/32	1-3/8	2
N/R	108-26[5]	N/R	N/R	N/R	N/R	1-1/16	1-3/16	1-3/8	1-7/16
N/R	108-26[5]	N/R	N/R	N/R	N/R	1-1/16	1-3/16	1-3/8	1-7/16
N/R	108-26[5]	N/R	N/R	N/R	N/R	1-1/16	1-3/16	1-3/8	1-7/16
Black	108-32	108-31	108-30	108-30	108-27	1-13/64	1-13/32	1-3/8	2
N/R	108-26[5]	N/R	N/R	N/R	N/R	1-1/16	1-3/16	1-3/8	1-7/16
N/R	108-26[5]	N/R	N/R	N/R	N/R	1-1/16	1-3/16	1-3/8	1-7/16
Plain	108-32	108-31	108-30	108-30	108-27	1-13/64	1-13/32	1-3/8	2
Black	108-32	108-31	108-30	108-30	108-27	1-13/64	1-13/32	1-3/8	2
N/R	108-26[6]	N/R	N/R	N/R	N/R	1-1/16	1-3/16	1-3/8	1-7/16
.028	108-33	108-29	108-33	108-29	N/R	1-9/16	1-9/16	1-11/16	1-11/16
N/R	108-26[5]	N/R	N/R	N/R	N/R	1-1/16	1-3/16	1-3/8	1-7/16
N/R	108-26[5]	N/R	N/R	N/R	N/R	1-1/16	1-3/16	1-3/8	1-7/16
N/R	108-26[5]	N/R	N/R	N/R	N/R	1-1/16	1-3/16	1-3/8	1-7/16
N/R	108-26[5]	N/R	N/R	N/R	N/R	1-1/16	1-3/16	1-3/8	1-7/16
Black	108-32	108-31	108-30	108-30	108-27	1-13/64	1-13/32	1-3/8	2
.035	108-33	108-36	108-33	108-36	N/R	1-11/16	1-11/16	2	2
N/R	108-26[5]	N/R	N/R	N/R	N/R	1-1/16	1-3/16	1-3/8	1-7/16
N/R	108-26[5]	N/R	N/R	N/R	N/R	1-1/16	1-3/16	1-3/8	1-7/16
Black	108-33	108-31	108-30	108-30	108-27	1-1/4	1-5/16	1-9/16	1-9/16
Plain	108-33	108-31	108-30	108-30	108-27	1-1/4	1-5/16	1-9/16	1-9/16
N/R	108-26[5]	N/R	N/R	N/R	N/R	1-1/16	1-3/16	1-3/8	1-7/16
N/R	108-26[5]	N/R	N/R	N/R	N/R	1-1/16	1-3/16	1-3/8	1-7/16
N/R	108-26[5]	N/R	N/R	N/R	N/R	1-1/16	1-3/16	1-3/8	1-7/16
N/R	108-26[5]	N/R	N/R	N/R	N/R	1-1/16	1-3/16	1-3/8	1-7/16
N/R	108-26[5]	N/R	N/R	N/R	N/R	1-1/16	1-3/16	1-3/8	1-7/16
Plain	108-33	108-29	108-33	108-29	N/R	1-3/8	1-7/16	1-11/16	1-11/16
N/R	108-26[5]	N/R	N/R	N/R	N/R	1-1/16	1-3/16	1-3/8	1-7/16
N/R	108-26[5]	N/R	N/R	N/R	N/R	1-1/16	1-3/16	1-3/8	1-7/16
Black	108-33	108-31	108-30	108-30	108-27	1-1/4	1-5/16	1-9/16	1-9/16
Black	108-33	108-31	108-30	108-30	108-27	1-1/4	1-5/16	1-9/16	1-9/16
N/R	N/R	N/R	N/R	N/R	N/R	1-1/25	1-1/16	1-7/25	1-7/16
Black	108-33	108-31	108-30	108-30	108-27	1-1/4	1-5/16	1-9/16	1-9/16
.035	108-33	108-36	108-33	108-36	N/R	1-11/16	1-11/16	2	2
.035	108-33	108-36	108-33	108-36	N/R	1-13/16	1-13/16	2	2
.031	108-33	108-29	108-33	108-29	N/R	1-3/8	1-3/8	1-11/16	1-11/16
.031	108-33	108-29	108-33	108-29	N/R	1-9/16	1-9/16	1-3/4	1-3/4
.028	108-33	108-29	108-33	108-29	N/R	1-9/16	1-9/16	1-11/16	1-11/16
N/R	N/R	N/R	N/R	N/R	N/R	1-1/25	1-1/16	1-7/25	1-7/16
N/R	N/R	N/R	N/R	N/R	N/R	1-1/25	1-1/16	1-7/25	1-7/16
N/R	N/R	N/R	N/R	N/R	N/R	1-1/25	1-1/16	1-7/25	1-7/16
N/R	N/R	N/R	N/R	N/R	N/R	1-1/25	1-1/16	1-7/25	1-7/16
N/R	N/R	N/R	N/R	N/R	N/R	1-1/25	1-1/16	1-7/25	1-7/16
Black	108-33	108-31	108-30	108-30	108-27	1-1/4	1-5/16	1-9/16	1-9/16
N/R	N/R	N/R	N/R	N/R	N/R	1-1/25	1-1/16	1-7/25	1-7/16
.045	108-33	108-29	108-33	108-29	N/R	1-3/8	1-3/8	1-11/16	1-11/16
.045	108-33	108-29	108-33	108-29	N/R	1-9/16	1-9/16	1-3/4	1-3/4
N/R	108-33	108-29	N/R	N/R	N/R	1-3/8	N/R	1-11/16	N/R
N/R	N/R	N/R	N/R	N/R	N/R	1-1/25	1-1/16	1-7/25	1-7/16
N/R	N/R	N/R	N/R	N/R	N/R	1-1/25	1-1/16	1-7/25	1-7/16
N/R	108-26	N/R	N/R	N/R	N/R	1-1/16	1-3/16	1-3/8	1-7/16
N/R	N/R	N/R	N/R	N/R	N/R	1-1/25	1-1/16	1-7/25	1-7/16
N/R	N/R	N/R	N/R	N/R	N/R	1-1/25	1-1/16	1-7/25	1-7/16
N/R	N/R	N/R	N/R	N/R	N/R	1-1/25	1-1/16	1-7/25	1-7/16
N/R	N/R	N/R	N/R	N/R	N/R	1-1/25	1-1/16	1-7/25	1-7/16
N/R	108-26	N/R	N/R	N/R	N/R	1-1/16	1-3/16	1-3/8	1-7/16
N/R	N/R	N/R	N/R	N/R	N/R	1-1/25	1-1/16	1-7/25	1-7/16
N/R	108-33	108-29	108-30	108-30	108-27	1-3/32	1-3/32	1-1/2	1-1/2
N/R	108-26[5]	N/R	N/R	N/R	N/R	1-1/16	1-3/16	1-3/8	1-7/16
N/R	N/R	N/R	N/R	N/R	N/R	1-1/25	1-1/16	1-7/25	1-7/16
N/R	N/R	N/R	N/R	N/R	N/R	1-1/25	1-1/16	1-7/25	1-7/16
N/R	N/R	N/R	N/R	N/R	N/R	1-1/25	1-1/16	1-7/25	1-7/16
Black	108-33	108-31	108-30	108-30	108-27	1-1/4	1-5/16	1-9/16	1-9/16
Black	108-33	108-31	108-30	108-30	108-27	1-1/4	1-5/16	1-9/16	1-9/16
Black	108-33	108-31	108-30	108-30	108-27	1-1/4	1-5/16	1-9/16	1-9/16
N/R	N/R	N/R	N/R	N/R	N/R	1-1/25	1-1/16	1-7/25	1-7/16
N/R	108-26	N/R	N/R	N/R	N/R	1-1/16	1-3/16	1-3/8	1-7/16
Black	108-32	108-31	108-30	108-30	108-27	1-13/64	1-13/32	1-3/8	2
N/R	N/R	N/R	N/R	N/R	N/R	1-1/25	1-1/16	1-7/25	1-7/16
N/R	N/R	N/R	N/R	N/R	N/R	1-1/25	1-1/16	1-7/25	1-7/16
Black	108-32	108-31	108-30	108-30	108-27	1-13/64	1-13/32	1-3/8	2
N/R	N/R	N/R	N/R	N/R	N/R	1-1/25	1-1/16	1-7/25	1-7/16
N/R	108-26	N/R	N/R	N/R	N/R	1-1/16	1-3/16	1-3/8	1-7/16
N/R	N/R	N/R	N/R	N/R	N/R	1-1/25	1-1/16	1-7/25	1-7/16
N/R	108-26	N/R	N/R	N/R	N/R	1-1/16	1-3/16	1-3/8	1-7/16

21. 125-65 Primary
22. 125-35 Secondary

24. 25R-475A-13

† Early versions must use 108-29 to seal pump passage.

9.1f Carburetor specifications (continued)

Carburetor Part No.	Carb. Model No.	CFM	Renew Kit	Trick Kit	Primary & Secondary Needle & Seat	Primary Main Jet	Secondary Main Jet or Plate	Primary Power Valve	Primary Discharge Nozzle Size
0-9948	4175	650	37-741	N/A	Notes 16, 17	122-563	134-50	125-211	.025
0-9973	4360	450	37-1138	N/A	6-514	124-171	124-330	125-203	.028
0-9976	4175	650	37-741	N/A	Notes 16, 17	122-582	134-49	125-211	.025
0-80054	5200	280	37-716	N/A	6-512	124-231	124-247	125-36	.023
0-80055	5200	280	37-716	N/A	6-512	124-231	124-247	125-36	.023
0-80056	5200	280	37-716	N/A	6-512	124-231	124-247	125-36	.023
0-80057	5200	280	37-716	N/A	6-512	124-132	124-135	125-36	.023
0-80073	4175	650	37-1421	N/A	Notes 16, 17	122-642	134-52	125-213	.037
0-80086	4360	450	37-1316	N/A	6-514	124-199	124-550	125-214	.028
0-80095	2305	500	37-749	N/A	6-504	122-55	122-73	125-85	.035
0-80098	4180	600	37-1346	N/A	6-517	122-612	34R-6153-61	125-215	.028
0-80099	4180	600	37-1346	N/A	6-517	122-622	34R-6153-61	125-218	.028
0-80111	4180	600	37-1346	N/A	6-517	122-612	34R-6153-30	125-216	.028
0-80112	4180	600	37-1346	N/A	6-517	122-622	34R-6153-30	125-217	.028
0-80120	2305	350	37-749	N/A	6-504	122-52	122-65	125-85	.035
0-80128	4175	650	37-741	—	6-510	122-582	134-50	125-211	.031
0-80133	4180	600	37-1346	—	6-517	122-611	34R-6153-30	125-216	.028
0-80134	4180	600	37-1346	—	6-517	122-612	34R-6153-30	25R-609-8A	.028
0-80135	4180	600	37-1346	—	6-517	122-612	34R-6153-30	25R-609-8A	.028
0-80136	4180	600	37-1346	—	6-517	122-612	34R-6153-30	25R-609-8A	.028
0-80137	4180	600	37-1346	—	6-517	122-612	34R-6153-30	25R-609-8A	.028
0-80139	4175	650	37-741	—	6-510	122-592	134-21	25R-475-13A	.037
0-80140	4175	650	37-1421	—	6-510	122-642	134-52	125-213	.037
0-80145	4150	600	37-1184	N/A	6-504	122-68	122-70	125-65	.031
0-80155	4175	650	37-741	—	6-510	122-632	134-21	25R-475-13A	.037
0-80163	4180	600	37-1346	—	6-517	122-622	34R-6153-30	25R-609-8A	.028
0-80164	4180	600	37-1422	—	6-517	122-612	34R-6153-30	25R-609-8A	.028
0-80165	4180	600	37-1346	—	6-517	122-612	34R-6153-30	25R-609-8A	.028
0-80166	4180	600	37-1346	—	6-517	122-612	34R-6153-30	25R-609-8A	.028
0-80169	4175	650	37-741	N/A	6-510	122-543	134-53	125-211	.025
0-80186	4500	750	37-487	37-920	6-504	122-70	122-70	125-65[15]	.028
0-80431	4160	550	37-119	37-912	6-506	N/A*	N/A*	N/A*	N/A*
0-80432	4160	550	37-119	37-912	6-506	N/A*	N/A*	N/A*	N/A*
0-80436	4150	850	37-1184	37-915	6-504	122-80	122-80	125-65[22]	.040
0-80450	4160	600	37-119	37-912	6-506	—	—	—	—
0-80451	4160	600	37-119	37-912	6-506	—	134-39	125-208	.031
0-80452	4160	600	37-119	37-912	6-506	—	—	—	—
0-80453	4160	600	37-119	37-912	6-506	—	—	—	—
0-80454	4160	600	37-119	37-912	6-506	—	—	—	—
0-80457	4160	600	37-119	37-912	6-506	—	—	—	—
0-80460	4160	600	37-119	37-192	6-506	—	134-39	125-208	.031
0-81850	4160	600	37-119	37-912	6-506	122-66	134-9	125-65	.025
0-82010	2010	350	37-1467	37-931	6-504	122-58	N/A	125-65	.035
0-82011	2010	500	37-1468	37-932	6-504	122-80	N/A	122-65	.035
0-82012	2010	560	37-1468	37-932	6-504	N/A*	N/A	125-65	.035
0-83310,-1	4160	750	37-754	37-921	6-504	122-72	134-21	125-65	.025
0-83311	4160	750	37-754	37-921	6-504	122-72	134-21	125-65	.025
0-83312	4160	750	37-754	37-921	6-504	122-72	134-21	125-65	.025
0-84010	4010	600	37-1445	37-927	6-504	122-67	122-75	125-65	.026
0-84010-1	4010	600	37-1445	37-927	6-504	122-67	122-75	125-65	.035
0-84011	4010	750	37-1445	37-927	6-504	122-75	122-75	125-65[15]	.026
0-84011-1	4010	750	37-1445	37-927	6-504	122-75	122-75	125-65[16]	.035
0-84012	4010	600	37-1446	37-928	6-504	122-67	122-77	125-65	.026
0-84012-1	4010	600	37-1446	37-928	6-504	122-67	122-77	125-65	.035
0-84013	4010	750	37-1446	37-928	6-504	122-79	122-79	125-65[15]	.026
0-84013-1	4010	750	37-1446	37-928	6-504	122-79	122-79	125-65[15]	.035
0-84014	4011	650	37-1447	37-929	6-504	122-60	122-66	125-65[15]	.026
0-84015	4011	800	37-1447	37-929	6-504	122-64	122-90	125-65[15]	.026
0-84016	4011	650	37-1448	37-930	6-504	122-64	122-64	125-65[15]	.026
0-84017	4011	800	37-1448	37-930	6-504	122-64	122-90	125-65 (15)	.026
0-84020	4010	600	37-1445	37-927	6-504	122-67	122-75	125-65	.026
0-84020-1	4010	600	37-1445	37-927	6-504	122-67	122-75	125-65	.035
0-84021	4011	650	37-1447	37-929	6-504	122-60	122-64	125-65 (15)	.026
0-84035	4010	600	37-1445	37-927	6-504	122-67	122-75	125-65	.035
0-84047	4010	750	37-1445	37-927	6-504	N/A	N/A	N/A	N/A
0-84412	2300	500	37-474	37-901	6-504	122-73	N/R	125-50	.028
0-84776	4150	600	37-485	37-910	6-504	122-66	122-73	125-65	.028
0-84777	4150	650	37-485	37-910	6-504	122-67	122-73	125-65	.028
0-84778	4150	700	37-485	37-910	6-504	122-69	122-78	125-65	.028
0-84779	4150	750	37-485	37-910	6-504	122-70	122-73	125-65	.028
0-84780	4150	800	37-485	37-910	6-504	122-71	122-85	125-65	.031
0-84781	4150	850	37-485	37-916	6-504	122-80	122-78	125-65	.031
0-87448	2300	350	37-749	37-901	6-504	122-61	N/A	125-85	.031
0-89834	4160	600	37-720	37-912	6-506	122-68	134-39	125-65	.031

*Indicates Center Inlet Dual Feed Bowls
**Indicates Dual Accel. Pump & Dual Feed Bowls

5. Main Body Gasket
12. 125-85 Secondary

13. 125-105 Primary
14. 125-85 Primary

15. 125-65 Secondary
16. 6-511 Primary

17. 6-510 Secondary

9.1g Carburetor specifications (by list number)

Secondary Nozzle Size or Spring Color	Primary Bowl Gasket	Primary Metering Block Gasket	Secondary Bowl Gasket	Secondary Metering Block Gasket	Secondary Metering Plate Gasket	Venturi Diameter Primary	Venturi Diameter Secondary	Throttle Bore Diameter Primary	Throttle Bore Diameter Secondary
Black	108-32	108-31	108-30	108-30	108-27	1-13/64	1-13/32	1-3/8	2
N/R	108-26	N/R	N/R	N/R	N/R	1-1/16	1-3/16	1-3/8	1-7/16
Black	108-32	108-31	108-30	108-30	108-27	1-13/64	1-13/32	1-3/8	2
N/R	108-39[5]	N/R	N/R	N/R	N/R	1-1/25	1-1/16	1-7/25	1-7/16
N/R	108-39[5]	N/R	N/R	N/R	N/R	1-1/25	1-1/16	1-7/25	1-7/16
N/R	108-39[5]	N/R	N/R	N/R	N/R	1-1/25	1-1/16	1-7/25	1-7/16
N/R	108-39[5]	N/R	N/R	N/R	N/R	1-1/25	1-1/16	1-7/25	1-7/16
Black	34-202	108-31	108-30	108-30	108-27	1-13/64	1-13/32	1-3/8	2
N/R	108-26	N/R	N/R	N/R	N/R	1-1/16	1-3/16	1-3/8	1-7/16
.028	108-33	108-29	N/R	N/R	N/R	1-3/8	1-3/8	1-11/16	1-11/16
Purple	108-56	108-55	108-30	108-30	108-13	1-1/4	1-5/16	1-9/16	1-9/16
Purple	108-56	108-55	108-30	108-30	108-13	1-1/4	1-5/16	1-9/16	1-9/16
Orange/Red	108-56	108-55	108-30	108-30	108-13	1-1/4	1-5/16	1-9/16	1-9/16
Orange/Red	108-56	108-55	108-30	108-30	108-13	1-1/4	1-5/16	1-9/16	1-9/16
.028	108-33	108-29	N/R	N/R	N/R	1-3/16	1-3/16	1-11/16	1-11/16
White	108-33	108-31	108-30	108-30	108-27	1-13/32	1-13/64	1-3/8	2
Orange-Red	108-56	108-55	108-30	108-30	108-13	1-1/4	1-5/16	1-9/16	1-9/16
Brown	108-56	108-55	108-30	108-30	108-13	1-1/4	1-5/16	1-9/16	1-9/16
Brown	108-56	108-55	108-30	108-30	108-13	1-1/4	1-5/16	1-9/16	1-9/16
Brown	108-56	108-55	108-30	108-30	108-13	1-1/4	1-5/16	1-9/16	1-9/16
Brown	108-56	108-55	108-30	108-30	108-13	1-1/4	1-5/16	1-9/16	1-9/16
Black	108-33	108-31	108-30	108-30	108-27	1-13/32	1-13/64	1-3/8	2
Black	34-202	108-31	108-30	108-30	108-27	1-13/32	1-13/64	1-3/8	2
Plain	108-33	108-35	108-33	108-29	N/R	1-1/4	1-5/16	1-9/16	1-9/16
Black	108-33	108-31	108-30	108-30	108-27	1-13/32	1-13/64	1-3/8	2
Purple	108-56	108-55	108-30	108-30	108-13	1-1/4	1-5/16	1-9/16	1-9/16
Brown	108-56	108-55	108-30	108-30	108-13	1-1/4	1-5/16	1-9/16	1-9/16
Pink	108-56	108-55	108-30	108-30	108-13	1-1/4	1-5/16	1-9/16	1-9/16
Pink	108-56	108-55	108-30	108-30	108-13	1-1/4	1-5/16	1-9/16	1-9/16
Black	108-32	108-35	108-30	108-30	108-27	1-13/32	1-13/64	1-3/8	2
.035	108-33	108-29	108-33	108-29	N/R	1-11/16	1-11/16	2	2
N/A*	108-33	108-29	108-30	108-30	108-27	1-3/16	1-1/4	1-1/2	1-1/2
N/A*	108-33	108-29	108-30	108-30	108-27	1-3/16	1-1/4	1-1/2	1-1/2
Pink	108-33	108-29	108-33	108-29	N/R	1-9/16	1-9/16	1-3/4	1-3/4
--	108-33	108-29	108-30	108-30	N/R	1-1/4	1-5/16	1-9/16	1-9/16
Black	108-33	108-29	108-30	108-30	N/R	1-1/4	1-5/16	1-9/16	1-9/16
--	108-33	108-29	108-30	108-30	N/R	1-1/4	1-5/16	1-9/16	1-9/16
--	108-33	108-29	108-30	108-30	N/R	1-1/4	1-5/16	1-9/16	1-9/16
--	108-33	108-29	108-30	108-30	N/R	1-1/4	1-5/16	1-9/16	1-9/16
--	108-33	108-29	108-30	108-30	N/R	1-1/4	1-5/16	1-9/16	1-9/16
Black	108-33	108-29	108-30	108-30	N/R	1-1/4	1-5/16	1-9/16	1-9/16
Plain	108-33-1	108-29-1	108-30-1	108-30-1	N/R	1-1/4	1-5/16	1-9/16	1-9/16
N/R	N/A	N/A	N/A	N/A	N/R	1-3/16	N/R	1-11/16	N/R
N/R	N/A	N/A	N/A	N/A	N/R	1-9/16	N/R	1-11/16	N/R
N/R	N/A	N/A	N/A	N/A	N/R	1-9/16	N/R	1-3/4	N/R
Plain	108-33	108-29	108-30	108-30	108-27	1-3/8	1-7/16	1-11/16	1-1/16
Plain	108-33-1	108-29-1	108-33-1	108-29-1	N/R	1-9/16	1-9/16	1-3/4	1-3/4
Plain	108-33-1	108-29-1	108-33-1	108-29-1	N/R	1-9/16	1-9/16	1-3/4	1-3/4
Purple	N/A	N/A	N/A	N/A	N/A	1-1/4	1-1/4	1-11/16	1-11/16
Purple	N/A	N/A	N/A	N/A	N/A	1-1/4	1-1/4	1-11/16	1-11/16
Purple	N/A	N/A	N/A	N/A	N/A	1-1/2	1-1/2	1-11/16	1-11/16
Purple	N/A	N/A	N/A	N/A	N/A	1-1/2	1-1/2	1-11/16	1-11/16
.026	N/A	N/A	N/A	N/A	N/A	1-1/4	1-1/4	1-11/16	1-11/16
.026	N/A	N/A	N/A	N/A	N/A	1-1/4	1-1/4	1-11/16	1-11/16
.026	N/A	N/A	N/A	N/A	N/A	1-1/2	1-1/2	1-11/16	1-11/16
.026	N/A	N/A	N/A	N/A	N/A	1-1/2	1-1/2	1-11/16	1-11/16
Plain	N/A	N/A	N/A	N/A	N/A	1-5/32	1-3/8	1-3/8	2
Yellow	N/A	N/A	N/A	N/A	N/A	1-5/32	1-23/32	1-3/8	2
.026	N/A	N/A	N/A	N/A	N/A	1-5/32	1-3/8	1-3/8	2
.026	N/A	N/A	N/A	N/A	N/A	1-5/32	1-23/32	1-3/8	2
Purple	N/A	N/A	N/A	N/A	N/A	1-1/4	1-1/4	1-11/16	1-11/16
Purple	N/A	N/A	N/A	N/A	N/A	1-1/4	1-1/4	1-11/16	1-11/16
Plain	N/A	N/A	N/A	N/A	N/A	1-5/32	1-3/8	1-3/8	2
Purple	N/A	N/A	N/A	N/A	N/A	1-1/4	1-1/4	1-11/16	1-11/16
Black	N/A	N/A	N/A	N/A	N/A	1-1/2	1-1/2	1-11/16	1-11/16
N/R	108-33-1	108-29-1	N/R	N/R	N/R	1-3/8	N/R	1-11/16	N/R
.032	108-33-1	108-29-1	108-33-1	108-29-1	N/R	1-1/14	1-5/16	1-9/16	1-9/16
.028	108-33-1	108-29-1	108-33-1	108-29-1	N/R	1-1/4	1-5/16	1-11/16	1-11/16
.031	108-33-1	108-29-1	108-33-1	108-29-1	N/R	1-5/16	1-3/8	1-11/16	1-11/16
.031	108-33-1	108-29-1	108-33-1	108-29-1	N/R	1-3/8	1-3/8	1-11/16	1-11/16
.031	108-33-1	108-29-1	108-33-1	108-29-1	N/R	1-3/8	1-7/16	1-11/16	1-11/16
.031	108-33-1	108-29-1	108-33-1	108-29-1	N/R	1-9/16	1-9/16	1-3/4	1-3/4
N/R	108-33-1	108-29-1	N/R	N/R	N/R	1-3/16	N/R	1-1/2	N/R
Black	108-33-1	108-31-1	108-30-1	108-30-1	108-27-1	1-1/4	1-5/16	1-9/16	1-9/16

21. 125-65 Primary   24. 25R-475A-13   N/A* = Not available at time of printing   † Early versions must use 108-29 to seal pump passage.
22. 125-35 Secondary

9.1h Carburetor specifications (continued)

NEEDLE AND SEAT ASSEMBLIES

STEEL INLET NEEDLES & SEATS FOR HI-PERFORMANCE FUELS

Needles and Seat	Seat Size
 3/8-32 thread	.097 .110 .120 .130

Most Holley performance carburetors come equipped with "viton" needle and seat. If you use exotic fuels or additives such as alcohol, benzine or acetone, use a Holley steel needle and seat .097" for small 4 barrels, .110" up to 735 CFM .120"-.130" for 750 CFM and above.

9.2 There are many different styles of needle and seat assemblies for various models. Also note that materials may vary based on standard versus performance usage

HOLLEY VITON INLET NEEDLES & SEATS

Needles and Seat	Seat Size
 3/8-32 thread	.097 .097 .097 .097 .110 Lemans
	2.00 MM
	.0785 .100 .110
	.110
	.097

FLOATS

SIDE HUNG FLOAT

Material
Nitrophyl
Brass
DURACON

CENTER HUNG FLOAT

Material
Brass
Nitrophyl
DURACON

MODEL 4360 FLOAT

Material
Brass

MODEL 5200/ 5210 FLOAT

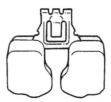

Material
Nitrophyl
DURACON

9.3 There are floats that are specific for the style of float bowl used on your carburetor, but even within the same float style there may be several selections on the type of material used

SECONDARY DIAPHRAGM SPRING KIT

CONTAINS

COLOR	RELATIVE LOAD
White	Lightest
Yellow	Lighter
Yellow	Light
Purple	Med. Light
Plain (No Color)	Medium
Brown	Med. Heavy
Black	Heavy

SECONDARY THROTTLE OPERATION RANGES

Diaphragm Secondary Springs From 20-13 Used In Model 4150, List 3310-1 Carburetor

Spring Color	350 CID Engine		402 CID Engine	
	RPM to Open	RPM at Full Open	RPM to Open	RPM at Full Open
Yellow (short spring)	1620	5680	1410	4960
Yellow	1635	5750	1420	5020
Purple	1915	6950	1680	6050
Plain (Std Spring)	2240	8160	1960	7130
Brown	2710	8750	2380	7650
Black	2720	Not fully open at maximum air flow	2390	Not fully open at maximum air flow

NOTE:
All data taken without air cleaner. An air cleaner would cause earlier opening in all cases. Values subject to change due to cleaner restrictions.

9.4 The vacuum secondary opening rate is determined by the vacuum your engine creates and the spring rate under the vacuum secondary diaphragm. Listed here are the color-coded springs and their relative opening rates. This will give you the ability to "dial-in" your own secondary opening rates through experimenting with the different springs

HOLLEY STANDARD MAIN JETS

Holley Main Jets are threaded for ease of installation and replacement. To order, use the basic part number 122- followed by a suffix "Jet Number." Example: 122-85 will have an approximate restriction size of .100" diameter.

Jet No.	Drill Size	Jet No.	Drill Size	Jet No.	Drill Size	Jet No.	Drill Size	Jet No.	Drill Size	Jet No.	Drill Size
40	.040	51	.050	61	.060	71	.076	81	.093	91	.105
41	.041	52	.052	62	.061	72	.079	82	.093	92	.105
42	.042	53	.052	63	.062	73	.079	83	.094	93	.105
43	.043	54	.053	64	.064	74	.081	84	.099	94	.108
44	.044	55	.054	65	.065	75	.082	85	.100	95	.118
45	.045	56	.055	66	.066	76	.084	86	.101	96	.118
47	.047	57	.056	67	.068	77	.086	87	.103	97	.125
48	.048	58	.057	68	.069	78	.089	88	.104	98	.125
49	.048	59	.058	69	.070	79	.091	89	.104	99	.125
50	.049	60	.060	70	.073	80	.093	90	.104	100	.128

*All sizes shown are for reference use only. Actual control of size is by flow check. In all instances, a higher stamped number indicates a greater average flow rate. Do not drill or damage the jet metering orifice. Change jet for your exact requirements.

CLOSE LIMIT MAIN JETS

These Main Jets were developed for OEM applications to obtain tighter carburetor emission flow limits. Their jet flow variation is reduced by approximately 60% compared to standard main jets. They can be a real asset for "fine tuning" a competition set-up.

NOTE: First two digits of jet No. determine size. A No. 602 jet is based on the flow of a No. 60 standard Holley main jet.

BASIC PART NO. 122-				
JET NO.				
352	432	512	592	672
362	442	522	602	682
372	452	532	612	692
382	462	542	622	702
392	472	552	632	712
402	482	562	642	722
412	492	572	652	742
422	502	582	662	

MODEL 4360, 5200 & 5210 MAIN JETS

Model 4360 main jets are interchangeable with Model 5200/5210 main jets.

To order use basic part No. 124- followed by a suffix "Jet number"

BASIC PART NO. 124-					
Jet No.	Drill Size	Jet No.	Drill Size	Jet No.	Drill Size
101	.035	219	.048	351	.061
104	.035	223	.048	357	.061
107	.036	227	.049	362	.062
110	.036	231	.050	368	.062
113	.037	235	.051	374	.062
116	.037	239	.052	380	.063
119	.038	243	.052	386	.063
123	.039	247	.052	392	.063
127	.039	251	.053	398	.064
131	.040	255	.053	404	.065
135	.040	259	.053	411	.066
139	.041	263	.054	417	.066
143	.041	267	.054	423	.067
147	.042	271	.055	429	.067
151	.043	275	.055	435	.068
155	.043	279	.055	441	.068
159	.043	283	.056	448	.068
163	.044	287	.056	455	.069
167	.044	291	.056	463	.070
171	.044	295	.056	470	.071
175	.045	299	.056	477	.071
179	.045	303	.056	485	.072
183	.046	307	.057	492	.072
187	.046	311	.057	500	.073
191	.046	315	.059	511	.074
195	.046	320	.059	524	.074
199	.047	325	.060	537	.075
203	.047	330	.060	550	.076
207	.047	335	.061	563	.076
211	.047	341	.061	576	.077
215	.048	346	.061	589	.077

MODEL 5200 AND 5210

IDLE JET ASSEMBLY

To order, use Basic Part No. 123- followed by a suffix "Idle Jet No."

BASIC PART NO. 123-	
Jet No.	Drill Size
45	.018
50	.020
55	.022
60	.024
65	.026
70	.028
80	.032

HIGH SPEED AIR BLEEDS

To order, use basic Part No. 123- followed by a suffix "airbleed No."

BASIC PART NO. 123-	
Air Bleed No.	Drill Size
150	.059
160	.063
170	.067
175	.069
180	.071
185	.073
190	.075
195	.077

9.5 There are different jets and air bleeds available, depending on carburetor model, and they have opening numbers related to a standard drill index numbering system. This makes checking, cleaning or modifying existing jets simple and accurate

HOLLEY SECONDARY METERING PLATES

Main Hole "A"	Idle Hole "B"	Part Stamped
.052	.026	7
.052	.029	34
.055	.026	3
.059	.026	4
.059	.029	32
.059	.035	40
.063	.026	5
.064	.028	18
.064	.029	30
.064	.031	13
.064	.043	33
.067	.026	8
.067	.028	23
.067	.029	16
.067	.031	9
.067	.035	36
.070	.026	6
.070	.028	19
.070	.031	20
.070	.033	41
.071	.029	35
.073	.029	39
.073	.031	37
.073	.040	17
.076	.026	10

Main Hole "A"	Idle Hole "B"	Part Stamped
.076	.028	22
.076	.029	43
.076	.031	12
.076	.035	3
.076	.040	28
.078	.029	38
.078	.040	52
.079	.031	11
.079	.035	24
.081	.029	44
.081	.033	49
.081	.040	21
.081	.052	31
.081	.063	29
.082	.031	46
.086	.043	25
.089	.031	47
.089	.037	5
.089	.040	27
.089	.043	26
.093	.040	4
.094	.070	15
.096	.031	50
.096	.040	45
.098	.070	14
.113	.026	42

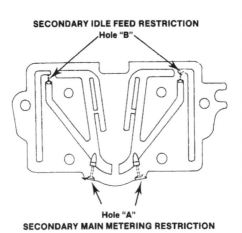

SECONDARY IDLE FEED RESTRICTION
Hole "B"

Hole "A"
SECONDARY MAIN METERING RESTRICTION

9.6 Secondary metering plates can come with more than one combination of Hole "A" and "B" sizes. Refer to the stamped number on your metering plate to determine which combination you have

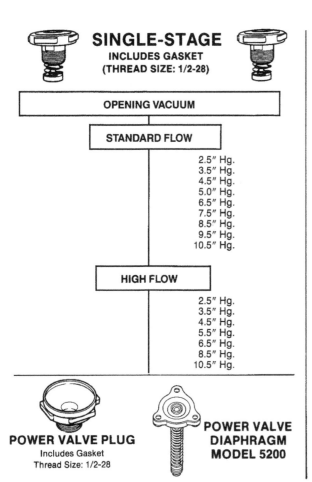

SINGLE-STAGE
INCLUDES GASKET
(THREAD SIZE: 1/2-28)

OPENING VACUUM

STANDARD FLOW

2.5" Hg.
3.5" Hg.
4.5" Hg.
5.0" Hg.
6.5" Hg.
7.5" Hg.
8.5" Hg.
9.5" Hg.
10.5" Hg.

HIGH FLOW

2.5" Hg.
3.5" Hg.
4.5" Hg.
5.5" Hg.
6.5" Hg.
8.5" Hg.
10.5" Hg.

POWER VALVE PLUG
Includes Gasket
Thread Size: 1/2-28

POWER VALVE DIAPHRAGM MODEL 5200

TWO-STAGE

INCLUDES GASKET   (THREAD SIZE: 1/2-28)

WARNING: Not recommended for Performance Applications.

1st Stage Opening	2nd Stage Opening
MODEL 4160	
12.5" Hg.	5.5" Hg.
10.5" Hg.	5" Hg.
10.5" Hg.	5.5" Hg.
11.5" Hg.	5" Hg.
MODEL 4175	
11" Hg.	6" Hg.
9" Hg.	2.5" Hg.
10.5" Hg.	5.5" Hg.
12" Hg.	6.5" Hg.
10" Hg.	6" Hg.
8" Hg.	1.5" Hg.
10" Hg.	4" Hg.
11" Hg.	5.5 Hg.
MODEL 4360	
9" Hg.	5" Hg.
8" Hg.	5" Hg.
8" Hg.	4" Hg.
8.5" Hg.	5.5" Hg.
8" Hg.	3" Hg.
9" Hg.	3" Hg.

SINGLE STAGE 4360 POWER VALVE
USED IN R-7454, R-7455, R-7456, R-7555, R-7556

9.7 Power valves, regardless of design (single-stage or two-stage), all are calibrated to open at a fixed vacuum value. Depending on the modifications done to your engine (which will determine the vacuum available in your engine), you will need to experiment to find the correct power valve for your specific needs

PUMP DISCHARGE NOZZLES
Each discharge nozzle comes with 2 gaskets.

ANTI-PULLOVER TYPE

Hole Size
.025
.028
.031
.037
.040
.042
.045
.047
.050
.052

TUBE TYPE

Hole Size
.025
.028
.031
.035
.037
.040
.042
.045

STRAIGHT TYPE

Hole Size
.025
.028
.031
.032
.035
.037
.040
.042
.045
.047
.050
.052

9.8 The three different types of accelerator pump discharge nozzles are available in the sizes listed here and they all have the size stamped on the nozzle itself. Again, as with main metering jets, the hole sizes have their numbering related directly to a standard drill index numbering system making checking, clean-up or modification very simple

CHEVROLET

Carb.	O.E. No.	Application	Transmission	Choke	C.F.M.
0-3124	3868826	1965½ Corvette 396/425	M/T	Auto Integral	780
0-3247	3886101	1966 Corvette 427/425	M/T	Auto Divorced	780
0-3367	3884505	1966 Corvette 327 (300, 350 H.P.)	Manual	Auto Divorced	585
0-3370	3882835	1966 Corvette 427/390	All	Auto Divorced	585
0-3418-1	3886091	1966-67 Corvette 427 (425 H.P.)	Manual	Auto Divorced	855
0-3613	3893229	1966 Chevelle 396 (375 H.P.)	Manual	Auto Divorced	770
0-3659	3902353	1967-69 Corvette 427 (3x2) Outboard Carb.	–	–	466
0-3660	3902355	1967 Corvette 427 (3x2) Center Carb.	Manual	Auto Divorced	350
0-3807	3903391	1967 Chevelle 327 (325 H.P.)	All	Auto Divorced	595
0-3810	3906631	1967 Corvette 327 (300, 350 H.P.)	Manual	Auto Divorced	585
0-3811	3906633	1967 Corvette 427 (390 H.P.)	All	Auto Divorced	585
0-3910	3916143	1967 Z-28 Chevelle 302, 396	All	Auto Divorced	780

CHRYSLER

Carb.	O.E. No.	Application	Transmission	Choke	C.F.M.
0-4144-1	3418550	1969-70 440 – Center Carb.	All	Auto Divorced	350
0-4365-1	3462373	1969-70 440 – Outboard Carbs.	N/A	N/A*	500
0-4670	3512835	1971 440 – Center Carb.	All	Auto Divorced	350
0-4672	3512837	1971 440 – Outboard Carbs.	N/A	N/A*	500
0-4790	3577185	1970-71 340 Outboard Carbs.	N/A	N/A*	500
0-4791	3577182	1970-71 340 – Center Carbs.	Manual	Auto Divorced	350
0-4792	3577183	1970-71 340 – Center Carbs.	Automatic	Auto Divorced	350
0-4235	2946263	1968 426 Hemi	All	N/A	770
0-4236	2946262	1968 426 Hemi	All	N/A	770

CHEVROLET (cont'd)

Carb.	O.E. No.	Application	Transmission	Choke	C.F.M.
0-4053	3923289	1968 Z-28/Chevelle 302, 396, 427	All	Auto Divorced	780
0-4055-1	3940929	1968-69 Corvette 427 (3x2) Center Carb.	Manual	Auto Divorced	350
0-4056-1	3940930	1968-69 Corvette 427 (3x2) Center Carb.	Automatic	Auto Divorced	350
0-4295	3957859	1969 Z-28 (2x4)	Manual	–	585
0-4296	3955205	1969 Corvette 427 (425 H.P.) L-88	Manual	Auto Divorced	850
0-4346	3959164	1969 Z-28, Camaro Chevelle 302, 396, 427	All	Auto Divorced	780
0-4490	3972120	1970 Z-28 350 (360 HP)	Automatic	Auto Divorced	780
0-4555	3972121	1970 Z-28 350 (360 HP)	Manual	Auto Divorced	780
0-4800-1	3989022	1971 Z-28, Corvette 350 LT-1	Automatic	Auto Divorced	780
0-4801-1	3989021	1971 Z-28, Corvette 350 LT-1	Manual	Auto Divorced	780
0-4802-1	3986196	1971 Chevelle 454 (LS-6, LS-7)	Automatic	Auto Divorced	780
0-4803-1	3986195	1971 Chevelle 454 (LS-6, LS-7)	Manual	Auto Divorced	780
0-6238-1	3997788	1972 Z-28, Corvette 350 (LT-1)	Automatic	Auto Divorced	780
0-6239-1	3999263	1972 Z-28, Corvette 350 (LT-1)	Manual	Auto Divorced	780

FORD

Carb.	O.E. No.	Application	Transmission	Choke	C.F.M.
0-3259-1	S2MS-9510A	1965-66 Shelby 289 (306 H.P.)	Manual	Manual	725
0-4514-1	DOZF-9510-AB	1970 428 Cobra Jet	Automatic	Manual	700
0-4609	C9AF-9510-U	1969 428 Cobra Jet	Automatic	Auto. Integral	730
0-4628	D0OF-9510-R	1970 429 Super Cobra Jet	Automatic	Auto. Integral	780
0-4647	D0OF-9510-S	1970 429 Boss	Manual	Manual	735
0-4653	DOZF-9510-ZA	1970 302 Boss	Manual	Manual	780
0-6129	D1ZF-9510-VA	1971 351 Boss	Manual	Manual	780

9.9 The muscle car era was a period of time from approximately the mid-1960's through the early 1970's and most of these so-called muscle cars were produced by the "Big 3" auto makers - General Motors, Chrysler and Ford. This table gives the most popular cars of that era and the Holley carburetor that came on it from the manufacturer

Conversion factors

Length (distance)

Inches (in)	X 25.4	= Millimetres (mm)	X 0.0394	= Inches (in)
Feet (ft)	X 0.305	= Metres (m)	X 3.281	= Feet (ft)
Miles	X 1.609	= Kilometres (km)	X 0.621	= Miles

Volume (capacity)

Cubic inches (cu in; in^3)	X 16.387	= Cubic centimetres (cc; cm^3)	X 0.061	= Cubic inches (cu in; in^3)
Imperial pints (Imp pt)	X 0.568	= Litres (l)	X 1.76	= Imperial pints (Imp pt)
Imperial quarts (Imp qt)	X 1.137	= Litres (l)	X 0.88	= Imperial quarts (Imp qt)
Imperial quarts (Imp qt)	X 1.201	= US quarts (US qt)	X 0.833	= Imperial quarts (Imp qt)
US quarts (US qt)	X 0.946	= Litres (l)	X 1.057	= US quarts (US qt)
Imperial gallons (Imp gal)	X 4.546	= Litres (l)	X 0.22	= Imperial gallons (Imp gal)
Imperial gallons (Imp gal)	X 1.201	= US gallons (US gal)	X 0.833	= Imperial gallons (Imp gal)
US gallons (US gal)	X 3.785	= Litres (l)	X 0.264	= US gallons (US gal)

Mass (weight)

Ounces (oz)	X 28.35	= Grams (g)	X 0.035	Ounces (oz)
Pounds (lb)	X 0.454	= Kilograms (kg)	X 2.205	= Pounds (lb)

Force

Ounces-force (ozf; oz)	X 0.278	= Newtons (N)	X 3.6	= Ounces-force (ozf; oz)
Pounds-force (lbf; lb)	X 4.448	= Newtons (N)	X 0.225	= Pounds-force (lbf; lb)
Newtons (N)	X 0.1	= Kilograms-force (kgf; kg)	X 9.81	= Newtons (N)

Pressure

Pounds-force per square inch (psi; lbf/in^2; lb/in^2)	X 0.070	= Kilograms-force per square centimetre (kgf/cm^2; kg/cm^2)	X 14.223	= Pounds-force per square inch (psi; lbf/in^2; lb/in^2)
Pounds-force per square inch (psi; lbf/in^2; lb/in^2)	X 0.068	= Atmospheres (atm)	X 14.696	= Pounds-force per square inch (psi; lbf/in^2; lb/in^2)
Pounds-force per square inch (psi; lbf/in^2; lb/in^2)	X 0.069	= Bars	X 14.5	= Pounds-force per square inch (psi; lbf/in^2; lb/in^2)
Pounds-force per square inch (psi; lbf/in^2; lb/in^2)	X 6.895	= Kilopascals (kPa)	X 0.145	= Pounds-force per square inch (psi; lbf/in^2; lb/in^2)
Kilopascals (kPa)	X 0.01	= Kilograms-force per square centimetre (kgf/cm^2; kg/cm^2)	X 98.1	= Kilopascals (kPa)

Torque (moment of force)

Pounds-force inches (lbf in; lb in)	X 1.152	= Kilograms-force centimetre (kgf cm; kg cm)	X 0.868	= Pounds-force inches (lbf in; lb in)
Pounds-force inches (lbf in; lb in)	X 0.113	= Newton metres (Nm)	X 8.85	= Pounds-force inches (lbf in; lb in)
Pounds-force inches (lbf in; lb in)	X 0.083	= Pounds-force feet (lbf ft; lb ft)	X 12	= Pounds-force inches (lbf in; lb in)
Pounds-force feet (lbf ft; lb ft)	X 0.138	= Kilograms-force metres (kgf m; kg m)	X 7.233	= Pounds-force feet (lbf ft; lb ft)
Pounds-force feet (lbf ft; lb ft)	X 1.356	= Newton metres (Nm)	X 0.738	= Pounds-force feet (lbf ft; lb ft)
Newton metres (Nm)	X 0.102	= Kilograms-force metres (kgf m; kg m)	X 9.804	= Newton metres (Nm)

Power

Horsepower (hp)	X 745.7	= Watts (W)	X 0.0013	= Horsepower (hp)

Velocity (speed)

Miles per hour (miles/hr; mph)	X 1.609	= Kilometres per hour (km/hr; kph)	X 0.621	= Miles per hour (miles/hr; mph)

Fuel consumption*

Miles per gallon, Imperial (mpg)	X 0.354	= Kilometres per litre (km/l)	X 2.825	= Miles per gallon, Imperial (mpg)
Miles per gallon, US (mpg)	X 0.425	= Kilometres per litre (km/l)	X 2.352	= Miles per gallon, US (mpg)

Temperature

Degrees Fahrenheit = (°C x 1.8) + 32   Degrees Celsius (Degrees Centigrade; °C) = (°F - 32) x 0.56

*It is common practice to convert from miles per gallon (mpg) to litres/100 kilometres (l/100km),
where mpg (Imperial) x l/100 km = 282 and mpg (US) x l/100 km = 235*

Glossary

A

Accelerator pump - A pump - usually some sort of piston or diaphragm - located in its own well in the carburetor float bowl, that adds a small amount of extra fuel under high pressure into the throttle bore to make up for the momentary shortage of fuel caused by the loss of vacuum that occurs when the throttle plates are suddenly opened.

Air bleed - A calibrated (precisely-drilled) hole in the carburetor body which allows air to enter a metering circuit.

Air density - The amount of air per unit measure, such as length, area or volume. The mass per unit volume of air under specified or standard conditions of pressure and temperature. Air density varies directly with pressure and temperature: The lower the pressure, or the higher the temperature, the less dense the air.

Air-fuel mixture - The air and fuel traveling to the combustion chamber after being mixed by the carburetor.

Air-fuel ratio - The proportions of air and fuel (by weight) supplied for combustion. See *stoichiometric*.

Air horn - The part of the carburetor body that the air enters first.

Air pressure - Atmospheric pressure (14.7 psi at sea level).

Annular-discharge booster venturi - A type of booster venturi with discharge holes located along the inner circumference of the trailing edge of the booster.

Anti-dieseling solenoid - Also referred to as an *idle-stop solenoid*. An electrically-operated two-position plunger used to provide a predetermined throttle setting. When activated, the idle stop solenoid allows the throttle plate(s) to close only so far; when deactivated, it allows the throttle(s) to close all the way, blocking the entry of any more air/fuel mixture into the engine to prevent dieseling (engine run-on) after the ignition key is switched off.

Anti-siphon bleeds - Small holes drilled into the cluster that prevent main-system fuel from continuing to flow when the throttle is closed, stopping airflow through the carburetor.

Anti-stall dashpot - A diaphragm unit mounted on the carburetor that allows air to escape slowly from its vacuum chamber to prevent throttle plate(s) in the carburetor from closing too suddenly - and stalling the engine - during deceleration.

Aspirator channel - This passage is plumbed from the point of greatest depression (lowest vacuum) on an upward angle to the main mixing well.

Atmospheric pressure - The weight of the atmosphere per unit area. Atmospheric pressure at sea level is 14.7 pounds per square inch (psi). Atmospheric pressure decreases as the altitude increases.

Atomization - The joining of fuel and air molecules into a spray of mist droplets.

Automatic choke - A device that positions the choke valve automatically in accordance with engine temperature, or in accordance with time.

Auxiliary air bleeds - Used on some idle systems; usually add air to idle system downstream from the regular idle air bleed; act in parallel with idle air bleed.

B

Bakelite - A synthetic resin used for insulating the automatic choke heating element and the carburetor spacer plate.

Bimetal - Two types of metal bonded into a strip and formed into a coil. Each type of metal has different thermal expansion characteristics, so the coil straightens when heated and coils up when cold. Bimetals are used to open and close choke plates.

Booster venturi - A small venturi located immediately above and concentric with the main venturi. Boosters are designed to amplify the weak venturi vacuum signal that occurs during low airflow conditions.

Bowl vent - Connects the float bowl to the carburetor's air inlet. Depressurizes the fuel being pumped into the float bowl by the fuel pump and acts as a vapor separator by allowing vapors in the float bowl to escape into the carburetor air inlet. Bowl vents are cut at a 45-degree angle and face incoming air so that reference pressure remains the same regardless of airflow.

C

Center-hung float - This type of float pivots on an axis that's parallel with the vehicle axles (assuming the carburetor isn't mounted sideways). It's a better float design than a side-hung float during high-speed cornering because the float isn't affected by centrifugal force or fuel sloshing from side-to-side, so it won't pull the inlet valve open in the middle of a corner.

cfm - Cubic feet per minute; a measurement of the air capacity through the carburetor venturi(s) determined by the cylinder displacement and engine rpm.

Charge density - Another term for *mass flow*. A dense charge has more air mass, so higher compression and burning pressures can be developed for higher power output. Least charge density is available at idle; highest density is at wide open throttle. See *air density*.

Choke - See *choke plate*.

Choke index - Automatic chokes have index marks. The factory setting closes the choke when the bimetal is about 70-degrees F. If you want less or more choke at this temperature, move the choke index one mark in the direction indicated by the arrows designating a leaner or richer mixture (you will seldom need to move the choke more than one mark).

Choke kick - A preset position for the choke valve set by manifold vacuum that is routed through a carburetor body passage to the choke diaphragm.

Choke plate - Or simply *choke*. A device installed in the air horn of the carburetor, used when starting a cold engine. When closed, the choke plate(s) chokes off airflow through the air horn, creating a pressure differential (partial vacuum) below it in the venturi area, which pulls fuel from the main-metering cir-

cuit, adding to and enrichening the air/fuel mixture provided by the idle circuit during warm-up.

Closed loop fuel control - The normal operating mode for a *feedback carburetor* system. Once the engine is warmed up, the computer can interpret an analog voltage signal from an exhaust gas oxygen sensor and alter the air/fuel ratio accordingly with a *duty-cycle solenoid* or *solenoid-controlled valve*.

Cluster assembly - A design which integrates several metering circuit functions into one removable casting; used only on Holley 4010 and 4011 carburetors. The cluster contains the pump shooters for the accelerator pump system, the annular discharge boosters for the main fuel system, the idle and high speed bleeds and the well tubes.

Condensation - Change of state during which a gas turns to liquid, usually because of temperature or pressure changes.

Cross-jetting - Re-jetting the carburetor jets from left to right to compensate for a left-to-right variation in performance. These tests are usually conducted using an engine dynamometer.

Cruising circuit - See *main metering system*.

Curb-idle stop screw - A screw which provides an adjustable stop for the throttle lever.

Curb-idle port - See *idle discharge hole*.

D

Dead-head pressure - A fuel pressure reading taken directly at the fuel pump outlet. Many systems use a fuel pressure regulator; dead-head pressure is an unregulated measurement.

Dashpot - A device consisting of a piston and cylinder with a restricted opening, used to slow down or delay the operation of some moving part. On carburetors, dashpots are used to prevent the throttle plates from slamming closed when the accelerator pedal is lifted suddenly.

Diaphragm - A thin dividing sheet or partition which separates a housing into two chambers, one of which is usually vented to vacuum while the other is not; used in vacuum-controlled secondaries, anti-stall dashpots and other carburetor control devices.

Discharge check ball - A small check ball or needle that lifts off its seat when the pump well is pressurized by the accelerator pump, which allows fuel to be discharged into the venturi through the shooter nozzle.

Discharge nozzle - The end of the main delivery tube that discharges fuel into the venturi area.

Divorced choke - Also known as a *remote* choke. Vacuum diaphragm is mounted on the carburetor, but the bimetal spring is mounted either on a pad on the intake manifold or in a heat well in the exhaust manifold. Choke lever is operated by a mechanical linkage rod from the bimetal spring.

Double-pumper - A carburetor equipped with two accelerator pumps.

Dry setting - The adjustment of the float with a graduated rule or drill bit while the carburetor is disassembled on the bench. Usually consists of setting a prescribed clearance between the top of the float and the top of the airhorn.

Duty-cycle solenoid - Also known as a *solenoid-controlled valve*. The duty-cycle solenoid is a computer-controlled device in a feedback carburetor that alters the mixture adjustment.

E

Economizer valve - An archaic term for the power valve.

Electric choke - Chokes can be operated by a bimetal spring heated by a solid-state heating unit (modern Holleys) or by an electric resistor nichrome-wire (no longer used by Holley). Both types increase temperature just like a coolant-controlled choke as engine warms up.

Emulsification - The process of making an emulsion.

Emulsion - Air/fuel mixture in the carburetor main circuit, before it's fully atomized.

Emulsion tube - A small tube with holes in it; protrudes down into the main or mixing well, introduces air from a bleed into the well to emulsify the fuel.

F

Fast idle cam - A cam or eccentric, attached to the choke plate by a linkage rod, that prevents a cold engine from stalling by holding the throttle plate partially open, which allows the engine to run at a faster-than-normal speed as long as the choke is applied.

Feedback carburetor - A modern emissions-era carb which responds not only to the amount of air moving through it, but also the demands of the engine operating conditions. Feedback carbs are just one part of a *closed-loop system* (an oxygen sensor, various other sensors, a computer, a duty-cycle solenoid or solenoid-controlled valve and a catalytic converter) which uses a microprocessor to monitor and adjust the air/fuel mixture.

Flame arrester - A device installed in the air filter housing; prevents flames from escaping into the engine compartment during a backfire.

Float - A vessel which floats in the fuel in the float bowl. The float is attached to and controls the inlet fuel valve (needle and seat). When the fuel level goes down, the float lowers and opens the inlet valve; when the fuel level rises to the "full" level, the float rises with it and closes the inlet valve. Floats are made of brass, nitrophyl or molded plastic.

Float bowl - The carburetor's reservoir for storing the fuel needed to supply all metering circuits.

Float bumper spring - A small spring installed under the float tang to minimize float bounce and vibration

Float level - The float position at which the float needle closes against its seat, shutting off the fuel inlet valve to prevent further delivery of fuel.

Float system - The circuit that controls the entry of fuel into - and the level of fuel in - the float bowl.

Flooding - A condition that occurs when the fuel level in the float bowl is too high. Flooding occurs when the float level is incorrectly adjusted, the float is rubbing on the side of the fuel bowl or the fuel inlet needle is held off the seat by some foreign matter or by a worn out seat. A flooded float bowl causes the idle and main circuits to deliver an air/fuel mixture that's too rich to burn.

Four-barrel carburetor - A carburetor with four venturis, four throttle plates, etc. A four-barrel is really a pair of two-barrel carbs in a single assembly.

G

Gradient power valve - A non-adjustable type of power valve that varies the volume of the fuel smoothly in accordance with manifold vacuum; under some conditions, a gradient power valve will supply a metered quantity of fuel that is less than its full output, in response to the residual manifold vacuum that remains for a brief period.

H

High-speed bleeds - Another Holley name for the main air bleeds; located in the cluster assembly or in the air horn.

High-speed circuit - See *main metering system*.

Hot idle compensator - A small air valve that allows fresh air to enter the manifold and lean the mixture when the engine is hot

Hot soak - Occurs when the engine is stopped during hot weather or after it has been run long enough to be fully warmed up; also the period during which the phenomenon known as *percolation* occurs (see *Percolation*).

I

Idle air bleed - A tube that allows air into the idle circuit to *emulsify* the fuel.

Idle channel restriction - Does the same thing as the *idle feed restriction* but is located in the idle passage just below the idle air bleed instead of the bottom of the idle tube. In the primary idle channel, the amount of fuel emitted by the idle discharge port is adjustable because there's a screw with a tapered tip extending into the port. On the secondary side of most carburetors, the idle discharge port uses a fixed idle channel restriction with no adjustment, although some carburetors use adjusting screw just like the primary side.

Idle discharge hole - Also known as the *curb-idle port*. The hole through which the idle mixture enters the airstream flowing past the throttle plate.

Idle feed restriction - Also known as an *idle orifice* or *idle jet*. A metering orifice that controls the amount of fuel that can can enter the idle tube.

Idle limiter - Any device that limits the maximum richness of the idle air/fuel mixture in the carburetor; also aids in preventing overly rich idle adjustments. Limiters take either of the two following forms: An external plastic cap (see *idle limiter cap*) or an internal-needle type located in the idle passages of the carburetor.

Idle limiter cap - An external plastic cap on the head of the idle mixture adjustment screw on older Holleys; limits the idle mixture screw adjustment to about 3/4-turn.

Idle mixture - The air/fuel mixture supplied to the engine during idling.

Idle mixture screw - The adjustment screw that can be turned in or out to lean out or enrich the idle mixture.

Idle orifice - See *idle restriction tube or idle jet*.

Idle solenoid - An electric solenoid designed to raise the curb-idle speed to compensate for an additional load, such as the air conditioner, on the engine.

Idle system - The circuit through which fuel is fed when the engine is idling.

Idle-transfer hole - See *idle transfer port*.

Idle transfer port/slot - A port drilled into the carburetor body slightly above the idle port to allow extra fuel/air emulsion into the airstream during the transition period when the throttle plate is opening from its idle (closed) position to a larger (cruising) opening angle.

Idle tube - A tube in the carburetor from the main well that allows fuel to pass through during idle

Inlet valve assembly - The needle and seat that control the admission of fuel into the float bowl. When the float level drops, the float lever arm pulls the needle off its seat and allows fuel to enter the float bowl; when the float rises, the arm pushes the needle against its seat and fuel flow into the float bowl is blocked.

Integral choke - A one-piece choke assembly that mounts both the bimetal assembly and the vacuum piston housing on the carburetor.

Internal bowl vent - A tube designed to vent excess fuel vapors from the fuel bowl back into the carburetor during acceleration and cruising conditions.

J

Jet - A calibrated passage through which fuel flows. Jets are usually machined out of solid brass and threaded so they can be replaced. However, some jets are actually drilled right into the carburetor body.

J-type vent tubes - Special tubes located on the top of marine carburetors that direct any overflow from the carburetor bowl(s) back into the throttle bores. United States Coast Guard-approved method of preventing fire in the engine compartment if the inlet valve sticks in an open position.

K

Kickdown linkage - On vehicles with an automatic transmission, the linkage on the carburetor that produces a downshift when the accelerator pedal is pushed down to the floorboard.

L

Lean surge - A change in engine rpm caused by an extremely lean setting.

List number - The part number stamped onto the body of the carburetor, used for identification; can be cross-referenced to obtain the model number.

M

Main air bleed - Reduces the signal from the discharge nozzle, which lowers pressure difference causing fuel flow and leans out the mixture. Decreasing bleed size increases pressure drop at main jet and gives a richer mixture. Main air bleed also acts as an anti-siphon, so fuel doesn't continue to dribble into venturi after airflow is reduced or stopped.

Main body - The center section of the carburetor through which air passes and is mixed with fuel. The main body houses the choke plate(s) and the venturi(s). On some models the main body also houses the float bowl and the metering circuits.

Main delivery tube - Also known as the *main nozzle*. The tube that delivers fuel from the main mixing well to the venturi.

Main jet - The removable orifice positioned at the bottom of the float bowl or in the metering block that allows fuel into the main well. Main jets are available in various sizes to alter the characteristics of the main metering circuit.

Main metering circuit - See *main metering system*.

Main metering system - Also known as the *main metering circuit*, the *cruising circuit* or the *high-speed circuit*. Supplies the correct air/fuel mixture to the engine during cruising and high-speed conditions.

Main mixing well - See *main well*.

Main nozzle - See *main delivery tube*.

Main venturi(s) - The large venturi(s) cast into the carburetor main body.

Main well - The reservoir in which fuel for the main system is stored. The main well can be in the main body or in the metering block.

Main-well tube - Also known as the *emulsion tube;* admits air from the air bleed into the main well to emulsify the main system fuel; extends from the top of the carburetor into the main well.

Manifold vacuum - The difference in air pressure, or *pressure drop*, between atmospheric pressure and air pressure in the intake manifold, that occurs just below the throttle plate(s); usually expressed in inches of Mercury (in-Hg). Manifold vacuum is proportional to the angle of the throttle plate(s): When the plates are closed or nearly closed, manifold vacuum is high; as they open farther, manifold vacuum decreases; by the time they're fully open, there's virtually no manifold vacuum.

Mass flow - In physics, mass is the measure of a body's resistance to acceleration. The mass of a body is different from, but proportional to, its weight.In automotive circles, mass flow refers to the amount of air that can be pumped into each cylinder. Mass flow is proportional to air density The greater the density, the greater the mass.

Mechanical secondary - A secondary throttle plate controlled by linkage connected to the primary throttle plate.

Metering block - Sometimes referred to as a metering *plate*, but it's not exactly the same thing (see *metering plate* below). A unique-to-Holley removable casting that's installed between the float bowl and the main body casting on the primary side of street performance Holleys such as the 4160, 4175, 4180, 4190, etc. and on the secondary side of all-out high-performance carbs such as the 4150 and 4500 (street performance carbs use a metering *plate* on the secondary side). The metering block contains discharge passages for such circuits as the the accelerator pump, timed spark, curb-idle, main metering, etc. and vacuum and vent passages for these same circuits. It also houses the main jets and the power valve. Metering blocks can be tailored to specific carburetor applications by changing the main jets and the power valves. See *metering plate*.

Metering plate - Sometimes referred to as a metering *block*, but it's not exactly the same thing (see *metering block* above). Think of the metering plate as an abbreviated metering block. It's used only on the secondary side of street performance Holleys such as the 4160, 4175, 4180, 4190, 3160, etc. It lacks only a power valve and power valve circuit, and has fixed main jet orifices drilled right into the plate instead of replaceable jets like the metering block.

Metering restriction - Precisely-machined orifice in an air bleed, jet or metering circuit; regulates amount of air, fuel or air/fuel mixture that can get through.

Metering signal - A (relative) vacuum signal generated by the pressure differential that occurs at the venturi. The strength of the metering signal determines how much fuel is pulled from the main circuit into the venturi: The smaller the venturi, the greater the pressure drop and the stronger the metering signal; the larger the venturi, the smaller the pressure drop and the weaker the metering signal.

Mile Dial - A special performance carburetor that uses a single duty-cycle solenoid in the primary float bowl. The solenoid, which is electronically controlled by a signal generator inside the car, varies the fuel flow into the power valve circuit. Flow is limited by the size of the power valve channel restrictions; once the power valve opens, the Mile Dial has no effect on the mixture.

Mixture adjusting screw - A tapered screw used to regulate the amount of air/fuel mixture through the curb-idle discharge port into the airstream. Usually located near the discharge hole.

N

Needle and seat - The components of the inlet valve assembly.The needle is attached to the float lever arm; when the float rises the lever arm closes the needle against its seat; preventing fuel from entering the float bowl; when the float sinks, the lever arm pulls the needle off its seat, allowing more fuel to enter the float bowl.

Nitrophyl - A closed cell material that's impervious to gasoline and fuel additives; used as a float material.

Non-staged carburetor - A four-barrel carburetor that has secondary throttle plates which open at the same time as the primary throttle plates, or a two-barrel carburetor with only one throttle shaft (both throttle valves open simultaneously.

Nozzle drip - Air rushing by the venturi can cause fuel to drip from the discharge nozzle for the main metering circuit.

O

Off-idle discharge ports - Another name for *transfer ports*.

Open-loop fuel control - A non-feedback mode of operation which a feedback system resorts to when the engine is started while it's still cold. During this period, the oxygen sensor isn't yet able to supply reliable data to the computer for controlling the air/fuel mixture ratio because the engine isn't yet warmed up. So mixture control is handled by a program stored in computer memory.

P

Part-throttle mixture - As the primary throttle plate opens, vacuum below the plate is reduced. Some of the fuel flow is from the idle port and some is from the idle transfer port or slot. Proportionally, more air is bypassed around the throttle plates, so the mixture gradually leans out. This transition phase is known as "part throttle mixture."

Percolation - Phenomenon that occurs during a *hot soak* period: Fuel in the main system between the float bowl and the main discharge nozzle boils over, vapor bubbles push liquid fuel out of the main system into the venturi, fuel falls onto the throttle plate and trickles into the manifold. Excess vapors from the fuel bowl and from bubbles escaping down the main well are heavier than air and drift down into the manifold. This makes the engine hard to start.

Phenolic spacer - A carburetor base gasket made from a thermosetting resin used specifically for heat insulation between the carburetor and the intake manifold.

"Picture-window" power valve - The standard performance power valve because its square holes offer better flow characteristics than conventional power valves with four or six drilled holes.

Ported vacuum - A slot-type port located right at the throttle plates, used for controlling various devices that must work in proportion to throttle plate opening, such as the EGR valve. When the throttle plates are closed at idle, there's virtually no vacuum signal at this slot. But as the throttle plates open during acceleration, they expose the slot to a progressively increasing amount of intake manifold vacuum (see *manifold vacuum*).

Power circuit - See *power system*.

Power system - Or *power circuit*. The circuit that supplies a richer mixture when the engine needs to produce extra power for acceleration or passing. The power circuit consists of a vacuum passage in the carburetor that supplies manifold vacuum to a spring-loaded power valve piston or diaphragm; when manifold vacuum drops below a certain level, spring pressure opens the power valve and extra fuel is admitted to the venturi.

Power valve - A spring loaded piston and/or diaphragm assembly that works like a "switch" operated by manifold vacuum. At a given load, the amount of manifold vacuum is no longer sufficient to overcome spring pressure, so the power valve is opened by the spring, which allows extra fuel from the power circuit to enter the venturi (see *staged power valve*).

Power valve channel restriction (PVCR) - A small, precisely-machined hole in the power valve circuit. PVCRs meter the flow of fuel into the main well.

Pressure differential - Also known as *pressure drop*. The drop in air pressure that occurs when air flowing through a tube meets a restriction such as a choke, venturi or throttle plate, causing a relative vacuum toward which air or fuel flows when acted upon by atmospheric pressure.

Pressure drop - See *pressure differential*.

Primary bore(s) - The throttle bore(s) and venturi(s) for the primary system.

Primary cluster - The cluster assembly mounted in the primary bores of 4010/1011 models. See *cluster assembly*.

Primary inlet system - The float circuit for the primary side of the carburetor.

Primary pull-off diaphragm - Device that partially opens the choke when vacuum develops (i.e. when the engine starts), allowing more air to pass through the carburetor, thinning out the excessively rich idle mixture.

Primary side - The choke and throttle plates, venturis and metering circuits associated with the *primary system*.

Primary system - The half of the carburetor that operates all the time.

Primary throttle - Linkage and throttle levers associated with the primary side of the carburetor.

Pull-off diaphragm - See *vacuum break diaphragm*.

Pump inlet ball - See *pump inlet check valve*.

Pump inlet check valve - A valve or steel ball located in the accelerator pump well. The pump inlet check valve prevents fuel from escaping from the well when the throttle is opened and pressure is exerted on the fuel in the pump well by the accelerator pump.

Pump sag - A hesitation in carburetor performance between the time the accelerator pump squirts fuel into the venturi and the point at which the main fuel circuit is activated.

Q

Quadrajet - A popular four-barrel carburetor made by Rochester designed with a spread-bore flange.

Qualifying diaphragm - See *vacuum-break diaphragm*.

Quarter Mile Dial - A special performance carburetor which uses a duty-cycle solenoid that's electronically controlled by a signal generator inside the car to control all fuel metering above idle. Quarter Mile Dial conversions have a solenoid-equipped bowl for both primary and secondary on carbs with mechanical secondaries; on carbs with vacuum-operated secondaries, there's only one solenoid-equipped bowl, on the primary side.

R

RAM tuning - A resonance phenomenon gained by changing the length of the intake passage from the carburetor to the intake valve thereby improving torque at one point on the rpm band depending upon the tuned length.

Reference pressure - The fuel bowl is vented to the outside air to maintain a constant (atmospheric) pressure on the fuel, thus maintaining a constant fuel level as a point of reference for the other systems in the carburetor.

Renew Kit - A carburetor overhaul kit from Holley that includes all gaskets, fuel needles and seats, pump diaphragms and the pump inlet and outlet checks.

Reverse-idle carburetor - A carburetor on which the idle mixture screw adjustment is "backwards," i.e. instead of turning the idle mixture screw counterclockwise to richen the mixture, you turn it clockwise. A reverse-idle carb is identified by a silver tag with an arrow and the word "lean" above the mixture screw.

S

Secondary side - The choke and throttle plates, venturis and metering circuits associated with the *secondary system*.

Secondary system - The half of a carburetor that's only used to supply extra fuel and air for increased power. On most carburetors, the secondary system doesn't open until the primary system has already opened a certain amount; then it's opened by mechanical linkage or by a vacuum-actuated diaphragm.

Secondary throttle - The linkage and throttle levers associated with the secondary system.

Secondary venturi(s) - The venturi(s) associated with the secondary system.

Shooter nozzle - Another name for the pump discharge nozzle that squirts extra fuel into the throttle bore when the accelerator pump circuit is pressurized y the pump piston.

Shooters - Small pump-discharge restrictions in the cluster assembly. These small cavities prevent accelerator pump pullover feeding from the pump system at high airflows.

Side-hung float - This float design has a pivot axis that's perpendicular to the vehicle axles (assuming the carburetor isn't mounted sideways). It has slightly better float control and fuel handling during acceleration and braking than its center-pivoted counterpart.

Signal - Another name for *vacuum* transmitted from one location in the carburetor to another. For example, a vacuum signal from the low-pressure venturi area to the float bowl, which is referenced to atmospheric pressure, initiates main circuit fuel flow.

Signal amplifier - Any device, such as the booster venturi, that amplifies a vacuum signal.

Solenoid-controlled valve - See *duty-cycle solenoid*.

Spillover point - Also known as the *pullover point*. The location of the main circuit discharge in the venturi, which is always higher than the fuel level in the bowl so fuel won't run into the venturi when it shouldn't. When spillover begins is determined by the size of the venturi and by the displacement of the engine pulling air through the carburetor.

Spread bore - A type of throttle body configuration. A spread-bore carburetor has smaller-diameter primary throttle bores than the secondaries, and the primaries are set on a wider centerline than the secondaries.

Square jetting - Same size jetting in all four holes or same size in primary barrels with a different "same size" in the secondaries).

Staged carburetors - Carburetors equipped with secondary throttle plates (or plate) which begin to open only after the primary throttle plates (or plate) has opened a specified amount.

Staged power valve - Power valve that flows a small amount of fuel at a certain manifold vacuum, then increases this flow as manifold vacuum drops further (see *power valve*).

Stagger jetting - See *cross-jetting.*

Stoichiometric - Stoichiometry is the methodology and technology by which the quantities of reactants and products in chemical reactions are determined. In automotive science, the ideal, or stoichiometric, air/fuel ratio has been determined to be 14.7 parts air to 1 part fuel, or 14.7:1.

T

Throat - The narrowest diameter of the venturi.

Throttle body - On carburetors, the casting which houses the throttle plate(s) and shaft(s) and the throttle linkage for the primary and, if equipped, the secondary bores. The throttle body is a separate component that's fastened to the underside of the carburetor main body.

Throttle bore - Each barrel-shaped opening cast into the carburetor main body and throttle body that directs air through the carburetor.

Throttle plate - Also known as a *throttle valve*. Round disc-shaped valve positioned at the bottom of the throttle bore to control airflow. The throttle plate does not control the volume of air drawn into the engine; the engine displacement never changes, so the volume of air pulled into the engine is constant for a given speed. The throttle controls the *density* or *mass flow* of air pumped in to the engine by the pistons. Least charge density is available at idle; highest density is available at wide open throttle. A dense charge has more air mass, so higher compression and burning pressures can be developed for higher power output. See *air density and mass flow*.

Throttle shaft - The shaft to which the throttle plate is affixed and on which it rotates. The throttle shaft is slightly offset (about 0.020-inch on primaries and about 0.060-inch on secondaries), so one side of the throttle plate has a larger area to induce self-closing.

Throttle stop screw - This screw sets the angle of the throttle plate opening at rest (or idle).

Throttle valve - See *throttle plate*.

Transfer ports - Also known as *off-idle discharge ports*. The holes that deliver fuel from the idle circuit during the transition from curb-idle to the main metering circuit. Located just above the throttle plates. At curb idle, off-idle ports function as an extra air bleed for further emulsification of the idle mixture; but as vacuum moves up the carb bore when the throttle plates are opened, they become fuel discharge ports. Either one or more holes, or a single slot (slots are usually used because they're cheaper to manufacture).

Trick Kit - This overhaul kit contains all the components of the Renew Kit along with a selection of power valves, pump cams and secondary diaphragm and springs to add performance

U

Unloader - If the engine doesn't start quickly, the extra fuel pulled into the carburetor by the choke plate creates an excessively rich mixture that floods the engine. To purge this excess fuel from the manifold, you push the accelerator pedal to the floor while cranking the engine. When the gas pedal is floored, a tang on the throttle lever contacts the fast idle cam, opening the choke enough to allow additional air through the carburetor and clear excess fuel from the manifold during engine start-up. This tang is known as the unloader.

V

Vacuum - In science, the absence of air. In automotive circles, any pressure that is lower than atmospheric pressure (14.7 psi). Vacuum is used extensively for controlling fuel flow within various metering circuits in a carburetor. Generally measured in inches of Mercury (in-Hg). There are three types of vacuum: manifold, ported and venturi. The strength of these vacuums depends on the throttle opening, engine speed and load. See *manifold vacuum, ported vacuum* and *venturi vacuum*.

Vacuum break diaphragm - Also known as a *pull-off diaphragm* or *qualifying diaphragm*. A choke mechanism on automatic chokes that partially opens the choke plate to a preset opening when vacuum develops (i.e. when the engine starts) to prevent an excessively-rich idle (though it's still about 20 to 50 percent richer than normal).

Vane type booster - A special type of venturi booster that delivers superior fuel distribution and atomization, at the expense of some airflow. Eight small spokes (vanes) connect the inner circle to the outer circumference of the booster, creating eight corresponding low pressure areas on their undersides.

Vaporization - A change of state from liquid to vapor or gas, by evaporation or boiling; a general term including both evaporation and boiling.

Vapor lock - The formation of gasoline vapor in the fuel lines; bubbles of gasoline vapor usually restrict or prevent fuel flow to the carburetor.

Velocity stacks - Add-on extensions of the carburetor throats mounted on top of the carburetor to clean up air entry so there is less flow-robbing turbulence.

Vena contracta - The point of lowest-pressure and highest velocity, located 0.030-inch below the venturi's throat (minimum diameter). The center of the discharge nozzle or the trailing edge of the booster venturi is placed at the vena contracta.

Vent - An opening through which air can leave an enclosed chamber.

Venturi - The part of the carburetor bore that necks down into a smooth-surface, bottleneck-shaped restriction.

Venturi effect - In an internal combustion engine, a relative vacuum is created in the cylinders by the downward strokes of the pistons. Because atmospheric pressure is higher than the low-pressure area in each cylinder, it rushes through the carburetor to equalize the low-pressure area in the cylinders. On its way to the cylinders, however, air must pass through the venturi. The venturi constricts the inrushing air column, then allows it to widen back out to the throttle bore diameter. This incoming air has a certain pressure. To get through the venturi, it must speed up, which lowers its pressure as it passes through the venturi. This change in pressure is known as the venturi effect.

Venturi vacuum - The low pressure area that occurs in the carburetor venturi area. Venturi vacuum is proportional to the speed of the air moving through the carburetor, which in turn is dependent on engine speed.

Viton-tipped needle - Special inlet valve needle with a hardened-rubber tip. Viton-tipped needles are resistant to dirt and conform to the seat even at low sealing pressures.

Volumetric efficiency - The ratio of the actual mass (weight) of air taken into the engine, to the mass of air the engine displacement would theoretically consume if there were no losses. The measure of how well an engine "breathes."

W

Wet setting - The adjustment of the float with the carburetor mounted on engine and the float bowl full of fuel.

WOT (wide open throttle) - The largest possible opening angle of the throttle plate(s).

Index

Notes

Notes

Haynes Automotive Manuals

NOTE: If you do not see a listing for your vehicle, please visit **haynes.com** for the latest product information and check out our **Online Manuals!**

ACURA
12020 Integra '86 thru '89 **& Legend** '86 thru '90
12021 Integra '90 thru '93 **& Legend** '91 thru '95
Integra '94 thru '00 - *see HONDA Civic (42025)*
MDX '01 thru '07 - *see HONDA Pilot (42037)*
12050 Acura TL all models '99 thru '08

AMC
14020 Mid-size models '70 thru '83
14025 (Renault) Alliance & Encore '83 thru '87

AUDI
15020 4000 all models '80 thru '87
15025 5000 all models '77 thru '83
15026 5000 all models '84 thru '88
Audi A4 '96 thru '01 - *see VW Passat (96023)*
15030 Audi A4 '02 thru '08

AUSTIN-HEALEY
Sprite - *see MG Midget (66015)*

BMW
18020 3/5 Series '82 thru '92
18021 3-Series incl. Z3 models '92 thru '98
18022 3-Series incl. Z4 models '99 thru '05
18023 3-Series '06 thru '14
18025 320i all 4-cylinder models '75 thru '83
18050 1500 thru 2002 except Turbo '59 thru '77

BUICK
19010 Buick Century '97 thru '05
Century (front-wheel drive) - *see GM (38005)*
19020 Buick, Oldsmobile & Pontiac Full-size (Front-wheel drive) '85 thru '05
Buick Electra, LeSabre and Park Avenue; **Oldsmobile** Delta 88 Royale, Ninety Eight and Regency; **Pontiac** Bonneville
19025 Buick, Oldsmobile & Pontiac Full-size (Rear wheel drive) '70 thru '90
Buick Estate, Electra, LeSabre, Limited, **Oldsmobile** Custom Cruiser, Delta 88, Ninety-eight, **Pontiac** Bonneville, Catalina, Grandville, Parisienne
19027 Buick LaCrosse '05 thru '13
Enclave - *see GENERAL MOTORS (38001)*
Rainier - *see CHEVROLET (24072)*
Regal - *see GENERAL MOTORS (38010)*
Riviera - *see GENERAL MOTORS (38030, 38031)*
Roadmaster - *see CHEVROLET (24046)*
Skyhawk - *see GENERAL MOTORS (38015)*
Skylark - *see GENERAL MOTORS (38020, 38025)*
Somerset - *see GENERAL MOTORS (38025)*

CADILLAC
21015 CTS & CTS-V '03 thru '14
21030 Cadillac Rear Wheel Drive '70 thru '93
Cimarron - *see GENERAL MOTORS (38015)*
DeVille - *see GENERAL MOTORS (38031 & 38032)*
Eldorado - *see GENERAL MOTORS (38030)*
Fleetwood - *see GENERAL MOTORS (38031)*
Seville - *see GM (38030, 38031 & 38032)*

CHEVROLET
10305 Chevrolet Engine Overhaul Manual
24010 Astro & GMC Safari Mini-vans '85 thru '05
24013 Aveo '04 thru '11
24015 Camaro V8 all models '70 thru '81
24016 Camaro all models '82 thru '92
24017 Camaro & Firebird '93 thru '02
Cavalier - *see GENERAL MOTORS (38016)*
Celebrity - *see GENERAL MOTORS (38005)*
24018 Camaro '10 thru '15
24020 Chevelle, Malibu & El Camino '69 thru '87
Cobalt - *see GENERAL MOTORS (38017)*
24024 Chevette & Pontiac T1000 '76 thru '87
Citation - *see GENERAL MOTORS (38020)*
24027 Colorado & GMC Canyon '04 thru '12
24032 Corsica & Beretta all models '87 thru '96
24040 Corvette all V8 models '68 thru '82
24041 Corvette all models '84 thru '96
24042 Corvette all models '97 thru '13
24044 Cruze '11 thru '19
24045 Full-size Sedans Caprice, Impala, Biscayne, Bel Air & Wagons '69 thru '90
24046 Impala SS & Caprice and Buick Roadmaster '91 thru '96
Impala '00 thru '05 - *see LUMINA (24048)*
24047 Impala & Monte Carlo all models '06 thru '11
Lumina '90 thru '94 - *see GM (38010)*
24048 Lumina & Monte Carlo '95 thru '05
Lumina APV - *see GM (38035)*
24050 Luv Pick-up all 2WD & 4WD '72 thru '82
24051 Malibu '13 thru '19
24055 Monte Carlo all models '70 thru '88
Monte Carlo '95 thru '01 - *see LUMINA (24048)*
24059 Nova all V8 models '69 thru '79

24060 Nova and Geo Prizm '85 thru '92
24064 Pick-ups '67 thru '87 - Chevrolet & GMC
24065 Pick-ups '88 thru '98 - Chevrolet & GMC
24066 Pick-ups '99 thru '06 - Chevrolet & GMC
24067 Chevrolet Silverado & GMC Sierra '07 thru '14
24068 Chevrolet Silverado & GMC Sierra '14 thru '19
24070 S-10 & S-15 Pick-ups '82 thru '93, Blazer & Jimmy '83 thru '94,
24071 S-10 & Sonoma Pick-ups '94 thru '04, including Blazer, Jimmy & Hombre
24072 Chevrolet TrailBlazer, GMC Envoy & Oldsmobile Bravada '02 thru '09
24075 Sprint '85 thru '88 & Geo Metro '89 thru '01
24080 Vans - Chevrolet & GMC '68 thru '96
24081 Chevrolet Express & GMC Savana Full-size Vans '96 thru '19

CHRYSLER
10310 Chrysler Engine Overhaul Manual
25015 Chrysler Cirrus, Dodge Stratus, Plymouth Breeze '95 thru '00
25020 Full-size Front-Wheel Drive '88 thru '93
K-Cars - *see DODGE Aries (30008)*
Laser - *see DODGE Daytona (30030)*
25025 Chrysler LHS, Concorde, New Yorker, Dodge Intrepid, Eagle Vision, '93 thru '97
25026 Chrysler LHS, Concorde, 300M, Dodge Intrepid, '98 thru '04
25027 Chrysler 300 '05 thru '18, Dodge Charger '06 thru '18, Magnum '05 thru '08 & Challenger '08 thru '18
25030 Chrysler & Plymouth Mid-size front wheel drive '82 thru '95
Rear-wheel Drive - *see Dodge (30050)*
25035 PT Cruiser all models '01 thru '10
25040 Chrysler Sebring '95 thru '06, Dodge Stratus '01 thru '06 & Dodge Avenger '95 thru '00
25041 Chrysler Sebring '07 thru '10, 200 '11 thru '17 Dodge Avenger '08 thru '14

DATSUN
28005 200SX all models '80 thru '83
28012 240Z, 260Z & 280Z Coupe '70 thru '78
28014 280ZX Coupe & 2+2 '79 thru '83
300ZX - *see NISSAN (72010)*
28018 510 & PL521 Pick-up '68 thru '73
28020 510 all models '78 thru '81
28022 620 Series Pick-up all models '73 thru '79
720 Series Pick-up - *see NISSAN (72030)*

DODGE
400 & 600 - *see CHRYSLER (25030)*
30008 Aries & Plymouth Reliant '81 thru '89
30010 Caravan & Plymouth Voyager '84 thru '95
30011 Caravan & Plymouth Voyager '96 thru '02
30012 Challenger & Plymouth Sapporro '78 thru '83
30013 Caravan, Chrysler Voyager & Town & Country '03 thru '07
30014 Grand Caravan & Chrysler Town & Country '08 thru '18
30016 Colt & Plymouth Champ '78 thru '87
30020 Dakota Pick-ups all models '87 thru '96
30021 Durango '98 & '99 & Dakota '97 thru '99
30022 Durango '00 thru '03 & Dakota '00 thru '04
30023 Durango '04 thru '09 & Dakota '05 thru '11
30025 Dart, Demon, Plymouth Barracuda, Duster & Valiant 6-cylinder models '67 thru '76
30030 Daytona & Chrysler Laser '84 thru '89
Intrepid - *see CHRYSLER (25025, 25026)*
30034 Neon all models '95 thru '99
30035 Omni & Plymouth Horizon '78 thru '90
30036 Dodge & Plymouth Neon '00 thru '05
30040 Pick-ups full-size '74 thru '93
30042 Pick-ups full-size models '94 thru '08
30043 Pick-ups full-size models '09 thru '18
30045 Ram 50/D50 Pick-ups & Raider and Plymouth Arrow Pick-ups '79 thru '93
30050 Dodge/Plymouth/Chrysler RWD '71 thru '89
30055 Shadow & Plymouth Sundance '87 thru '94
30060 Spirit & Plymouth Acclaim '89 thru '95
30065 Vans - Dodge & Plymouth '71 thru '03

EAGLE
Talon - *see MITSUBISHI (68030, 68031)*
Vision - *see CHRYSLER (25025)*

FIAT
34010 124 Sport Coupe & Spider '68 thru '78
34025 X1/9 all models '74 thru '80

FORD
10320 Ford Engine Overhaul Manual
10355 Ford Automatic Transmission Overhaul
11500 Mustang '64-1/2 thru '70 Restoration Guide
36004 Aerostar Mini-vans all models '86 thru '97
36006 Contour & Mercury Mystique '95 thru '00
36008 Courier Pick-up all models '72 thru '82

36012 Crown Victoria & Mercury Grand Marquis '88 thru '11
36014 Edge '07 thru '19 & Lincoln MKX '07 thru '18
36016 Escort & Mercury Lynx all models '81 thru '90
36020 Escort & Mercury Tracer '91 thru '02
36022 Escape '01 thru '17, Mazda Tribute '01 thru '11, & Mercury Mariner '05 thru '11
36024 Explorer & Mazda Navajo '91 thru '01
36025 Explorer & Mercury Mountaineer '02 thru '10
36026 Explorer '11 thru '17
36028 Fairmont & Mercury Zephyr '78 thru '83
36030 Festiva & Aspire '88 thru '97
36032 Fiesta all models '77 thru '80
36034 Focus all models '00 thru '11
36035 Focus '12 thru '14
36045 Fusion '06 thru '14 & Mercury Milan '06 thru '11
36048 Mustang V8 all models '64-1/2 thru '73
36049 Mustang II 4-cylinder, V6 & V8 models '74 thru '78
36050 Mustang & Mercury Capri '79 thru '93
36051 Mustang all models '94 thru '04
36052 Mustang '05 thru '14
36054 Pick-ups & Bronco '73 thru '79
36058 Pick-ups & Bronco '80 thru '96
36059 F-150 '97 thru '03, Expedition '97 thru '17, F-250 '97 thru '99, F-150 Heritage '04 & Lincoln Navigator '98 thru '17
36060 Super Duty Pick-ups & Excursion '99 thru '10
36061 F-150 full-size '04 thru '14
36062 Pinto & Mercury Bobcat '75 thru '80
36063 F-150 full-size '15 thru '17
36064 Super Duty Pick-ups '11 thru '16
36066 Probe all models '89 thru '92
Probe '93 thru '97 - *see MAZDA 626 (61042)*
36070 Ranger & Bronco II gas models '83 thru '92
36071 Ranger '93 thru '11 & Mazda Pick-ups '94 thru '09
36074 Taurus & Mercury Sable '86 thru '95
36075 Taurus & Mercury Sable '96 thru '07
36076 Taurus '08 thru '14, Five Hundred '05 thru '07, Mercury Montego '05 thru '07 & Sable '08 thru '09
36078 Tempo & Mercury Topaz '84 thru '94
36082 Thunderbird & Mercury Cougar '83 thru '88
36086 Thunderbird & Mercury Cougar '89 thru '97
36090 Vans all V8 Econoline models '69 thru '91
36094 Vans full size '92 thru '14
36097 Windstar '95 thru '03, Freestar & Mercury Monterey Mini-van '04 thru '07

GENERAL MOTORS
10360 GM Automatic Transmission Overhaul
38001 GMC Acadia '07 thru '16, Buick Enclave '08 thru '17, Saturn Outlook '07 thru '10 & Chevrolet Traverse '09 thru '17
38005 Buick Century, Chevrolet Celebrity, Oldsmobile Cutlass Ciera & Pontiac 6000 all models '82 thru '96
38010 Buick Regal '88 thru '04, Chevrolet Lumina '88 thru '04, Oldsmobile Cutlass Supreme '88 thru '97 & Pontiac Grand Prix '88 thru '07
38015 Buick Skyhawk, Cadillac Cimarron, Chevrolet Cavalier, Oldsmobile Firenza, Pontiac J-2000 & Sunbird '82 thru '94
38016 Chevrolet Cavalier & Pontiac Sunfire '95 thru '05
38017 Chevrolet Cobalt '05 thru '10, HHR '06 thru '11, Pontiac G5 '07 thru '09, Pursuit '05 thru '06 & Saturn ION '03 thru '07
38020 Buick Skylark, Chevrolet Citation, Oldsmobile Omega, Pontiac Phoenix '80 thru '85
38025 Buick Skylark '86 thru '98, Somerset '85 thru '87, Oldsmobile Achieva '92 thru '98, Calais '85 thru '91, & Pontiac Grand Am all models '85 thru '98
38026 Chevrolet Malibu '97 thru '03, Classic '04 thru '05, Oldsmobile Alero '99 thru '03, Cutlass '97 thru '00, & Pontiac Grand Am '99 thru '03
38027 Chevrolet Malibu '04 thru '12, Pontiac G6 '05 thru '10 & Saturn Aura '07 thru '10
38030 Cadillac Eldorado, Seville, Oldsmobile Toronado & Buick Riviera '71 thru '85
38031 Cadillac Eldorado, Seville, DeVille, Fleetwood, Oldsmobile Toronado & Buick Riviera '86 thru '93
38032 Cadillac DeVille '94 thru '05, Seville '92 thru '04 & Cadillac DTS '06 thru '10
38035 Chevrolet Lumina APV, Oldsmobile Silhouette & Pontiac Trans Sport all models '90 thru '96
38036 Chevrolet Venture '97 thru '05, Oldsmobile Silhouette '97 thru '04, Pontiac Trans Sport '97 thru '98 & Montana '99 thru '05
38040 Chevrolet Equinox '05 thru '17, GMC Terrain '10 thru '17 & Pontiac Torrent '06 thru '09

GEO
Metro - *see CHEVROLET Sprint (24075)*
Prizm - '85 thru '92 see CHEVY (24060), '93 thru '02 see TOYOTA Corolla (92036)
40030 Storm all models '90 thru '93
Tracker - *see SUZUKI Samurai (90010)*

(Continued on other side)

NOTE: If you do not see a listing for your vehicle, please visit **haynes.com** for the latest product information and check out our **Online Manuals!**

GMC
Acadia - see *GENERAL MOTORS (38001)*
Pick-ups - see *CHEVROLET (24027, 24068)*
Vans - see *CHEVROLET (24081)*

HONDA
42010 **Accord CVCC** all models '76 thru '83
42011 **Accord** all models '84 thru '89
42012 **Accord** all models '90 thru '93
42013 **Accord** all models '94 thru '97
42014 **Accord** all models '98 thru '02
42015 **Accord** '03 thru '12 **& Crosstour** '10 thru '14
42016 **Accord** '13 thru '17
42020 **Civic 1200** all models '73 thru '79
42021 **Civic 1300 & 1500 CVCC** '80 thru '83
42022 **Civic 1500 CVCC** all models '75 thru '79
42023 **Civic** all models '84 thru '91
42024 **Civic & del Sol** '92 thru '95
42025 **Civic** '96 thru '00, **CR-V** '97 thru '01
     & Acura Integra '94 thru '00
42026 **Civic** '01 thru '11 **& CR-V** '02 thru '11
42027 **Civic** '12 thru '15 **& CR-V** '12 thru '16
42030 **Fit** '07 thru '13
42035 **Odyssey** all models '99 thru '10
     Passport - see *ISUZU Rodeo (47017)*
42037 **Honda Pilot** '03 thru '08, **Ridgeline** '06 thru '14
     & Acura MDX '01 thru '07
42040 **Prelude CVCC** all models '79 thru '89

HYUNDAI
43010 **Elantra** all models '96 thru '19
43015 **Excel & Accent** all models '86 thru '13
43050 **Santa Fe** all models '01 thru '12
43055 **Sonata** all models '99 thru '14

INFINITI
G35 '03 thru '08 - see *NISSAN 350Z (72011)*

ISUZU
Hombre - see *CHEVROLET S-10 (24071)*
47017 **Rodeo** '91 thru '02, **Amigo** '89 thru '94 & '98 thru '02
     & Honda Passport '95 thru '02
47020 **Trooper** '84 thru '91 **& Pick-up** '81 thru '93

JAGUAR
49010 **XJ6** all 6-cylinder models '68 thru '86
49011 **XJ6** all models '88 thru '94
49015 **XJ12 & XJS** all 12-cylinder models '72 thru '85

JEEP
50010 **Cherokee, Comanche & Wagoneer Limited**
     all models '84 thru '01
50011 **Cherokee** '14 thru '19
50020 **CJ** all models '49 thru '86
50025 **Grand Cherokee** all models '93 thru '04
50026 **Grand Cherokee** '05 thru '19
     & Dodge Durango '11 thru '19
50029 **Grand Wagoneer & Pick-up** '72 thru '91
     Grand Wagoneer '84 thru '91, Cherokee &
     Wagoneer '72 thru '83, Pick-up '72 thru '88
50030 **Wrangler** all models '87 thru '17
50035 **Liberty** '02 thru '12 **& Dodge Nitro** '07 thru '11
50050 **Patriot & Compass** '07 thru '17

KIA
54050 **Optima** '01 thru '10
54060 **Sedona** '02 thru '14
54070 **Sephia** '94 thru '01, **Spectra** '00 thru '09,
     Sportage '05 thru '20
54077 **Sorento** '03 thru '13

LEXUS
ES 300/330 - see *TOYOTA Camry (92007, 92008)*
ES 350 - see *TOYOTA Camry (92009)*
RX 300/330/350 - see *TOYOTA Highlander (92095)*

LINCOLN
MKX - see *FORD (36014)*
Navigator - see *FORD Pick-up (36059)*
59010 **Rear-Wheel Drive Continental** '70 thru '87,
     Mark Series '70 thru '92 **& Town Car** '81 thru '10

MAZDA
61010 **GLC (rear-wheel drive)** '77 thru '83
61011 **GLC (front-wheel drive)** '81 thru '85
61012 **Mazda3** '04 thru '11
61015 **323 & Protegé** '90 thru '03
61016 **MX-5 Miata** '90 thru '14
61020 **MPV** all models '89 thru '98
     Navajo - see *Ford Explorer (36024)*
61030 **Pick-ups** '72 thru '93
     Pick-ups '94 thru '09 - see *Ford Ranger (36071)*
61035 **RX-7** all models '79 thru '85
61036 **RX-7** all models '86 thru '91
61040 **626 (rear-wheel drive)** all models '79 thru '82
61041 **626 & MX-6 (front-wheel drive)** '83 thru '92
61042 **626** '93 thru '01 **& MX-6/Ford Probe** '93 thru '01
61043 **Mazda6** '03 thru '13

MERCEDES-BENZ
63012 **123 Series Diesel** '76 thru '85
63015 **190 Series** 4-cylinder gas models '84 thru '88
63020 **230/250/280** 6-cylinder SOHC models '68 thru '72
63025 **280 123 Series** gas models '77 thru '81
63030 **350 & 450** all models '71 thru '80
63040 **C-Class:** C230/C240/C280/C320/C350 '01 thru '07

MERCURY
64200 **Villager & Nissan Quest** '93 thru '01
     All other titles, see FORD Listing.

MG
66010 **MGB** Roadster & GT Coupe '62 thru '80
66015 **MG Midget, Austin Healey Sprite** '58 thru '80

MINI
67020 **Mini** '02 thru '13

MITSUBISHI
68020 **Cordia, Tredia, Galant, Precis & Mirage** '83 thru '93
68030 **Eclipse, Eagle Talon & Plymouth Laser** '90 thru '94
68031 **Eclipse** '95 thru '05 **& Eagle Talon** '95 thru '98
68035 **Galant** '94 thru '12
68040 **Pick-up** '83 thru '96 **& Montero** '83 thru '93

NISSAN
72010 **300ZX** all models including Turbo '84 thru '89
72011 **350Z & Infiniti G35** all models '03 thru '08
72015 **Altima** all models '93 thru '06
72016 **Altima** '07 thru '12
72020 **Maxima** all models '85 thru '92
72021 **Maxima** all models '93 thru '08
72025 **Murano** '03 thru '14
72030 **Pick-ups** '80 thru '97 **& Pathfinder** '87 thru '95
72031 **Frontier** '98 thru '04, **Xterra** '00 thru '04,
     & Pathfinder '96 thru '04
72032 **Frontier & Xterra** '05 thru '14
72037 **Pathfinder** '05 thru '14
72040 **Pulsar** all models '83 thru '86
72042 **Roque** all models '08 thru '20
72050 **Sentra** all models '82 thru '94
72051 **Sentra & 200SX** all models '95 thru '06
72060 **Stanza** all models '82 thru '90
72070 **Titan pick-ups** '04 thru '10, **Armada** '05 thru '10
     & Pathfinder Armada '04
72080 **Versa** all models '07 thru '19

OLDSMOBILE
73015 **Cutlass** V6 & V8 gas models '74 thru '88
     For other OLDSMOBILE titles, see BUICK,
     CHEVROLET or GENERAL MOTORS listings.

PLYMOUTH
For PLYMOUTH titles, see DODGE listing.

PONTIAC
79008 **Fiero** all models '84 thru '88
79018 **Firebird** V8 models except Turbo '70 thru '81
79019 **Firebird** all models '82 thru '92
79025 **G6** all models '05 thru '09
79040 **Mid-size Rear-wheel Drive** '70 thru '87
     Vibe '03 thru '10 - see *TOYOTA Corolla (92037)*
     For other PONTIAC titles, see BUICK,
     CHEVROLET or GENERAL MOTORS listings.

PORSCHE
80020 **911** Coupe & Targa models '65 thru '89
80025 **914** all 4-cylinder models '69 thru '76
80030 **924** all models including Turbo '76 thru '82
80035 **944** all models including Turbo '83 thru '89

RENAULT
Alliance & Encore - see *AMC (14025)*

SAAB
84010 **900** all models including Turbo '79 thru '88

SATURN
87010 **Saturn** all S-series models '91 thru '02
     Saturn Ion '03 thru '07- see *GM (38017)*
     Saturn Outlook - see *GM (38001)*
87020 **Saturn L-series** all models '00 thru '04
87040 **Saturn VUE** '02 thru '09

SUBARU
89002 **1100, 1300, 1400 & 1600** '71 thru '79
89003 **1600 & 1800** 2WD & 4WD '80 thru '94
89080 **Impreza** '02 thru '11, **WRX** '02 thru '14,
     & WRX STI '04 thru '14
89100 **Legacy** all models '90 thru '99
89101 **Legacy & Forester** '00 thru '09
89102 **Legacy** '10 thru '16 **& Forester** '12 thru '16

SUZUKI
90010 **Samurai/Sidekick & Geo Tracker** '86 thru '01

TOYOTA
92005 **Camry** all models '83 thru '91
92006 **Camry** '92 thru '96 **& Avalon** '95 thru '96
92007 **Camry, Avalon, Solara, Lexus ES 300** '97 thru '01

TOYOTA (continued)
92008 **Camry, Avalon, Lexus ES 300/330** '02 thru '06
     & Solara '02 thru '08
92009 **Camry, Avalon & Lexus ES 350** '07 thru '17
92015 **Celica Rear-wheel Drive** '71 thru '85
92020 **Celica Front-wheel Drive** '86 thru '99
92025 **Celica Supra** all models '79 thru '92
92030 **Corolla** all models '75 thru '79
92032 **Corolla** all rear-wheel drive models '80 thru '87
92035 **Corolla** all front-wheel drive models '84 thru '92
92036 **Corolla & Geo/Chevrolet Prizm** '93 thru '02
92037 **Corolla** '03 thru '19, **Matrix** '03 thru '14,
     & Pontiac Vibe '03 thru '10
92040 **Corolla Tercel** all models '80 thru '82
92045 **Corona** all models '74 thru '82
92050 **Cressida** all models '78 thru '82
92055 **Land Cruiser FJ40, 43, 45, 55** '68 thru '82
92056 **Land Cruiser FJ60, 62, 80, FZJ80** '80 thru '96
92060 **Matrix** '03 thru '11 **& Pontiac Vibe** '03 thru '10
92065 **MR2** all models '85 thru '87
92070 **Pick-up** all models '69 thru '78
92075 **Pick-up** all models '79 thru '95
92076 **Tacoma** '95 thru '04, **4Runner** '96 thru '02
     & T100 '93 thru '08
92077 **Tacoma** all models '05 thru '18
92078 **Tundra** '00 thru '06 **& Sequoia** '01 thru '07
92079 **4Runner** all models '03 thru '09
92080 **Previa** all models '91 thru '95
92081 **Prius** all models '01 thru '12
92082 **RAV4** all models '96 thru '12
92085 **Tercel** all models '87 thru '94
92090 **Sienna** all models '98 thru '10
92095 **Highlander** '01 thru '19
     & Lexus RX330/330/350 '99 thru '19
92179 **Tundra** '07 thru '19 **& Sequoia** '08 thru '19

TRIUMPH
94007 **Spitfire** all models '62 thru '81
94010 **TR7** all models '75 thru '81

VW
96008 **Beetle & Karmann Ghia** '54 thru '79
96009 **New Beetle** '98 thru '10
96016 **Rabbit, Jetta, Scirocco & Pick-up**
     gas models '75 thru '92 & Convertible '80 thru '92
96017 **Golf, GTI & Jetta** '93 thru '98, **Cabrio** '95 thru '02
96018 **Golf, GTI, Jetta** '99 thru '05
96019 **Jetta, Rabbit, GLI, GTI & Golf** '05 thru '11
96020 **Rabbit, Jetta & Pick-up** diesel '77 thru '84
96021 **Jetta** '11 thru '18 **& Golf** '15 thru '19
96023 **Passat** '98 thru '05 **& Audi A4** '96 thru '01
96030 **Transporter 1600** all models '68 thru '79
96035 **Transporter 1700, 1800 & 2000** '72 thru '79
96040 **Type 3 1500 & 1600** '63 thru '73
96045 **Vanagon Air-Cooled** all models '80 thru '83

VOLVO
97010 **120, 130 Series & 1800 Sports** '61 thru '73
97015 **140 Series** all models '66 thru '74
97020 **240 Series** all models '76 thru '93
97040 **740 & 760 Series** all models '82 thru '88
97050 **850 Series** all models '93 thru '97

TECHBOOK MANUALS
10205 **Automotive Computer Codes**
10206 **OBD-II & Electronic Engine Management**
10210 **Automotive Emissions Control Manual**
10215 **Fuel Injection Manual** '78 thru '85
10225 **Holley Carburetor Manual**
10230 **Rochester Carburetor Manual**
10305 **Chevrolet Engine Overhaul Manual**
10320 **Ford Engine Overhaul Manual**
10330 **GM and Ford Diesel Engine Repair Manual**
10331 **Duramax Diesel Engines** '01 thru '19
10332 **Cummins Diesel Engine Performance Manual**
10333 **GM, Ford & Chrysler Engine Performance Manual**
10334 **GM Engine Performance Manual**
10340 **Small Engine Repair Manual,** 5 HP & Less
10341 **Small Engine Repair Manual,** 5.5 thru 20 HP
10345 **Suspension, Steering & Driveline Manual**
10355 **Ford Automatic Transmission Overhaul**
10360 **GM Automatic Transmission Overhaul**
10405 **Automotive Body Repair & Painting**
10410 **Automotive Brake Manual**
10411 **Automotive Anti-lock Brake (ABS) Systems**
10420 **Automotive Electrical Manual**
10425 **Automotive Heating & Air Conditioning**
10435 **Automotive Tools Manual**
10445 **Welding Manual**
10450 **ATV Basics**

Over a 100 Haynes
motorcycle manuals
also available

10/22

Haynes North America, Inc. • (805) 498-6703 • www.haynes.com